AF581886

Novel Methods and Applications for Mineral Exploration

Novel Methods and Applications for Mineral Exploration

Special Issue Editor

Paul Alexandre

MDPI • Basel • Beijing • Wuhan • Barcelona • Belgrade • Manchester • Tokyo • Cluj • Tianjin

Special Issue Editor
Paul Alexandre
Department of Geology,
Brandon University
Canada

Editorial Office
MDPI
St. Alban-Anlage 66
4052 Basel, Switzerland

This is a reprint of articles from the Special Issue published online in the open access journal *Minerals* (ISSN 2075-163X) (available at: https://www.mdpi.com/journal/minerals/special_issues/exploration).

For citation purposes, cite each article independently as indicated on the article page online and as indicated below:

LastName, A.A.; LastName, B.B.; LastName, C.C. Article Title. *Journal Name* **Year**, *Article Number*, Page Range.

ISBN 978-3-03928-943-1 (Pbk)
ISBN 978-3-03928-944-8 (PDF)

Cover image courtesy of Paul Alexandre.

Contents

About the Special Issue Editor

Paul Alexandre is an experienced mineral exploration and metallogeny scientist. He obtained his Ph.D. in France, working on emerald, gold, and rare earth element (REE) deposits, before moving to Canada, where he studies such commodities as uranium, REE, base metals, and gold. His main methods include geochemistry, petrology and mineralogy, geochronology, and geostatistics. He is a dedicated educator who teaches and supervises graduate students at Brandon University, Canada.

Preface to "Novel Methods and Applications for Mineral Exploration"

Exploration geologists, both industry- and academia-based, have always been highly innovative in their approach to exploration. This mindset has never been as visible as in these exciting times, with the advent of advanced computer technologies such as big data, artificial intelligence, and machine learning. These tools lend their considerable power to the re-imagination and re-invention of some of the classical and most common mineral exploration methods such as geophysics (e.g., electromagnetics, magnetotellurics, and gravity) and geochemistry (e.g., trace elements), but also to the emergence of novel conceptual frameworks, for example, hyperspectral exploration.

The present Special Issue was envisioned with the desire to capture, in one volume, the latest and most inventive applications of these methods in the field of mineral exploration. In that regard, it has been successful beyond our expectations: all articles, summarized in the Editorial, are highly innovative and make full use of the abovementioned computational developments. This is particularly true in the re-imagination of geophysical data interpretation and specifically addressing the thorny problem of non-unique interpretation. Several papers also use enhanced computation for integrated 3D geochemistry-geophysics modelling, resulting in effective examples of what exploration will look like in the future. One can imagine that in the future, far from battling with several, and sometimes contradictory, interpretations of the raw data, exploration geologists will rely on trustworthy, dependable, and robust models that will greatly reduce the risks inherent in mineral exploration.

In conclusion, the present volume provides a timely and accurate snapshot of the forefront of mineral exploration research and provides insights into what mineral exploration will look like in the future. Specifically, all articles in this Special Issue make full use of the most modern computational tools and modeling concepts, underlining the significance of the intimate collaboration between academia-based scientists and the exploration industry.

Finally, I would like to personally acknowledge all those who worked tirelessly to make this volume a great success. Chief among those are our authors who contributed some wonderful groundbreaking research. They are closely followed in importance by the countless and selfless reviewers who, for the love of science and animated by altruism, worked hard not only to provide thoughtful and constructive reviews, but to significantly improve the quality of the papers. Finally, this volume would not have been possible without the dedicated MDPI staff and editors, in particular Managing Editor Irwin Liang, who has spared no effort to make this volume a success. To all these, my heartfelt and deepest thank you!

Paul Alexandre
Special Issue Editor

Editorial

Editorial for Special Issue "Novel Methods and Applications for Mineral Exploration"

Paul Alexandre

Department of Geology, Brandon University, John R. Brodie Science Centre, 270–18th Street, Brandon, MB R7A 6A9, Canada; AlexandreP@BrandonU.CA

Received: 15 March 2020; Accepted: 23 March 2020; Published: 25 March 2020

1. Introduction

The mineral exploration industry is undergoing a profound transformation, reflecting not only the presence of some novel societal, economic, and environmental considerations, but also reflecting the changes in the deposits themselves, which tend to be deeper, with lower grades, and in more remote regions. On the other hand, recent technological advances, not only in geophysics and geochemistry, but in fields such as in artificial intelligence, computational methods, and hyperspectral exploration, to name but a few, have profoundly changed the way exploration is now conducted. This special volume is a representation of these cutting-edge and pioneering ways to consider and conduct exploration and should serve both as a valuable compendium of the most innovative exploration methodologies available and as a fore-shadowing of what form the future of exploration will likely take. As such, this volume is of significant importance and would be useful to any exploration geologist and company.

2. Review of the Papers in the Special Issue

The papers published in this Special Issue are diverse, with contributions in the fields of geophysics (four papers), computational methods (three papers), geochemistry (two papers), and one review paper on a specific deposit type. These distinctions are, of course, somewhat artificial, as modern exploration geophysics and geochemistry heavily rely on computation, data treatment, and interpretation. The individual contributions will be briefly reviewed here.

2.1. Geophysics

The contribution by Zhang et al. [1] provides an example of successful deep-seated deposit exploration, where the geological background was interpreted in combination with geophysical methods such as gravity, aeromagnetic, and controlled source audio-frequency magnetotellurics (CSAMT). The method was applied to one of the largest Ni–Cu–(PGE) deposits in the world: The Jinchuan Cu–Ni sulfide deposit in the North China Craton. The authors found that medium-low resistivity, high density, and high magnetic anomaly areas near the structural belt tend to correspond to the known ore-bearing rocks in the area, thus providing an exploration tool for this type of deposits.

The paper by Guo et al. [2] reports the application of electromagnetics (EM) combined with controlled source audio-frequency magnetotellurics (CSAMT) to the exploration of the Eagle's Nest lead–zinc deposits in Jianshui, SW China. Importantly, the authors report several specific optimizations of the methods, based on previously obtained dual- frequency induced polarization data, allowing them to infer that the Pb–Zn ore-bodies correlate with high induced polarization and low resistivity, suggesting that EM and CSAMT can be used for similar deposits in the area.

The contribution by Zhang et al. [3] reports the development of a geophysical exploration method based on the joint inversion of 2D gravity, gradiometry, and magnetotelluric data, based on data-space and normalized cross-gradient constraints. Both a synthetic example and a real-world example (from the Haigou gold mine, Jilin, Northern China) are provided to test the method, allowing the authors to conclude that the method can be applied with relative ease and can be useful, in particular in geologically complicated terrains.

The next paper, by Zhang et al. [4], also deals with the joint 2D inversion of gravity and magnetotelluric data that are structurally constrained in this study. A synthetic and a real-world (Linjiang Cu mine, Jilin, Northern China) example are used to test the method, allowing the authors to conclude that the elastic-net regularization method and the cross-gradient constraints help to provide a more meaningful, integrated interpretation of the subsurface. The method results in more detailed and sharp boundary models leading to the less ambiguous distinction of geologic units and materials.

2.2. Geochemistry

The contribution by Steiner et al. [5] provides an elegant example of a combined stream sediment geochemistry and automated mineralogy approach to the exploration of the Kagenfels and Natzwiller fractionated granites, Vosges Mountains, NE France. Characteristic geochemical fractionation and principal component analysis trends are combined with mineralogical evidence from a series of stream sediment samples to suggest that the fractionated granite suites in the northern Vosges Mountains contain rare metal mineralization indicators and are therefore highly prospective for further exploration.

An intriguing paper by Harmon et al. [6] describes a novel analytical tool that has high potential in mineral exploration: laser-induced breakdown spectroscopy (LIBS). A review of previously published research and new data demonstrates the high usefulness of this method in geochemical fingerprinting, sample classification and discrimination, quantitative geochemical analysis, rock characterization by grain size analysis, and in situ geochemical imaging. Given that LIBS data can be obtained in the field by a hand-held instrument, LIBS has high potential in mineral exploration.

2.3. Computational Methods

The contribution by Mao et al. [7] reports the results of mineral prospectivity modeling, involving a combination of 3D geological modeling, 3D spatial analysis, and prospectivity modeling, applied to the Axi low-sulfidation epithermal gold deposit NW China. The results suggest that genetic algorithm optimized support vector regression (GA-SVR) outperforms multiple nonlinear regression or fuzzy weights-of-evidence in complicated nonlinear and high-dimensional cases of prospectivity modeling.

The contribution by Battalgazy et al. [8] focuses on the use of complex bi-variate plots and provides an algorithm for combining projection pursuit multivariate transform (PPMT) with a conventional (co)-simulation. The proposed algorithm is applied to geochemical exploration data from a real-world case: a deposit in south Kazakhstan.

A valuable contribution by Chen et al. [9] provides a novel method for the use of a one-class support vector machine (OCSVM) algorithm by combining it with the bat algorithm. This combination results in the automatic optimization of the initialization parameters of the OCSVM. The bat-optimized OCSVM is then applied to the mineral prospectivity of the Helong district, Jilin Province, China.

2.4. Review of a Deposit Type

The paper by Steiner [10] provides a comprehensive review of the main controls for the formation of Li–Cs–Ta pegmatite deposits. The review recommends an optimized grassroots exploration workflow and suggests the methods that can be used in this exploration. It also provides specific case studies from the Vosges Mountains in northeast France and the Kaustinen pegmatite field in west Finland. It is a compendium that is very valuable as a "cookbook" to guide exploration for Li–Cs–Ta pegmatite deposits.

3. Summary

Exploration geologists have always been very innovative and have always strived to develop and utilize the most advanced exploration techniques. This has never been as visible as today, when some very significant technological advances, specifically in computational power, artificial intelligence, and machine learning, have opened completely new perspectives and vistas allowing not only to extract much more and more detailed and specific information from the raw observational data, but also to develop completely new and exciting exploration methods and techniques. The present volume provides a snapshot of the fore-front of exploration research, underlining the significance of the collaboration between academia-based scientists and the exploration industry.

References

1. Zhang, J.; Zeng, Z.; Zhao, X.; Li, J.; Zhou, Y.; Gong, M. Deep Mineral Exploration of the Jinchuan Cu–Ni Sulfide Deposit Based on Aeromagnetic, Gravity, and CSAMT Methods. *Minerals* **2020**, *10*, 168. [CrossRef]
2. Guo, Z.; Hu, L.; Liu, C.; Cao, C.; Liu, J.; Liu, R. Application of the CSAMT Method to Pb–Zn Mineral Deposits: A Case Study in Jianshui, China. *Minerals* **2019**, *9*, 726. [CrossRef]
3. Zhang, R.; Li, T. Joint Inversion of 2D Gravity Gradiometry and Magnetotelluric Data in Mineral Exploration. *Minerals* **2019**, *9*, 541. [CrossRef]
4. Zhang, R.; Li, T.; Zhou, S.; Deng, X. Joint MT and Gravity Inversion Using Structural Constraints: A Case Study from the Linjiang Copper Mining Area, Jilin, China. *Minerals* **2019**, *9*, 407. [CrossRef]
5. Steiner, B.M.; Rollinson, G.K.; Condron, J.M. An Exploration Study of the Kagenfels and Natzwiller Granites, Northern Vosges Mountains, France: A Combined Approach of Stream Sediment Geochemistry and Automated Mineralogy. *Minerals* **2019**, *9*, 750. [CrossRef]
6. Harmon, R.S.; Lawley, C.J.; Watts, J.; Harraden, C.L.; Somers, A.M.; Hark, R.R. Laser-Induced Breakdown Spectroscopy—An Emerging Analytical Tool for Mineral Exploration. *Minerals* **2019**, *9*, 718. [CrossRef]
7. Mao, X.; Zhang, W.; Liu, Z.; Ren, J.; Bayless, R.C.; Deng, H. 3D Mineral Prospectivity Modeling for the Low-Sulfidation Epithermal Gold Deposit: A Case Study of the Axi Gold Deposit, Western Tianshan, NW China. *Minerals* **2020**, *10*, 233. [CrossRef]
8. Battalgazy, N.; Madani, N. Stochastic Modeling of Chemical Compounds in a Limestone Deposit by Unlocking the Complexity in Bivariate Relationships. *Minerals* **2019**, *9*, 683. [CrossRef]
9. Chen, Y.; Wu, W.; Zhao, Q. A Bat-Optimized One-Class Support Vector Machine for Mineral Prospectivity Mapping. *Minerals* **2019**, *9*, 317. [CrossRef]
10. Steiner, B.M. Tools and Workflows for Grassroots Li–Cs–Ta (LCT) Pegmatite Exploration. *Minerals* **2019**, *9*, 499. [CrossRef]

Article

Deep Mineral Exploration of the Jinchuan Cu–Ni Sulfide Deposit Based on Aeromagnetic, Gravity, and CSAMT Methods

Jianmin Zhang [1,2], Zhaofa Zeng [1,2,*], Xueyu Zhao [1,2], Jing Li [1,2,*], Yue Zhou [1,2] and Mingxu Gong [1,2]

1 College of Geo-Exploration Science and Technology, Jilin University, Changchun 130026, China; zjm16@mails.jlu.edu.cn (J.Z.); zhaoxueyu035035@163.com (X.Z.); zhouyue17@mails.jlu.edu.cn (Y.Z.); gongmx18@mails.jlu.edu.cn (M.G.)

2 Key Laboratory of Applied Geophysics, Ministry of Natural Resources of PRC, Changchun 130026, China

* Correspondence: zengzf@jlu.edu.cn (Z.Z.); inter.lijing@gmail.com (J.L.)

Received: 31 December 2019; Accepted: 11 February 2020; Published: 13 February 2020

Abstract: The exploration of deep mineral resources is an important prerequisite for meeting the continuous demand of resources. The geophysical method is one of the most effective means of exploring the deep mineral resources with a large depth and a high resolution. Based on the study of the geological background, petrophysical properties, and aeromagnetic anomaly characteristics of the Jinchuan Cu–Ni sulfide deposit, which is famous throughout the world, this paper uses the widely used gravity, aeromagnetic, and CSAMT (controlled source audio-frequency magnetotellurics) methods with a complementary resolution to reveal the favorable prospecting position. In order to obtain better inversion results, the SL0 norm tight support focusing regularization inversion method is introduced to process the section gravity and aeromagnetic data of the mining area. By combining the results with CSAMT, it is found that the medium-low resistivity, high density, and the high magnetic anomaly areas near the structural belt can nicely correspond with the known ore-bearing rock masses in the mining area. At the same time, according to the geophysical exploration model and geological and physical property data, four favorable ore-forming prospect areas are delineated in the deep part of the known mining area.

Keywords: Jinchuan Cu–Ni sulfide deposit; deep mineral exploration; CSAMT; inversion

1. Introduction

Mineral resources are the material basis of human science and technology progress and social and economic development [1]. In recent years, with the continuous and stable growth of China's economy, the contradiction between the supply and demand of mineral resources has become increasingly prominent, among which all kinds of metal mineral resources are in short supply. The effective way to alleviate the shortage of resource supply is to carry out deep prospecting and strengthen the resource reserve. The Jinchuan Cu–Ni sulfide deposit is one of the three largest Ni–Cu–(PGE) deposits in the world. It is of great significance to carry out deep prospecting in this area to ensure the supply of copper and nickel resources. Moreover, the available geological data show that the deep parts of the known mining area and the surrounding area have a good prospecting potential, mainly based on the following: according to the metallogenic model, structural characteristics, and the spatial location of ore-bearing magma emplacement, there may be more ore bodies in the deep part of the mining area; the lower end of the main rock mass in the first, second, and third mining areas is not completely revealed; the new ore body is indeed found in the deep of the II mining area [2]. An independent ore body has been found in the surrounding rock at the bottom of No.24 ore body, with an increase of more than 600,000 tons of nickel metal. It is believed that there may be a relatively rich copper–nickel sulfide ore bodies in the surrounding rocks in the footwall direction of the western rock bodies of Jinchuan, and there may be a large number of sulfide residues in the deep magma chamber corresponding to the eastern rock body [3]. For the surrounding area of the mining area, the previous data show that the rock mass of the mining area shows the characteristics of echelon arrangement in space, and there are new ore-bearing rock masses near the main ore body, such as No. 58 ore body of the third mining area [4].

The deep prospecting methods mainly include geological, geophysical, geochemical, and drilling methods. The geological prospecting method relies on the observation and analysis of surface outcrop or drilling core to infer the underground geological conditions, which has limitations for deep prospecting, especially for the concealed characteristics. Geochemical exploration is an important technical support condition for deep prospecting, and drilling engineering is the realization condition. However, the depth and scope that these two methods can reach is also limited. Geophysical exploration is the basic means to obtain the information of concealed parts beyond other prospecting methods, which has a large detection depth, a high resolution, and various means [5–8]. It can carry out multi-scale detection in the target area and provide rich information for deep ore prospecting [9,10]. Kheyrollahi et al. [11] discovered and predicted the distribution pattern of porphyry copper deposits in the tertiary magmatic belt by the upward extension and boundary enhancement of magnetic anomalies. Xiao and Wang [12] used Bouguer gravity and aeromagnetic data to further understand the geological and mineral resources near the porphyry copper molybdenum polymetallic mineralization in the Tianshan area of China. Hu et al. [13] explored potential iron and polymetallic lead–zinc–copper deposits in the Longmen area by the CSAMT method and found high-grade lead–zinc–silver–titanium ore through drilling based on inversion results. Guo et al. [14] applied the CSAMT method to the exploration of the Jianshui lead–zinc mine and drew the underground resistivity distribution map through data processing and inversion. According to the CSAMT results, the location of the ore body is inferred, and the results are verified by drilling. The lead–zinc ore body is 373.70–407.35 m in the well. Shah et al. [15] comprehensively used aeromagnetic induced polarization, magnetotelluric and borehole geological alteration, magnetic susceptibility, and density data to explore the copper–gold molybdenum Pebble porphyry deposit, and achieved good prospecting results. In order to accurately detect the underground structure of complex deposits and solve the problems of uniqueness and inconsistency in the single parameter inversion model, Zhang and Li [16] proposed a two-dimensional gravity gradient and a magnetotelluric joint inversion method based on data space and normalized cross gradient constraints. Melo et al. [17] proposed a geological characterization method that can identify copper deposits based on geophysical inversion. This method can use geophysical data and sparse geological information to evaluate the target quickly, especially for the first stage of deep target or concealed target exploration. The success of this method is verified by the inversion of magnetic data and the direct current resistivity data of Cristalino iron oxide copper gold deposit in northern Brazil. Lee et al. [18] obtained the resistivity model consistent with the regional geology through the 3D joint inversion of magnetotelluric and Z-axis tipper electromagnetic data, which not only shows the mineralization belt interpreted for the Morrison porphyry Cu–Au–Mo deposit but is also conducive to the exploration of the disseminated sulfide of other porphyry deposits.

Some researchers also used geophysical methods to carry out prospecting work in the Jinchuan copper–nickel mining area and its surrounding areas and obtained some knowledge or achievements. Through the joint interpretation of gravity and magnetic data, it is considered that the M-15 anomaly is caused by ultrabasic rocks with a buried depth of more than 1200 m, which has a positive effect on the indication of deep Cu–Ni deposits [19]. According to the comprehensive prospecting model of geology, geophysics, and geochemistry, Wen and Luo [20] carried out prospecting and prediction work in the deep and edges of the Jinchuan copper-nickel mining area and found five potential target areas. In 2006, Fu and Li [21] established a comprehensive geological geophysical prospecting model based on the characteristics of the geophysical geochemical field of rock masses and ore deposits in different mining areas. On the basis of a systematic analysis of metallogenic geological conditions and comprehensive geophysical, geochemical, and remote sensing information, Gao [2] believed that the joint area of the first and second mining areas, the joint area of the I and II mining areas, the joint area of No.1 and No.2 ore bodies in the II mining area, and the overlap areas of geophysical and geochemical anomalies in the III mining area are important locations for ore body tracing.

On the basis of systematically summarizing the geological background of mineralization, this paper uses the aeromagnetic and gravity methods with low exploration cost and high efficiency, and the CSAMT method with a large exploration depth and a high vertical resolution to indicate the favorable metallogenic locations in the deep of the Jinchuan Cu–Ni deposit and its surrounding area. In order to overcome the weakness of deep signal in deep gravity and magnetic exploration, the processing method of potential field data is studied. Additionally, the focus inversion method based on the SL0 norm with a good convergence effect is used to process the aeromagnetic or gravity data of the profile, and, based on these results and the CSAMT inversion results, geological and physical properties data and geophysical profile data are interpreted. The results show that the range of the abnormal bodies corresponds well with the ore-bearing bodies of the known mining areas, and four favorable metallogenic targets are delineated according to the results.

2. Geological Background of the Survey Area

The Jinchuan Cu–Ni sulfide deposit is a part of the Longshoushan metallogenic belt, which is located in Longshoushan terrane in the southwest of the Alxa block of the North China Craton (NCC) [22,23]. The NCC is one of the three major Precambrian blocks in China, which formed from an amalgamation of micro blocks [24–26]. The Alxa block is located in the westernmost part of the NCC and is in fault contact with the Tarim craton to the west, bounded by the North Qilian orogenic belt in the south and the Central Asia orogenic belt in the north (Figure 1a) [27,28]. The Longshoushan terrane is a long narrow northwest trending uplift. It is 195 km long and 30–35 km wide, which is controlled by deep faults on both sides of the north and the south (F_1 and F_2). Its north side is adjacent to the Chaoshui basin, and its south side is separated from the Qilian orogenic belt (Figure 1b). In the Longshoushan terrance, the main outcropping strata are the Paleoproterozoic Longshoushan group, the late Mesoproterozoic Dunzigou group [29], and the Neoproterozoic–Cambrian Hanmushan group [30,31]. The Longshoushan group is the oldest metamorphic basement in the Longshoushan terrane [32]. Strong metamorphism and deformation in multiple periods [33] make the strata fragmented and it is difficult to judge the original stratigraphic sequence [34]. It mainly consists of schlieren and homogenic migmatites, marble, biotite-plagioclase gneiss, granulites, quartz schist, amphibolite, and pyroclastic rock [27,35]. The Dunzigou group is the earliest sedimentary overlying strata in the Longshoushan terrane, which is in angular unconformity contact with the lower Longshoushan group. In the geological history, the Longshoushan terrane has experienced multiple structural changes. The present NW trending faults and folds are mainly Caledonian and later structures. Faults in the Longshoushan terrane are mainly NW- and NE-trending structures, cutting the metamorphic formation [23]. Magmatic rocks are widely developed in the region, and magmatism occurred from Paleoproterozoic to Neoproterozoic, mainly in the Paleoproterozoic and the Paleozoic [36–39]. With regard to the formation age of Jinchuan intrusion, different researchers have adopted different methods to obtain a large number of isotopic age data, which can be roughly divided into two ranges: 1400–1600 Ma and 800–1000 Ma, representing Mesoproterozoic and Neoproterozoic, respectively [40,41].

The ore bearing ultrabasic rock bodies unconformity intrudes into the Baijiazuizi formation of pre-Great Wall system, which is in direct contact with gneiss, marble and banded migmatite, in the form of wall. It is about 6500 m long, 20–527 m wide on the surface (Figure 2a) and has a southwestern downward extension of more than 1000 m from the ground surface (Figure 2b) [28]. The overall strike is N50° W, inclined to the SW, with a dip angle of 50°–80°. The rock mass area is about 1.34 km^2. The ultramafic intrusion is divided into four sections by F_8, $F_{16\text{-}1}$, and F_{23} and is numbered as III, I, II, and IV from west to east (Figure 2a), which are corresponding to the four mining areas, respectively.

The three main ore bodies with proved reserves of great economic value are respectively hosted in No.1 and No.2 ore bodies of Segment II and No.24 ore body of Segment I, and No.58 ore body is hosted in an independent rock body in the southwest side of Segment III (Figure 2b).

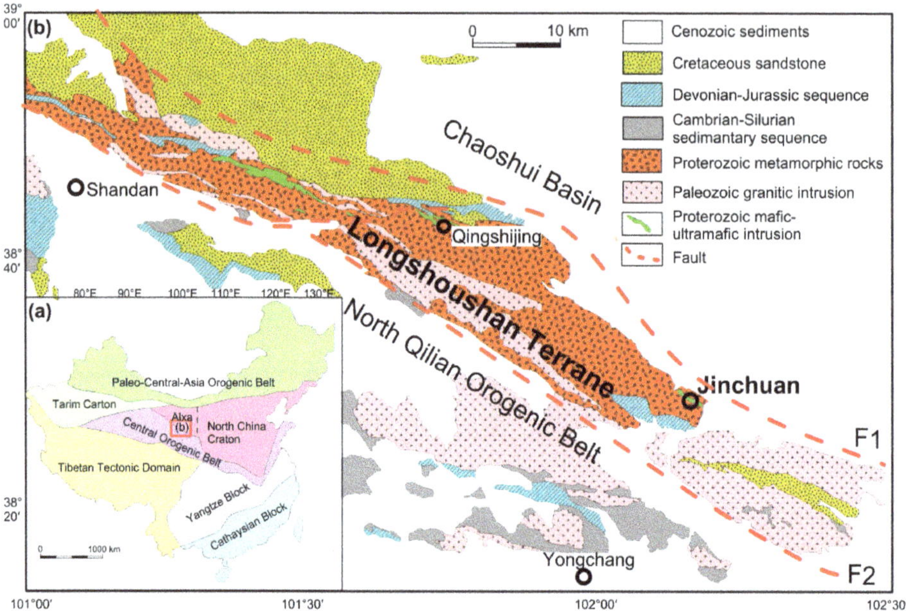

Figure 1. (**a**) The location of the study area and (**b**) a simplified geological map of the Longshoushan terrane. Both subfigures are based on [27,28].

The main strata in the mining area are the Baijiazuizi formation and the Quaternary system of the Longshoushan group of the pre-Great Wall system. The Quaternary sediments are mainly distributed in the east and west ends of the Jinchuan intrusion and the north of F_1. Baijiazuizi formation is the direct surrounding rock of the Jinchuan deposit, which is distributed in a NW–SE direction, consistent with the regional structural direction. Baijiazuizi formation underwent multiple magmatic intrusions and multiple metamorphisms, forming a series of rocks mainly composed of migmatite, gneiss, and marble (Figure 2a) [42].Among them, gneiss with stable chemical properties and poor water permeability is a good barrier layer, which enables the ore-forming materials to fully crystallize and differentiate in ultrabasic magma. Marble is active in chemical properties. It is favorable for the formation of contact metasomatic mineralization [43]. Therefore, Baijiazuizi formation is an important prospecting indicator.

The mining area has experienced structural activities many times. The structures of different periods and different directions superposed each other, making the mining area fold and fracture developed. The axial near the EW fold group includes the anticline where the deposit is located and a large syncline in the south of the mining area. The axial near the NE fold group is nearly vertical to the NW direction main structural lines of the mining area, among which the NE direction fold group across the ultramafic rock mass is the most significant, which plays an important role in the shape change and mineralization re-enrichment of the ore body, and the rich ore body is obviously thickened at the turning part of the fold. The NW trending faults are the most developed, followed by the NE and nearly EW trending faults. As one of the most important ore-controlling factors, faults not only control the emplacement of an ore bearing rock but also control the re-enrichment of mineralization and the spatial position of ore body. The NW ore-controlling faults are related to the spatial distribution of ore bodies. The NE or near EW ore controlling faults mainly cut the rock and ore bodies. The intersecting parts of faults in different directions can form irregular columnar ore bodies.

The ore-forming materials in the mining area mainly come from ultramafic magma. Ultrabasic rock is the ore-forming parent rock and the surrounding rock of the main ore body. The relationship between ultrabasic rock and mineralization is mainly reflected in the spatial change of rock mass and the relationship between lithofacies and mineralization. Only the spatial relationship between the two is introduced here. The shape of ore-bearing rock mass is irregular, and the development of the ore body is closely related to the floor. Generally, the concave part of the floor is favorable for the accumulation of ore-forming materials, and the ore body is thick. The occurrence of ore-bearing rock mass controls the occurrence of the stratoid ore body, the ore body is the same as the rock mass, and the strike is NW. The thickness of the ore-bearing rock mass is related to the thickness of the rich ore body. At the bottom of the thick rock mass, the rich ore body is also thick, which can be seen in the first and second mining areas.

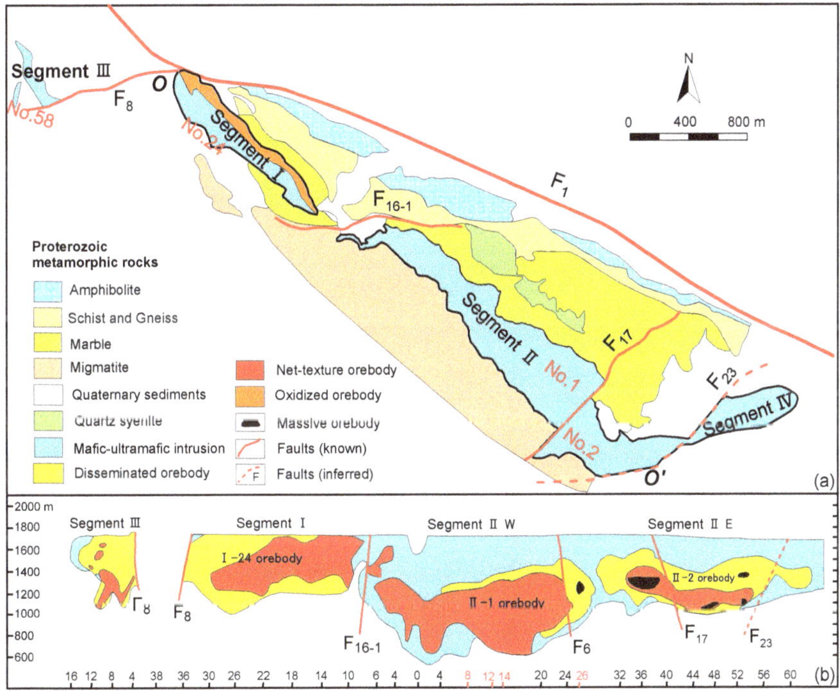

Figure 2. Geological map (**a**) and a cross section (**b**) of the Jinchuan intrusion, both subfigures are based on [28,42].

3. Aeromagnetic, Gravity, and CSAMT Surveys

Based on the characteristics of density, magnetism, and resistivity (Table 1), the rocks and ores in the mining area can be roughly divided into three categories. The first category is copper–nickel ore, showing the characteristics of high density, high magnetism, and low resistivity; the second category is ultrabasic rock, showing the characteristics of high density, strong magnetism, and medium resistivity; the third category is the rock surrounding the ultrabasic rock, with the characteristics of low density, weak magnetism, and high resistivity. The differences of these physical properties provide a precondition for geophysical exploration work such as gravity, magnetic, and electrical methods in the study area [21].

Table 1. Petrophysical properties in the Jinchuan Cu–Ni sulfide deposit [43], *Jr* represents remanent magnetization.

Rocks	Susceptibility $(k)/4\pi{*}10^{-6}$ SI	$Jr/10^{-3}$ A·m^{-1}	Density $(\sigma)/10^3$kg·m^{-3}	Resistivity$(\rho)/\Omega$·m
	Regular Value	**Regular Value**	**Regular Value**	**Regular Value**
Lherzolite	3900	900	2.72	320
Peridotite	2300	600	2.72	300
Migmatite	600	200		200
Granite	0	0	2.54~2.90	700
Gneissic granite	400	100	2.5	600
Biotite gneiss	0	0		
Marble	0	0	2.6	500
Amphibolite	200	200		376–1501
Tiny spotted ores	4300	800	2.73	62
Spotted ores	6100	500		90
Spongy ores	6600	1900	2.92	20

Convenient and efficient aeromagnetic exploration has been carried out in the main mining area. The aeromagnetic work uses the power glider as the carrier, and the measuring instrument is the helium optical-pumping magnetometer with a sensitivity of 0.001 nT. The average flight height is 93 m, the measurement scale is 1:10,000, the distance between survey lines is 100 m, and the distance between points is 2.7–3 m. The maximum dynamic noise level of the survey line in the survey area is 0.052 nT, most of which is less than 0.04 nT, and the average value is 0.027 nT, meeting the measurement requirements. Reduction to the pole can eliminate the asymmetry of the magnetic anomaly position caused by the declination and inclination of the magnetization field. After reducing magnetic anomaly to the pole, the anomaly information is more abundant, including the anomalies of different properties, scales, and depths. It is the basic data for anomaly interpretation. The induced magnetization of ultrabasic rocks and ores with high magnetic susceptibility in the Jinchuan Cu–Ni mining area is obviously greater than the residual magnetization [19], so reduction to the pole can be carried out, and the result is shown in Figure 3. The negative aeromagnetic anomaly of the mining area is located in the northeast, the isoline is relatively disordered, and the minimum negative anomaly is less than-140 nT. The positive aeromagnetic anomaly is mainly located in the southwest and central part, showing a significant northwest distribution. The known mining areas III, I, II, and IV (magenta curve range in Figure 3) are all in the high positive aeromagnetic anomaly area. The high positive anomaly in the III mining area is nearly circular, with a diameter of about 750 m, an area of about 0.4 km^2, and a maximum anomaly intensity of more than 350 nT. The high positive anomaly in the I mining area extends northwestward in a belt, with a length of about 1400 m, a width of about 600 m, and an area of about 0.75 km^2. The maximum anomaly value is located in the southeast end of the mining area, and the maximum anomaly value is more than 600 nT. The two ends of the high normal abnormal morphology in the II mining area have obvious distortion, but generally it is a strip extending northwestward, with a length of about 3000 m, a width of about 900 m in the west section, a width of about 650 m in the east section, an area of about 2.23 km^2, and the maximum abnormal value is more than 1500 nT. In addition, the cascade zones on both sides of the I and II mining areas are relatively steep, showing the characteristics of steepness in the northeast and slowness in the southwest, suggesting that the abnormal body is steeply inclined to the southwest. The corresponding high positive anomaly of the IV mining area has a irregular ellipse shape, with long axis in east-west direction, about 1000 m long, 850 m wide, and an area of about 0.75 km^2. Combined with the geological map and the characteristics of physical parameters of rocks and ores, the high positive aeromagnetic anomaly in the mining area is mainly caused by the ore bearing ultrabasic rocks, so the high positive aeromagnetic anomaly is an important indicator of ultrabasic rocks.

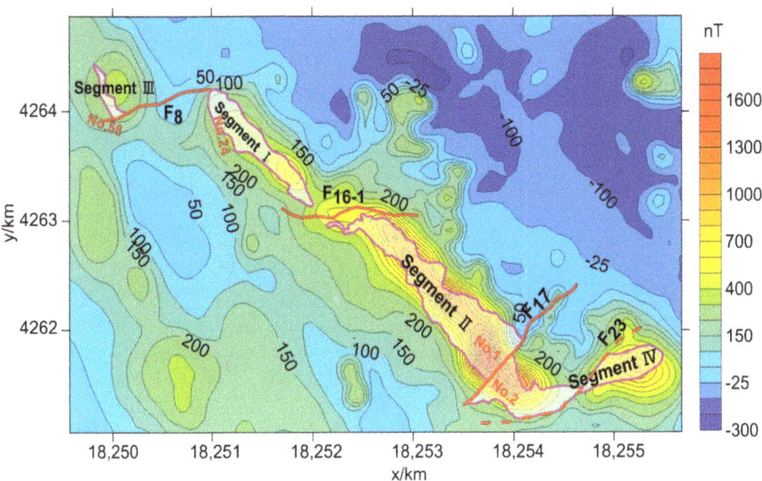

Figure 3. Aeromagnetic anomaly map of the Jinchuan Cu–Ni sulfide deposit after reduction to the pole.

In order to understand the scope of deep ore-bearing rock masses in the mining area, we use the boundary enhancement method presented by Zhang et al. [44] to detect the boundaries of aeromagnetic anomaly in the mining area. It can be seen from the results that the scope of the boundaries of the underground rock masses is determined (Figure 4), especially the edge positions of the deep rock masses of the four mining areas completely covered by the Quaternary are delineated, which provides the exploration scope for the prospecting of the deep Cu–Ni deposit. In order to better understand the characteristics of the deep rock anomalies, we first introduce the dual-tree complex wavelet into the multi-scale anomaly separation of aeromagnetic anomalies. The dual-tree complex wavelet not only has the advantages of wavelet transformation but also the characteristics of approximate translation invariance, more directional selectivity, and limited data redundancy. The results show that with the increase of decomposition scale, the detail information in the shallow part decreases gradually (Figure 5). The range of the high normal anomaly in the second mining area in the southwest side is gradually expanding, indicating that with the increase of burial depth, the range of ore bearing ultrabasic rock masses is gradually expanding, which shows that the deep mining area has a good prospecting potential.

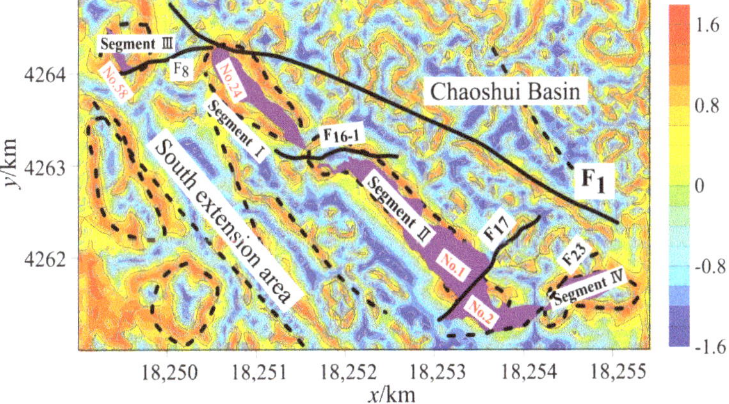

Figure 4. Boundaries detection results of aeromagnetic anomalies in Figure 3.

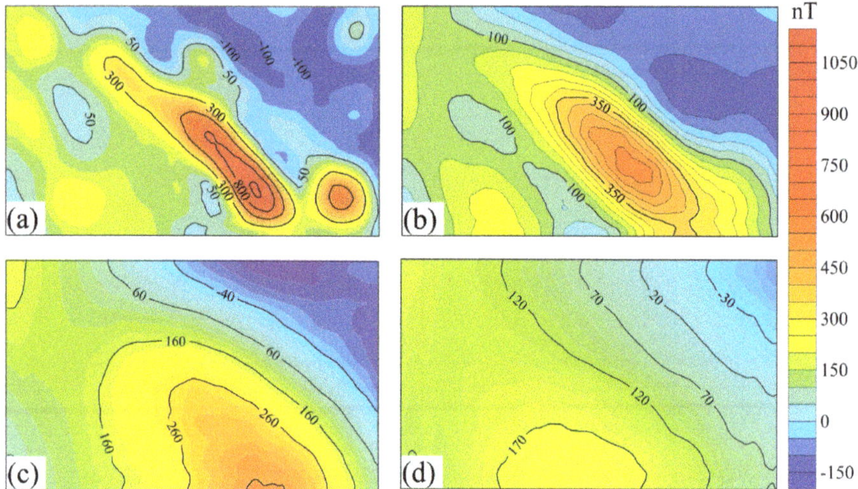

Figure 5. Multiscale separation results (**a–d**) of the aeromagnetic anomaly in Figure 3 using a dual-tree complex wavelet.

On this basis, we use the gravity and CSAMT data to acquire the locations of the deep potential ore-bearing rock masses in the II and IV mining areas, and the locations of the survey lines are shown in Figure 6. The profiles of gravity and CSAMT data are designed according to the characteristics of magnetic anomalies and the structure of the mining area. The direction of profile Lmg1~3 is 38.38° and that of Lmg4 is 0°. The parameters of CSAMT measurement of the four profiles are determined according to the proposed exploration depth. The minimum receiving and transmitting distance is 12 km, and the maximum is about 16 km. The power supply electrodes are arranged parallel to the survey line, the electrodes distance is 2 km, and the azimuth error is less than 3°. The sampling frequency is 1–9600 Hz, and there are 41 sampling frequency points. The station distance is 50 m. The quality inspection of the CSAMT measurement adopts the method of data observation on the inspection point again. The data quality evaluation is to calculate the mean square relative error of the resistivity of the inspection point. The mean square relative error of the resistivity of the single point in this work is 1.3–4.9%, which meets the specification and design requirements of less than or equal to 5%. The scale of gravity profile (Lmg-1 and 4) work is 1:5000, and the distance between profile points is 20 m. The instrument used is a CG-5 high-precision gravimeter with a reading resolution of 1 μgal. Before field operation, in addition to various checks and adjustments, static and dynamic tests are carried out to ensure the good performance of the gravimeter in use. The total mean square error of the Bouguer gravity anomaly is $\pm 0.079 \times 10^5$ m/s^2, which meets the design requirements and has reliable quality.

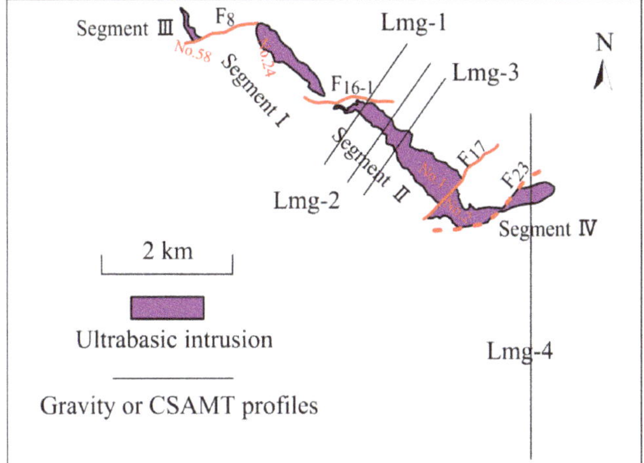

Figure 6. Locations of controlled source audio-frequency magnetotellurics (CSAMT) or gravity exploration lines in the study area.

4. Inversion Methods

In this part, the section gravity, corresponding aeromagnetic data and CSAMT data are inverted to understand the distribution of different geological bodies in the depth of the second and fourth mining areas, so as to predict the favorable prospecting prospect combining with geological data. In this paper, the constraint inversion method based on the SL0 norm tight support is used to process the section gravity and aeromagnetic data.

The following is the principle of tightly supported focused inversion based on the SL0 norm.

If **m** is model space and **d** is data space, the relationship between the two is **F**, and the forward process is expressed as:

$$\mathbf{d} = \mathbf{F}\mathbf{m} \tag{1}$$

The inversion is expressed as:

$$\mathbf{m} = \mathbf{F}^{-1}\mathbf{d} \tag{2}$$

Potential field data inversion is an underdetermined problem. In order to reduce multiple solutions, the Tikhonov regularization method is commonly used. The inversion process can be written as:

$$P^{\alpha}(\mathbf{m}) = \varphi(\mathbf{m}) + \alpha s(\mathbf{m}) \tag{3}$$

Among them, $\varphi(\mathbf{m})$ is the two norm of the difference between the observed data and the theoretical forward data, α is the regularization parameter, $s(\mathbf{m})$ is the stabilizer, which represents the model objective function based on the prior information constraint. In this paper, the minimum compactly supported functional is used as the stabilizer, which can make the inversion result have a better focusing effect [45].

The integral equation is used to express the minimum support functional stabilizers as follows:

$$s_{MS}(\mathbf{m}) = \left\{ \frac{\mathbf{m} - \mathbf{m}_{apr}}{\left[\left(\mathbf{m} - \mathbf{m}_{apr}\right)^2 + e^2\right]^{\frac{1}{2}}}, \frac{\mathbf{m} - \mathbf{m}_{apr}}{\left[\left(\mathbf{m} - \mathbf{m}_{apr}\right)^2 + e^2\right]^{\frac{1}{2}}} \right\} = \min \tag{4}$$

Among them, $\{\boldsymbol{a}, \boldsymbol{b}\}$ represents the internal product of $\boldsymbol{a}$ and $\boldsymbol{b}$, which is the focusing factor, and e is the focus factor, which is related to the focus effect of the inversion output. $\mathbf{m}_{apr}$ is a model based on prior information.

To simplify the above formula, variable weight functional $\omega_e(\mathbf{m})$ is introduced, which is expressed as follows:

$$\omega_e(\mathbf{m}) = \frac{1}{\left[\left(\mathbf{m} - \mathbf{m}_{apr}\right)^2 + e^2\right]^{\frac{1}{2}}} \tag{5}$$

Equation (4) can be changed to:

$$s_{MS}(\mathbf{m}) = \left\{\omega_e(\mathbf{m})\left(\mathbf{m} - \mathbf{m}_{apr}\right), \ \omega_e(\mathbf{m})\left(\mathbf{m} - \mathbf{m}_{apr}\right)\right\} = ||\mathbf{m} - \mathbf{m}_{apr}||^2_{\omega_e} \tag{6}$$

Then, the objective function can be written as:

$$\begin{aligned} P^{\alpha}(\mathbf{m}, \mathbf{d}) &= ||\mathbf{W}_d\mathbf{A}(\mathbf{m}) - \mathbf{W}_d\mathbf{d}||^2 + \alpha||\mathbf{W}_m\mathbf{m} - \mathbf{W}_m\mathbf{m}_{apr}||^2_{\omega_e} \\ &= (\mathbf{W}_d\mathbf{A}(\mathbf{m}) - \mathbf{W}_d\mathbf{d})^T(\mathbf{W}_d\mathbf{A}(\mathbf{m}) - \mathbf{W}_d\mathbf{d}) + \alpha\left(\mathbf{W}_e\mathbf{W}_m\mathbf{m} - \mathbf{W}_e\mathbf{W}_m\mathbf{m}_{apr}\right)^T\left(\mathbf{W}_e\mathbf{W}_m\mathbf{m} - \mathbf{W}_e\mathbf{W}_m\mathbf{m}_{apr}\right) \end{aligned} \tag{7}$$

where $||\mathbf{W}_d\mathbf{A}(\mathbf{m}) - \mathbf{W}_d\mathbf{d}||^2$ is the fitting difference, $||\mathbf{W}_m\mathbf{m} - \mathbf{W}_m\mathbf{m}_{apr}||^2_{\omega_e}$ is the stabilizer, $\mathbf{W}_e$ is the change matrix, which depends on $\mathbf{m}$, $\mathbf{W}_d$ and $\mathbf{W}_m$ are the weighting matrix of the traditional data space and the model space, respectively. In this paper, $\mathbf{W}_d$ and $\mathbf{W}_m$ are, respectively, as follows:

$$\mathbf{W}_m = \mathbf{diag}\left(\mathbf{A}^T\mathbf{A}\right)^{1/2} \tag{8}$$

$$\mathbf{W}_d = \mathbf{diag}\left(\mathbf{A}\mathbf{A}^T\right)^{1/2} \tag{9}$$

The objective function given by Equation (7) is similar to the traditional objective function form. The difference is that a variable weight matrix needs to be introduced into the model parameters of Equation (7). This paper uses the conjugate gradient method to solve the problem of parameter functional minimization given by Equation (7). In Equation (5), $\omega_e(\mathbf{m})$ can be regarded as a regularization parameter α, and since it also has a focusing effect, it can be called a regularization-focusing factor. As this parameter becomes smaller, the corresponding stabilizer can minimize the non-zero deviation of the model parameters from the prior information.

The smooth L0 algorithm (SL0 algorithm) comes from the sparse signal recovery theory and is used to solve the problem of how to accurately solve $\mathbf{m}$ in the inverse problem. It uses a suitable smooth continuous function to approximate the discontinuous L0 norm and minimizes it by using a minimization algorithm on the smooth function, thereby obtaining the minimum L0 norm and obtaining a sparse solution. In this paper, a Gaussian function with an expected value of 0 is selected to approximate the smooth function of the L0 norm. The continuous function be expressed as:

$$f_{\sigma}(m) = \frac{\sigma^2}{(m^2 + \sigma^2)} \tag{10}$$

Among them, σ represents the approximate degree of continuous and discontinuous L0 norm. Then there are:

$$\lim_{\sigma \to 0} f_{\sigma}(m) = \begin{cases} 1, m = 0 \\ 0, m \neq 0 \end{cases} \tag{11}$$

Or approximately:

$$f_{\sigma}(m) \approx \begin{cases} 1, |m| \ll \sigma \\ 0, |m| \gg \sigma \end{cases} \tag{12}$$

Define a new function:

$$E_\sigma(m) = \sum_{i=1}^{M} f_\sigma(m_i) \tag{13}$$

Then:

$$\lim_{\sigma \to 0} E_\sigma(m) = M - \|m\|_0 \tag{14}$$

The above formula shows that $\|m\|_0 \approx M - F_\sigma$ is true when σ is small, and, when $\sigma \to 0$, this approximate relationship tends to be equal. Therefore, in order to find the solution with the smallest L0 norm, we can take a small value of σ, and make $F_\sigma(m)$ the maximum. For small values of σ, F_σ is highly uneven and contains many local maxima, so it is difficult to maximize it. For a large value of σ, F_σ is smooth and contains fewer local maxima, so it is easier to maximize it. In order to have the largest F_σ for any value of σ, this paper uses a decreasing sequence of σ to maximize F_σ. For each σ with a large front value, the initial value of the maximization algorithm of F_σ is the maximum value of the corresponding F_σ. When σ gradually decreases, the initial value of F_σ corresponding to each σ starts from the maximum value close to the actual F_σ. Therefore, the SL0 algorithm does not fall into the local maximum problem and can find the actual maximum value of F_σ for a small value of σ, and give the solution of the smallest L0 norm. Compared with a tightly-supported focused inversion, SL0 norm-constrained tightly-supported focused inversion continuously adjusts $\mathbf{W}_e$ based on a priori information in the form of a weighting function, making the inversion results more accessible to actual physical parameter models.

Therefore, the objective function of the inversion method based on SL0 norm tight support focus can be expressed as follows:

$$\begin{aligned} P^{\alpha}_{SL0}(\mathbf{m},\mathbf{d}) &= \|\mathbf{W}_d\mathbf{A}(\mathbf{m}) - \mathbf{W}_d\mathbf{d}\|^2 + \alpha\|\mathbf{W}_m\mathbf{m} - \mathbf{W}_m\mathbf{m}^{SL0}_{apr}\|^2_{\omega_e} \\ &= (\mathbf{W}_d\mathbf{A}(\mathbf{m}) - \mathbf{W}_d\mathbf{d})^T(\mathbf{W}_d\mathbf{A}(\mathbf{m}) - \mathbf{W}_d\mathbf{d}) + \alpha(\mathbf{W}_m\mathbf{m} - \mathbf{W}_m\mathbf{m}^{SL0}_{apr})^T(\mathbf{W}_m\mathbf{m} - \mathbf{W}_m\mathbf{m}^{SL0}_{apr}) \end{aligned} \tag{15}$$

For the CSAMT data, this paper uses the widely used conventional SCS2D software for inversion. This program is an active audio magnetotelluric data processing program developed on the basis of magnetotelluric data processing. Its development level is relatively mature. The suitability selection of SCS2D software inversion parameters is an important part of data processing. The correct selection of inversion parameters will directly affect the accuracy of subsequent data interpretation. The initial background model selected in this paper is a moving average model.

5. Inversion Results and Interpretation

In this section, the section aeromagnetic and gravity data of the II and IV mining areas are inversed based on the SL0 tight support focus inversion method. At the same time, combined with the CSAMT inversion results and geological and rocks' physical properties, the corresponding structures of the survey lines are inferred and interpreted, and the favorable positions of deep mineralization are delineated.

5.1. Survey Results of the II Mining Area

The II mining area is located in the southeast of the $F_{16\text{-}1}$ fault and the northwest of the No.56 exploration line, with the largest copper–nickel ore body developed in the Jinchuan Cu–Ni deposit. The ore bearing strata are mainly pre-Sinian Baijiazuizi formations. Faults are developed, mainly including three groups of faults in the NW, NE, and nearly EW directions. The ultrabasic rock body is the ore-bearing parent rock. Under the control of the project, the rock body is in the shape of a rock wall, trending to the northwest, inclining to the southwest, with a length of more than 3000 m. the horizontal thickness is shown as thin at both ends and thick in the middle, up to 1550 m. West of line 26 of the mining area, the occurrence of the ore body is relatively steep, plate like and lens like, and is in the form of a completely intrusive contact or mixed gradual intrusive contact with the early lithofacies.

The ore body to the east of line 26 is in lenticular or stratoid shape. For the deep prospecting of the mining area, the previous study shows that the deep part of No.2 ore body in the II mining area has the possibility of a branch compound and a pinch-out reappearance of ore body. The main reason for this is that the extension of No.2 ore body below the 1000 m level is not revealed, and the geological sections of line 28–30 show that the ore body below the 1100 m level has not been pinched out [2]. Based on the multiple geophysical data, this paper investigates the deep of No.1 ore body so as to find out whether there is the possibility of new ore body in the deep.

The three survey lines of Lmg-1–3 arranged in No.1 ore body of the II mining area coincide with the No.8, 12, and 14 exploration lines, respectively, which are close to each other and are arranged in parallel along the southeast direction. It is of great significance for indicating the deep resistivity, density, and magnetic variations in the profiles. Gravity and CSAMT explorations have been carried out along the Lmg-1 line, respectively. Figure 7 shows the inversion results of gravity and corresponding aeromagnetic data by using the SL0 method. The density and magnetism of the media under the line are obviously different. There are two obvious high-density abnormal areas in the profile, which extend to the deep of the southwest part. The high-magnetism abnormal areas correspond to the high-density abnormal areas and show similar changing characteristics towards the deep. The inversion result of CSAMT shows that the high resistivity areas are distributed on both sides of the profile, the medium-low resistivity areas are mainly located in the middle, tend to the southwest, the dip angle changes from steep to slow, and the extension is large, which is consistent with the high density and high magnetic areas in Figure 7 (Figure 8). CSAMT exploration was also carried out along the Lmg-2 and Lmg-3 lines. The resistivity inversion results and the corresponding aeromagnetic anomaly inversion results show similar resistivity and magnetism distribution characteristics to Lmg-1 (Figure 9).

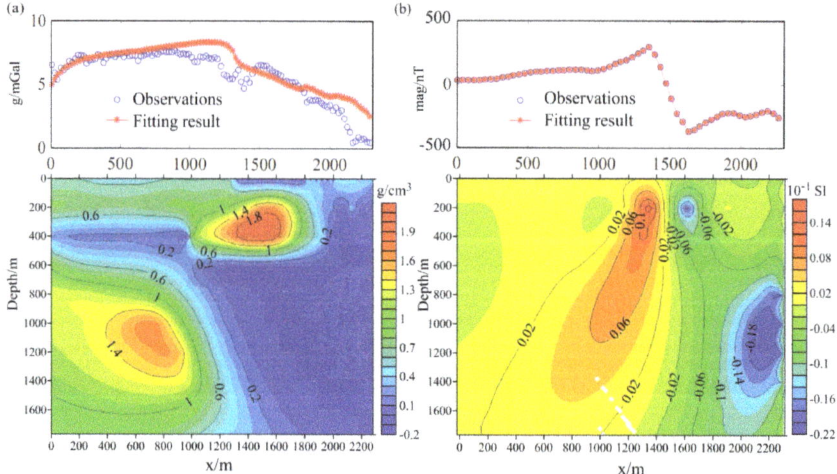

Figure 7. Inversion results of the gravity (**a**) and aeromagnetic (**b**) data of Lmg-1.

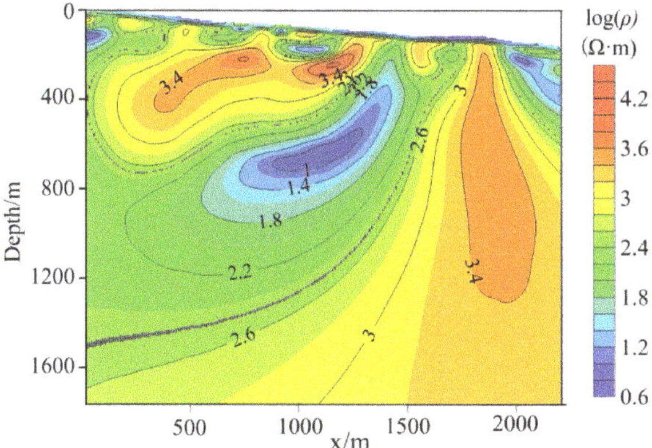

Figure 8. CSAMT data inversion result of the Lmg-1 profile.

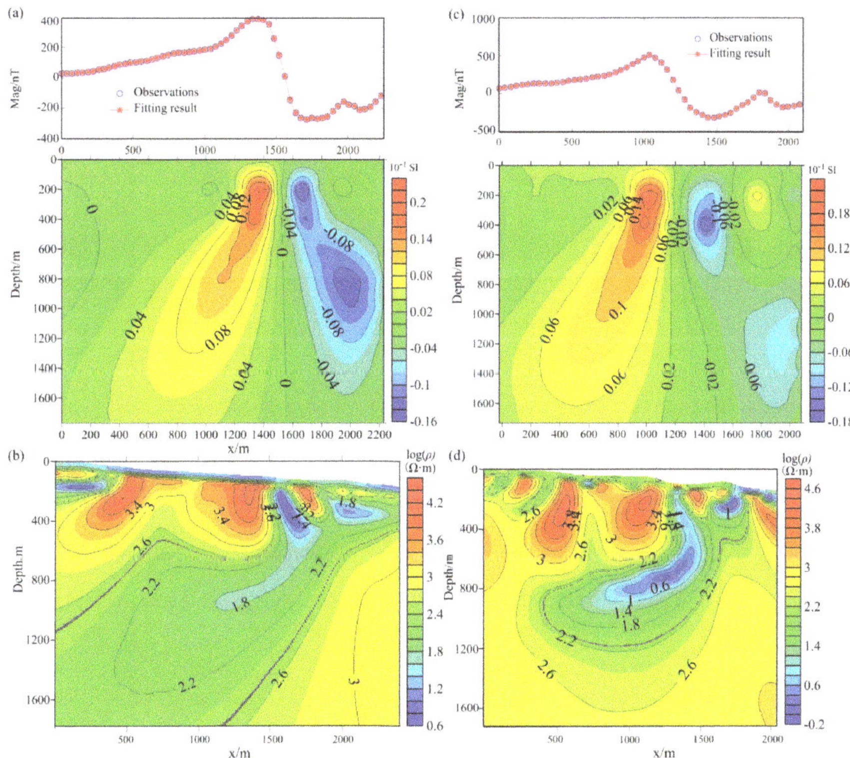

Figure 9. Inversion results of the aeromagnetic (**a**) and CSAMT (**b**) data of Lmg-2, and the inversion results of the aeromagnetic (**c**) and CSAMT (**d**) data of Lmg-3.

Combined with the geological map, the petrophysical properties of the study area and the results of the known exploration profiles, it is obvious that the high resistivity, low density, and low magnetism areas on both sides of the profiles are caused by the Baijiazuizi formation, the shallow medium-low resistivity, high density, and high magnetism areas are caused by the ore-bearing ultrabasic rock masses, and the transition zone on both sides of the middle abnormal area are the fault zones where ultrabasic rocks intrude into Baijiazuizi formation, as shown by the red dotted line in Figure 10. At the same time, we can see that the single geophysical method has limitations. For example, the resolution of CSAMT in the deep of the profile is insufficient, and it cannot clearly indicate the location of the deep abnormal target body. Therefore, we have roughly determined the target locations based on the high-density, high-magnetism, medium-low-resistivity geophysical exploration model, and the metallogenic law that is easy to form ore at a low-lying structure place, and we have inferred that the favorable metallogenic location of the line Lmg-1 is about 400–900 m and the burial depth is about 1100–1500 m, the favorable metallogenic area of the line Lmg-2 is about 800–1300 m and the burial depth is about 1200–1700 m, and the favorable metallogenic area of the line Lmg-3 is about 500–1000 m and the burial depth is about 1000–1500 m. The approximate location of the target body is shown by the black dotted line in Figure 10, and three drilling verification locations are designed.

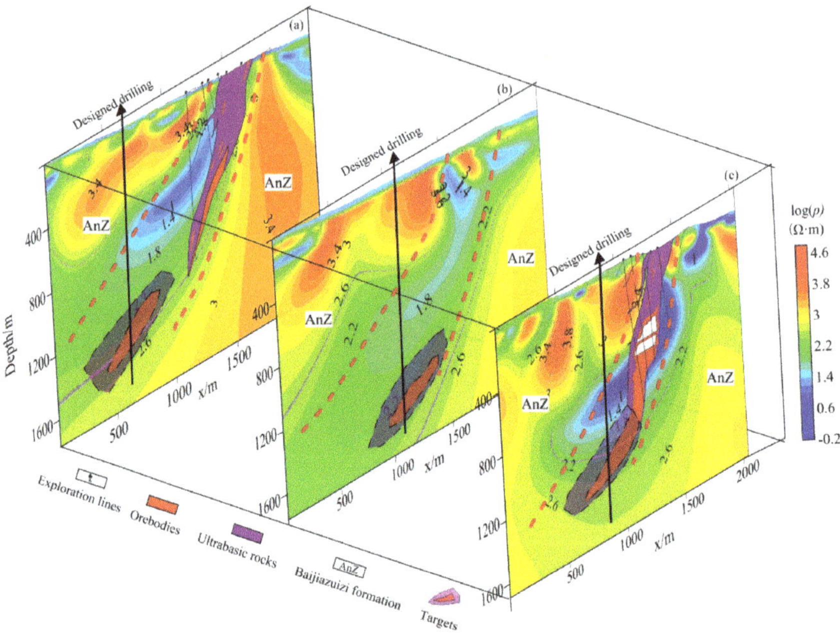

Figure 10. Interpretation results of gravity, aeromagnetic, and CSAMT data and the prediction of deep targets; the targets were indirectly determined based on the inferred locations of the ultrabasic rocks. (**a**) Lmg-1; (**b**) Lmg-2; (**c**) Lmg-3.

5.2. Survey Results of the IV Mining Area

The known ultramafic rock mass in the fourth mining area is 1160 m long and completely covered by the Quaternary. The thickness of the cover layer is 60–140 m. The Baijiazuizi formation and ore-bearing ultrabasic rock mass are developed below the cover layer. The ore body is dominated by low grade ore, and No.1 ore body is the main ore body. It is produced in the concave section of the bottom of the rock body and is lenticular. The upper and lower parts are small and the middle is large. The strike of the Lmg-4 exploration line in the mining area is just south, passing through No.1 ore body, close to the No.10 exploration line. Gravity and CSAMT surveys were carried out along this line, their lengths are slightly different.

Figure 11 is the inversion results of gravity and aeromagnetic anomalies based on the SL0 algorithm. The corresponding positions in the middle and lower parts of the profile have obvious large high-density and high-magnetic anomaly areas, and they all have the characteristics of extending to the upper left. The inversion result of CSAMT shows that the resistivity on the right side of the profile is relatively high, the resistivity in the middle is relatively low, and that their contact zone changes from steep to slow towards the deep (Figure 12). Compared with the inversion results in Figure 11, it was found that the high-density and high-magnetic area is consistent with the middle medium-low resistivity area. The inversion results are interpreted in combination with a geological map, the results of a nearby geological exploration line, and the physical parameters of the rocks. It can be seen that the shallow low-density, low-magnetic and low-resistivity areas of the profile are mainly caused by the Quaternary, while the high-resistivity area on the right side is the reflection of Baijiazuizi formation, the high-density, high-magnetic, and low-resistivity area in the upper part of the contact zone corresponds to the known ore-bearing ultrabasic rocks in the four mining areas. At the same time, it is inferred that the transition zone is a fault zone (red dotted line on the right side of Figure 12), which is the channel for ultrabasic magma to intrude into Baijiazuizi formation. Based on the geophysical prospecting model and metallogenic law, we speculate that the deep high density, high magnetism, and medium-low resistivity area in a large range near the channel is the favorable target area for prospecting, and the location is roughly within the survey line 2600–3200 m and its buried depth is 1200–1700 m. At the same time, the drilling hole location is designed, as shown in Figure 12.

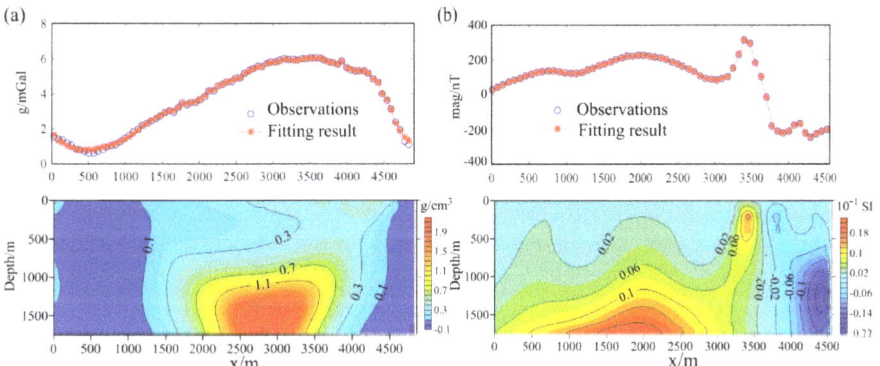

Figure 11. Inversion results of the gravity (**a**) and aeromagnetic (**b**) data of Lmg-4.

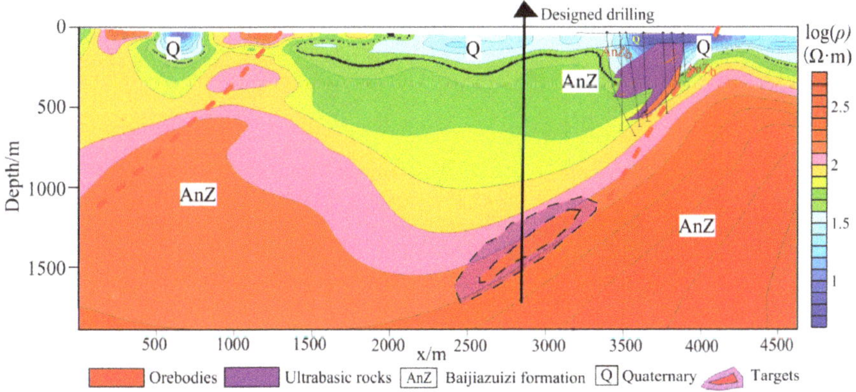

Figure 12. CSAMT data inversion result and interpretation of Lmg-4.

6. Discussion and Conclusions

The geophysical method is an important supporting technology for the exploration of deep metal mineral resources. In this paper, the deep exploration of the Jinchuan Cu–Ni sulfide deposit is carried out based on the gravity, aeromagnetic, and CSAMT methods with a complementary resolution. Among them, to overcome the ill-posed problem of inversion and reduce the multiplicity of solutions, the focus inversion method based on the SL0 norm is introduced to the inversion of gravity and aeromagnetic data, and the widely used SCS2D inversion software is used for the inversion of CSAMT data, both of which have achieved good inversion results. The medium-low resistivity, high density, and high magnetic areas shown in the inversion results can correspond to the known ore-bearing rocks in the shallow part well. In addition, four favorable target areas are delineated in the deep part of the mining area.

However, these methods are indirect and have some limitations. Because the ultrabasic rock is the parent rock and the surrounding rock of the ore body, the density and magnetism of the two are similar, and the ore body is often located in the middle or lower part of the ultrabasic rock body, so the geophysical signal produced by the ore body is easy to be covered by the ultrabasic rock body, meaning that it is difficult to distinguish the ultrabasic rock body and the ore body. Therefore, more geological data are needed to increase the reliability of target area prediction. In addition, the geophysical data in this paper are limited. It is expected to carry out seismic exploration in the study area and to obtain the velocity structure in the deep part of the mining area, so as to further supplement the supporting evidence of favorable prospective areas. At the same time, it is expected to carry out drilling work at the predicted favorable prospective areas, so as to further verify the results of geophysical deep exploration in this paper.

In general, it is of great significance to study the deep exploration of the Cu–Ni deposit in Jinchuan based on multiple geophysical methods. CSAMT, gravity, and aeromagnetic data not only indicate the known ore locations of the II and IV mining areas but also indicate the favorable ore locations in the deep. At the same time, this achievement provides a good reference for further deep exploration of the mining area and its surrounding areas, as well as a good demonstration of the feasibility and effectiveness of the geophysical methods used to detect deep metal mines.

Author Contributions: Conceptualization, Z.Z.; Methodology, Z.Z. and J.Z.; data curation, Z.Z. and J.Z.; software, X.Z.; validation, Z.Z., X.Z. and M.G.; formal analysis, J.L. and Y.Z.; writing—original draft preparation, J.Z.; writing—review & editing, Z.Z. and J.L.; project administration, Z.Z.; funding acquisition, Z.Z. and J.L. All authors have read and agreed to the published version of the manuscript.

Funding: This work is supported by the National Key Research and Development Program of China (2016YFC0600505), the National Natural Science Foundation of China (41874134) and Jilin Excellent Youth Fund (20190103142JH).

Acknowledgments: We would like to thank the editor and reviewers for their reviews that improved the content of this paper, and thank the Jinchuan Group Co., Ltd. for the providing of geological and geophysical data.

Conflicts of Interest: The authors declare no conflict of interest.

References

1. Teng, J.; Liu, J.; Liu, C.; Yao, J.; Han, L.; Zhang, Y. Prospecting for metal ore deposits in second deep space of crustal interior, the building of strategy reserve base of northeast China. *J. Jilin Univ. (Earth Sci. Ed.)* **2007**, *37*, 633–651.
2. Gao, Y. Study on Geological Characteristics, Temporal and Spatial Evolution, Prospecting in the Depth and Border of Jinchuan Deposit. Ph.D. Thesis, Lanzhou University, Lanzhou, China, 2009.
3. Song, X.; Hu, R.; Chen, L. Characteristics and inspirations of the Ni-Cu sulfide deposits in China. *J. Nanjing Univ. (Nat. Sci.)* **2018**, *54*, 221–235.
4. Wei, Z. Analysis on the Genetic Type and Metallegenic Prognosis about Orebody No. 58 of Jinchuan Copper–Nickel Deposit, Gansu Province. Master's Thesis, Central South University, Changsha, China, 2009.
5. Li, J.; Hanafy, S.; Schuster, G. Dispersion inversion of guided P-waves in a waveguide of arbitrary geometry. *J. Geophys. Res. Solid Earth* **2018**, *123*, 7760–7774. [CrossRef]
6. Zeng, Z.; Zhao, X.; Li, Z.; Li, J.; Wang, K.; Ma, L. Geophysical characteristics of the Shuanghu District in the Lungmu Co-Shuanghu-Lancangriver suture zone. *Chin. J. Geophys.* **2016**, *59*, 4594–4602.

7. Zeng, Z.; Huai, N.; Li, J.; Zhao, Z.; Liu, C.; Hu, Y.; Zhang, L.; Hu, Z.; Yang, H. Stochastic inversion of cross-borehole radar data from metalliferous vein detection. *J. Geophys. Eng.* **2017**, *14*, 1327. [CrossRef]
8. Li, J.; Sun, Z.; Weng, A.; Fan, Q. Determination of metallogenic tectonic environment by rock resistivity value—with Xia Dian Gold deposit as example. *Gold* **2008**, *29*, 15–19. (In Chinese)
9. Thmos, M.D.; Ford, K.L.; Keating, P. Review paper: Exploration geophysics for intrusion-hosted rare metals. *Geophys. Prospect.* **2016**, *64*, 1275–1304. [CrossRef]
10. Zhao, Z.; Zhou, X.; Guo, N.; Zhang, H.; Liu, Z.; Zheng, Y.; Zeng, Z.; Chen, Y. Superimposed W and Ag-Pb-Zn (-Cu-Au) mineralization and deep prospecting: Insight from a geophysical investigation of the Yinkeng orefield, South China. *Ore Geol. Rev.* **2018**, *93*, 404–412. [CrossRef]
11. Kheyrollahi, H.; Alinia, F.; Ghods, A. Regional magnetic lithologies and structures as controls on porphyry copper deposits: Evidence from Iran. *Explor. Geophys.* **2016**, *49*, 98–110. [CrossRef]
12. Xiao, F.; Wang, Z. Geological interpretation of Bouguer gravity and aeromagnetic data from the Gobi-desert covered area, Eastern Tianshan, China: Implications for porphyry Cu-Mo polymetallic deposits exploration. *Ore Geol. Rev.* **2017**, *80*, 1042–1055. [CrossRef]
13. Hu, X.; Peng, R.; Wu, G.; Wang, W.; Huo, G.; Han, B. Mineral exploration using CSAMT data: Application to Longmen region metallogenic belt, Guangdong Province, China. *Geophysics* **2013**, *78*, B111–B119. [CrossRef]
14. Guo, Z.; Hu, L.; Liu, C.; Cao, C.; Liu, J.; Liu, R. Application of the CSAMT method to Pb–Zn mineral deposits: A case study in Jianshui, China. *Minerals* **2019**, *9*, 726. [CrossRef]
15. Shah, A.K.; Bedrosian, P.A.; Anderson, E.D.; Kelley, K.D.; Lang, J. Integrated geophysical imaging of a concealed mineral deposit: A case study of the world-class Pebble porphyry deposit in southwestern Alaska. *Geophysics* **2013**, *78*, B317–B328. [CrossRef]
16. Zhang, R.; Li, T. Joint inversion of 2D gravity gradiometry and magnetotelluric data in mineral exploration. *Minerals* **2019**, *9*, 541. [CrossRef]
17. Melo, A.T.; Sun, J.; Li, Y. Geophysical inversions applied to 3D geology characterization of an iron oxide copper-gold deposit in Brazil. *Geophysics* **2017**, *82*, K1–K13. [CrossRef]
18. Lee, B.M.; Unsworth, M.J.; Hübert, J.; Richards, J.P.; Legault, J.M. 3D joint inversion of magnetotelluric and airborne tipper data: A casestudy from the Morrison porphyry Cu–Au–Mo deposit, British Columbia, Canada. *Geophys. Prospect.* **2018**, *66*, 397–421. [CrossRef]
19. Zhang, X.; Zhao, X.; Xie, Z. The application of the gravity and magnetic method to the exploration of the eastward extending M 15 anomaly of the Jinchuan copper–nickel deposit. *Geophys. Geochem. Explor.* **2010**, *34*, 139–143.
20. Wen, M.; Luo, X. A study of the ore-prospecting work based on multiple geosciences information in the Jinchuan Cu–Ni deposit. *Geol. China* **2013**, *40*, 594–601.
21. Fu, K.; Li, B. Geological and geophysical composite exploration model of Jinchuan copper–nickel sulfide deposit in Gansu province. *Gansu Geol.* **2006**, *15*, 62–67.
22. Tang, Z.L.; Li, W.Y. *Mineralisation Model and Geology of the Jinchuan Ni–Cu Sulfide Deposit Bearing PGE*; Geological Publishing House: Beijing, China, 1995. (In Chinese)
23. Song, X.; Danyushevsky, L.V.; Keays, R.R.; Chen, L.; Wang, Y.; Tian, Y.; Xiao, J. Structural, lithological, and geochemical constraints on the dynamic magma plumbing system of the Jinchuan Ni-Cu sulfide deposit, NW China. *Min. Depos.* **2012**, *47*, 277–297. [CrossRef]
24. Zhao, G.; Cawood, P.A. Precambrian geology of China. *Precambr. Res.* **2012**, *222–223*, 13–54. [CrossRef]
25. Zhao, G.C.; Sun, M.; Wilde, S.A.; Li, S.Z. Late Archean to Paleoproterozoic evolution of the North China Craton: Key issues revisited. *Precambr. Res.* **2005**, *136*, 177–202. [CrossRef]
26. Zhai, M.G.; Santosh, M. The early Precambrian odyssey of the North China Craton: A synoptic overview. *Gondwana Res.* **2011**, *20*, 6–25. [CrossRef]
27. Song, X.Y.; Keays, R.R.; Zhou, M.F.; Qi, L.; Ihlenfeld, C.; Xiao, J.F. Siderophile and chalcophile elemental constraints on the origin of the Jinchuan Ni-Cu-(PGE) sulfide deposit, NW China. *Geochim. Cosmochim. Acta* **2009**, *73*, 404–424. [CrossRef]
28. Mao, X.; Li, L.; Liu, Z.; Zeng, R.; Dick, J.M.; Yue, B.; Ai, Q. Multiple Magma Conduits Model of the Jinchuan Ni-Cu-(PGE) Deposit, Northwestern China: Constraints from the Geochemistry of Platinum-Group Elements. *Minerals* **2019**, *9*, 187. [CrossRef]
29. Xu, A.D.; Jiang, X.D. Characteristics and geological significance of the Dunzigou Group of the mesoproterozoic in the western edge of the North China Platform. *J. Earth Sci. Environ.* **2003**, *25*, 27–31. (In Chinese)

30. Xiao, P.X.; You, W.F.; Cao, X.D. Redefining of the Hanmushan Group in Longshoushan, central-western Gansu Province. *Geol. Bull. China* **2011**, *30*, 1228–1232. (In Chinese)
31. Xie, C.R.; Xiao, P.X.; Yang, Z.Z.; Cao, X.D.; Hu, Y.X. Progress in the studying of the Hanmushan Group in the Longshou mountains of Gansu province. *J. Stratigr.* **2013**, *37*, 54–57. (In Chinese)
32. Barnes, S.J.; Naldrett, A.J.; Gorton, M.P. The origin of the fractionation of platinum-group elements in terrestrial magmas. *Chem. Geol.* **1985**, *53*, 303–323. [CrossRef]
33. Tung, K.A.; Yang, H.Y.; Liu, D.Y.; Zhang, J.X.; Tseng, C.Y.; Wan, Y.S. SHRIMP U–Pb geochronology of the detrital zircons from the Longshoushan Group and its tectonic significance. *Chin. Sci. Bull.* **2007**, *52*, 1414–1425. [CrossRef]
34. Tang, Z.L.; Bai, Y.L. Geotectonic framework and metallogenic system in the southwest margin of north China paleocontinent. *Geosci. Front.* **1999**, *6*, 78–90. (In Chinese)
35. Zeng, R.Y.; Lai, J.Q.; Mao, X.C.; Li, B.; Zhang, J.D.; Bayless, R.; Yang, L.Z. Paleoproterozoic Multiple Tectonothermal Events in the Longshoushan Area, Western North China Craton and Their Geological Implication: Evidence from Geochemistry, Zircon U–Pb Geochronology and Hf Isotopes. *Minerals* **2018**, *8*, 361. [CrossRef]
36. Gong, J.H.; Zhang, J.X.; Wang, Z.Q.; Yu, S.Y.; Li, H.K.; Li, Y.S. Origin of the Alxa Block, western China: New evidence from zircon U–Pb geochronology and Hf isotopes of the Longshoushan Complex. *Gondwana Res.* **2016**, *36*, 359–375. [CrossRef]
37. Xiu, Q.Y.; Lu, S.N.; Yu, H.F.; Yang, C.L. The isotopic age evidence for main Longshoushan Group contributing to Palaeoproterozoic. *Prog. Precambr. Res.* **2002**, *25*, 93–96. (In Chinese)
38. Duan, J.; Li, C.S.; Qian, Z.Z.; Jiao, J.G. Geochronological and geochemical constraints on the petrogenesis and tectonic significance of Paleozoic dolerite dykes in the southern margin of Alxa Block, North China Craton. *J. Asian Earth Sci.* **2015**, *111*, 244–253. [CrossRef]
39. Zeng, R.Y.; Lai, J.Q.; Mao, X.C.; Li, B.; Ju, P.J.; Tao, S.L. Geochemistry, zircon U–Pb dating and Hf isotopies composition of Paleozoic granitoids in Jinchuan, NW China: Constraints on their petrogenesis, source characteristics and tectonic implication. *J. Asian Earth Sci.* **2016**, *121*, 20–33. [CrossRef]
40. Li, X.H.; Su, L.; Chung, S.L.; Li, Z.X.; Liu, Y.; Song, B.; Liu, D.Y. Formation of the Jinchuan ultramafic intrusion and the world's third largest Ni-Cu sulfide deposit: Associated with the ~825 Ma south China mantle plume? *Geochem. Geophys. Geosyst.* **2005**, *6*, 1–16. [CrossRef]
41. Zhang, M.J.; Kamo, S.L.; Li, C.S.; Hu, P.Q.; Ripley, E.M. Precise U–Pb zircon-baddeleyite age of the Jinchuan sulfide ore-bearing ultramafic intrusion, western China. *Miner. Depos.* **2010**, *45*, 3–9. [CrossRef]
42. Jiang, J.J.; Chen, L.M.; Song, X.Y.; Fu, Z.Q.; Wang, L.; Lu, J.Q.; Ai, Q.X.; Li, H. Siderophile and chalcophile element geochemistry of No. 58 ore body in Jinchuan Cu–Ni sulfide deposit and its geological significance. *Miner. Depos.* **2013**, *32*, 941–953. (In Chinese)
43. Li, T.H.; Luo, X.R.; Peng, Q.L.; Wang, W.; Luo, X.P.; Song, Z.B.; Wen, X.Q. Geological-geoelectrochemical-geophysical multifactor information ore prognosis in the depth and on the edge of No.1 mining area of the Jinchuan copper–nickel sulfide ore deposit, Gansu Province. *Geol. Bull. China* **2012**, *31*, 1192–1200. (In Chinese)
44. Zhang, J.M.; Zeng, Z.F.; Wu, Y.G.; Du, W.; Wang, Y.Z. Balanced morphological filters for horizontal boundaries enhancement of the potential field sources. *Appl. Geophys.* **2020**, 1–10. [CrossRef]
45. Zhao, X. The Study and Application on the Gravity and Magnetic Joint Inversion Based on the Smoothed L0 Norm (SL0) Constraint of Compactly Supported Conjugate Gradient Algorithm. Master's Thesis, Jilin University, Changchun, China, 2017.

Article

Application of the CSAMT Method to Pb–Zn Mineral Deposits: A Case Study in Jianshui, China

Zhenwei Guo [1,2,3], Longyun Hu [1,2,3], Chunming Liu [1,2,3], Chuanghua Cao [4], Jianxin Liu [1,2,3] and Rong Liu [1,2,3,*]

1 School of Geosciences and Info-Physics, Central South University, Changsha 410083, China; guozhenwei@csu.edu.cn (Z.G.); hulongyun@csu.edu.cn (L.H.); liuchunming@csu.edu.cn (C.L.); ljx6666@126.com (J.L.)
2 Hunan Key Laboratory of Nonferrous Resources and Geological Hazard Exploration, Changsha 410083, China
3 Key Laboratory of Metallogenic Prediction of Nonferrous Metals and Geological Environment Monitoring (Central South University) Ministry of Education, Changsha 410083, China
4 Geology Survey Institute of Hunan Province, Changsha 410116, China; caochuanghua@csu.edu.cn
* Correspondence: liurongkaoyan@126.com; Tel.: +86-731-8887-7151

Received: 4 October 2019; Accepted: 21 November 2019; Published: 25 November 2019

Abstract: The electromagnetic (EM) method is commonly used in mineral exploration due to the method's sensitivity to conductive targets. Controlled source audio-frequency magnetotellurics (CSAMT) is developed from magnetotelluric (MT) method with an artificial EM source to improve the signal amplitude. It has been used for mineral exploration for many years. In this study, we performed a case study of the CSAMT application for the Eagles-Nest lead–zinc (Pb–Zn) ore deposits in Jianshui, China. The Eagles-Nest deposit is located in southwest in China in forest-covered complex terrain, making it difficult to acquire the geophysical data. Based on the previous dual-frequency induced polarization (IP) results, we designed four profiles for the CSAMT data acquisition. After data processing and inversion, we mapped the subsurface resistivity distribution. From the CSAMT results, we inferred the location of the ore body, which was verified by the drilling wells. The Pb–Zn ore body was found at a depth between 373.70 m to 407.35 m in the well.

Keywords: CSAMT; dual-frequency IP; mineral exploration

1. Introduction

The Eagles-Nest lead–zinc (Pb–Zn) ore deposit is located in Jianshui, in Southwest China, in an area which is 90% covered by forests. The northeast is relatively flat; however, the western and southern mountains are cut by narrow and deep valleys. The elevation of the area ranges between 1271 m and 2503 m. It is difficult to carry out the ground geophysical surveys in this area because of the forest and steep mountains. During previous research 50 years ago, geologists investigated three profiles. Unfortunately, the reports of that investigation were lost.

During the 1950s, the magnetotelluric method (MT) was introduced for electromagnetic (EM) exploration by Cagniard [1]. In order to acquire strong signal, the controlled-source audio-frequency magnetotellurics (CSAMT) method was proposed by Goldstein [2]. Both the MT and CSAMT methods are applied in frequency domain to detect mineral deposits.

In mineral exploration, CSAMT method is one of the most important tools. Although electrical resistivity tomography (ERT) can describe the subsurface resistivity distribution, CSAMT can describe the subsurface resistivity distribution clearly. Some successful case studies have been conducted in geothermal [3,4], mineral deposits exploration [5] and groundwater [6]. Normally, the mineral deposits are shallower than geothermal sources. Chen et al. [7] successfully detected the Longtoushan Ag–Pb–Zn deposit using CSAMT in inner Mongolia, China. This method has also been applied with the iron and polymetallic (Pb–Zn–Cu) deposits in the Longmen region by Hu et al. [8]. The CSAMT method was also applied to explain geological structures, which were distinguished by the difference between Tamusu rock and surrounding rock [9].

Induced polarization, another geophysical method, is commonly used to delineate potential target zones and estimate the deposit projected on the surface. Schlumberger introduced the induced polarization (IP) method for geophysical surveys in the 1920s [10]. The IP method has high sensitivity to mineral deposits. Moreover, it was used to test the detectability on the sensitivity in geothermal systems [11,12], thus, combining the IP and CSAMT method is usually applied for the mineral exploration, for instance, in the case of massive chalcopyrite exploration [13].

Dual-frequency IP method is a kind of frequency domain IP method that utilizes information from two frequencies [14]. Bao and He (He Jishan) introduced the dual-frequency and multi-parameter IP method, which could find the anomaly of Percent Frequency Effect (PFE) and phase, and also provided the property information of the IP anomaly resource [14]. In this paper, we combine dual-frequency IP and CSAMT method to explore the Pb–Zn deposit. We used the dual-frequency IP method to estimate the location of the deposit, then we inverted and analyzed the CSAMT data to determine the depth of the orebody. The final result was analyzed and interpreted.

2. Geological Setting in Jianshui Area

The Eagles-Nest region lies in the south-western part of the Gejiu-Shiping faulted fold. At the southwest part of Honghe deep fault, it is connected with the Ailaoshan metamorphic block, and at the northeast part it is connected with Mile-Shizong fault.

The main exposed carbonate rocks belong to the Triassic Gejiu Formation in survey area (Figure 1). The third stratum (T_2g^3) of Gejiu Formation is located in the center of the survey area. The upper part lithology is thick layered dolomite of Triassic Gejiu Formation; the middle part is medium-thick layered fine-grained marble; the lower part is medium-thick layered dolomite sandwiched with thin layered limestone. This stratum goes through the whole area from the north to east with a strike of 130°, and dips 25° to 40°. The area is about 1.5 km^2. From south to north, the layers become thinner with an average thickness of 252.42 m.

The second stratum (T_2g^2) of Gejiu Formation is distributed in the northwestern part of the survey area. The lithology is light yellow-gray and medium-thick layered argillaceous limestone-bearing mudstone. The former is often metamorphosed into light yellow fine-grained marble. The latter is developed horizontally with a small amount of sea lily stem fragments. It stretches in a northeast direction with a strike of 160–170°, and dips 25° to 30°. The lithology of this section varies from carbonate to clastic rock with marble from south to north. The rock layer is generally metamorphosed where the rock is in direct contact with the granite body.

The Yanshanian granite ($\gamma_5{}^2$) is exposed as the main magnetic rock in the survey area, which is distributed in the south-central part of this area with east–west direction. It is plaque-like black cloud monzonitic granite with a semi-automorphic granular structure. The phenocrysts in granite are mainly single crystals of light flesh red potassium feldspar. The mineral composition includes biotite, quartz, plagioclase, and potassium feldspar. The granite body is rock-based and the carbonate rocks in contact with it are all marbled.

Tectonic movement has occurred many times; layers were cut by faults in south–north direction. The two main faults in the survey area are the south–north fault (F1) in early phase and east–west fault (F2) in late phase. Fault F1 is located in the central of the area with around 2031 m length and 5–10 width. Irregular and angular structured breccias are found in the fault zone. Along the fault zone, there are structural fracture zones within the second stratum (T_2g^2), the third stratum (T_2g^3) of the Gejiu Formation, and the Yanshanian granite ($\gamma_5{}^2$). The second fault (F2) went through the survey area in an east–west direction with 3154 m length and 5–20 m width. The F2 is a reverse fault with a strike of 160–170° and a north–east inclination. The small folds in survey area are well developed, and the axial direction is consistent with the direction of the large tectonic line in northeast direction.

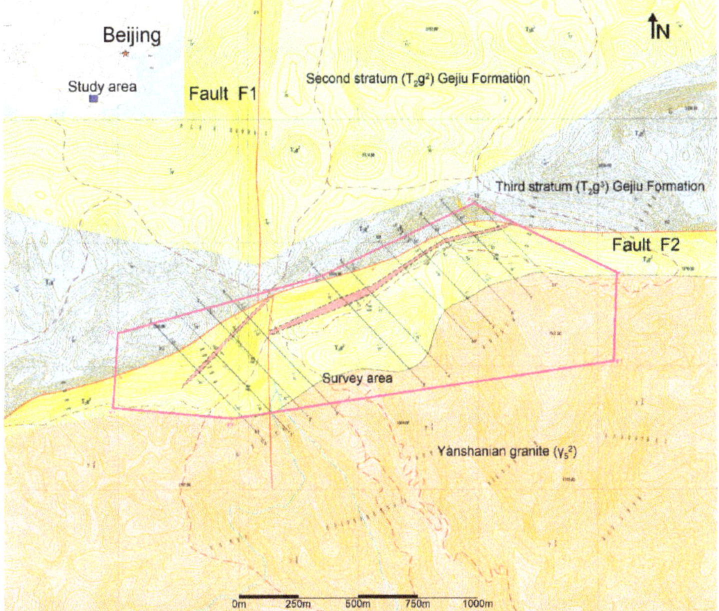

Figure 1. Simplified geological map of Eagles-Nest region. Fault F1 in N–S direction was interrupted by fault F2 in NE–SW direction. The study area is located in the pink area.

3. Overview of the CSAMT Method

The CSAMT method is a commonly-used, surface-based geophysical method which provides resistivity information of the subsurface. A horizontal dipole is used to transmit the EM signal. Signals from near-zone and transition-zone always result in distortions of Cagniard resistivity. Therefore, electrical and magnetic fields are measured on the ground with a distance at the far-zone, where the useful signal can be measured.

The three components of electrical (Equations (1)–(3)) and magnetic (Equations (4)–(6)) fields could be computed by the following equations:

$$E_x = \frac{I{\cdot}AB{\cdot}\rho_1}{2\pi r^3}{\cdot}\left(3cos^2\theta - 2\right), \tag{1}$$

$$E_y = \frac{3{\cdot}I{\cdot}AB{\cdot}\rho_1}{4\pi r^3}{\cdot}sin2\theta, \tag{2}$$

$$E_z = (i-1)\frac{I{\cdot}AB{\cdot}\rho_1}{2\pi r^3}{\cdot}\sqrt{\frac{\mu_0\omega}{2\rho_1}}{\cdot}cos\theta, \tag{3}$$

$$H_x = -(1+i)\frac{3I{\cdot}AB}{4\pi r^3}{\cdot}\sqrt{\frac{2\rho_1}{\mu_0\omega}}{\cdot}cos\theta{\cdot}sin\theta, \tag{4}$$

$$H_y = (1+i)\frac{I{\cdot}AB}{4\pi r^3}{\cdot}\sqrt{\frac{2\rho_1}{\mu_0\omega}}{\cdot}(3cos^2\theta - 2), \tag{5}$$

$$H_z = i\frac{3I \cdot AB \cdot \rho_1}{2\pi\mu_0\omega r^4} \cdot sin\theta \tag{6}$$

where E is electrical field; H is magnetic field; I is current; AB is the length of the transmitter source; ρ is resistivity; μ_0 is magnetic permeability in air; ω is angular frequency; (r, θ) is the coordinate of the observation point. The Cagniard resistivity [1] can be calculated by using the ratio of the orthogonal horizontal components Ex/Hy from the equations above.

$$\rho_a = \frac{1}{5f}\left|\frac{E_x}{H_y}\right|^2 \tag{7}$$

Finally, the impedance phase is given by

$$P = E_{\text{phase}} - H_{\text{phase}} \tag{8}$$

4. Geophysical Survey

4.1. IP Interpretation and CSAMT Data Acquisition Design

In 2015, we did a dual-frequency induced polarization (IP) investigation on the survey area (Figure 2). Based on the results of that study we set up the CSAMT data acquisition design. In Figure 2, the area with red line bounded is the dual-frequency IP investigation region. The grid of dual-frequency IP data is 100 m × 20 m, all line spacing and station spacing are 100 m and 20 m respectively, with red dots marking measurement points. A total of 31 profiles of dual-frequency IP survey described a distribution of IP and resistivity anomalies.

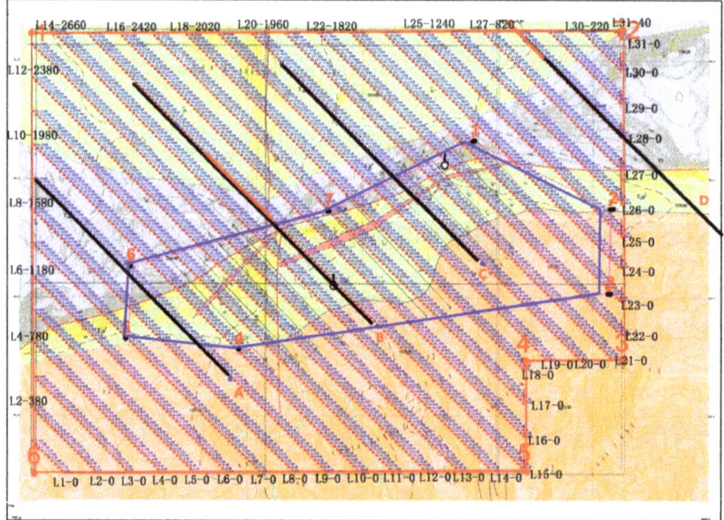

Figure 2. Geophysical survey history and controlled-source audio-frequency magnetotellurics (CSAMT) survey design. Outer red solid lines (1–6) delineate the survey area, induced polarization (IP) profiles are symbolized as 31 red dotted lines (L1–L31) near which point numbers are identified, 4 black lines (A–D) are CSAMT profiles and blue area in the center means mining available range.

Figures 3 and 4 show the IP and resistivity anomalies, respectively. The amplitude of an IP (Fs, %) anomaly was defined with the value larger than 2.4. Table 1 shows the resistivity and polarizability of characteristic rocks from this mining area. From the Table 1, the amplitude frequency background formation was around 2.0. Based on the physical difference between the rocks and ores, it was easy to infer the IP anomaly zones due to the Pb–Zn ore body or mineralization. Compared to the unmineralized rock formations, the Pb–Zn mineralization might reflect a low resistivity anomaly in the background. In order to interpret clearly, we drew the resistivity contour map (Figure 4) with the amplitude frequency anomaly. Integrated the amplitude frequency and resistivity map, the interesting area was selected for the CSAMT measurement to detect the deep target. We studied the subsurface resistivity distribution by using CSAMT method. The profiles were designed to span all of interesting area in a time-saving way as Figure 2 shows.

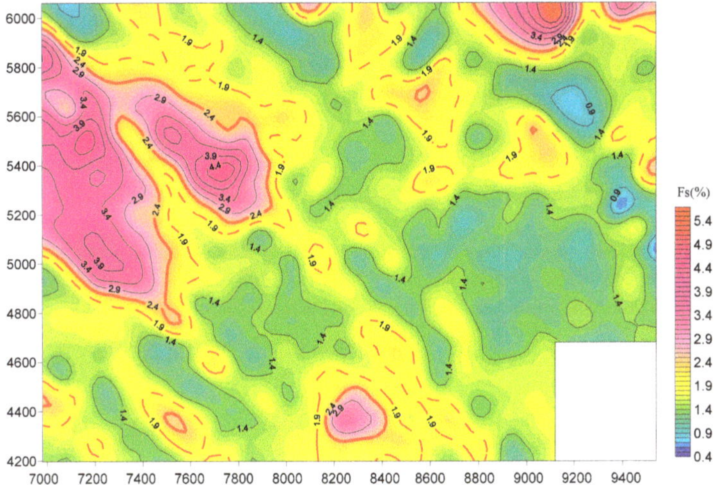

Figure 3. Amplitude of induced polarization (Fs) contour map.

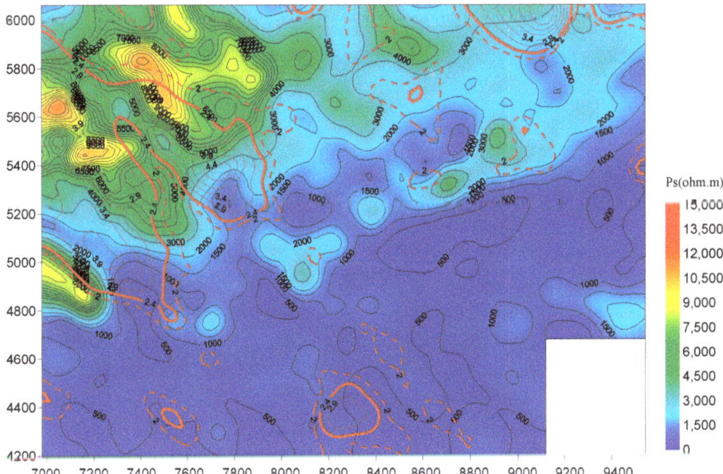

Figure 4. Resistivity (Ps) contour map with amplitude frequency anomaly. The range enclosed by red lines (dotted and solid) corresponds to interesting area in Fs analysis.

Table 1. Geophysical properties of Eagles-Nest deposit.

Rocks	Fs Range (%)	Fs Average	Resistivity Range (Ωm)	Resistivity Average
Pb–Zn	1.6–6.4	4.23	121.5–336.2	257.68
Limestone	1.6–2.7	2.08	325.7–336.5	676.5
Dolomites	1.09–1.38	1.26	547.4–978.5	734.98
Granite	1.1–1.7	1.27	294.5–597.4	430.46

4.2. Data Acquisition, Processing and Inversion

In order to describe the subsurface resistivity distribution, four CSAMT data acquisition profiles were designed with 1200 m length and 20 m station distance. Before the CSAMT survey, we tested the offset with 8 km and 12 km. Based on the results of the test, the offset was chosen as 11 km, and the horizontal current dipole length was 1200 m. The current was 9 A at low frequency. The CSAMT profiles A–D location are shown as black lines in Figure 2.

The quality evaluation of this CSAMT measurement was determined by calculating the mean square error (MSE) (<±5%) of Cagniard resistivity in two surveys at the same station. Data consistency was checked by comparing Cagniard resistivity data in two surveys at 1700 station 20 line as shown in Figure 5. Based on the same conditions, the relative error of the data collected twice was 2.76%. The low-frequency data reflected a slight error, which was the allowable range of normal error, and did not affect the subsequent inversion processing. The CSAMT data collection in this area was reliable, which provided guarantee for data processing and data interpretation.

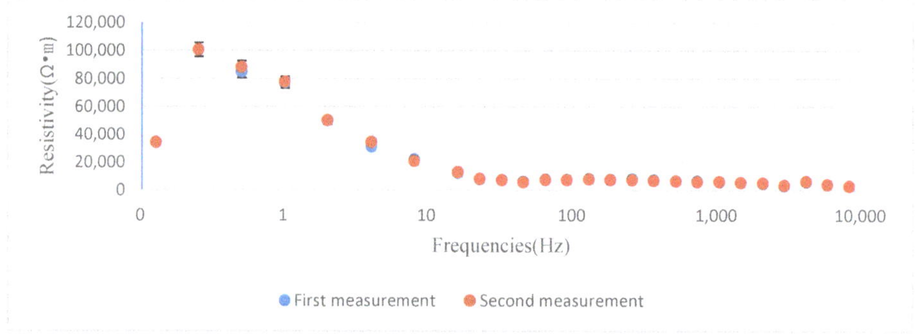

Figure 5. Quality evaluation of CSAMT data. Black and red line illustrate twice measurements.

Occam's inversion was introduced to find a smooth model to satisfy the geophysical data by Constable et al. [15]. It is a simple way to map the subsurface resistivity structure. The CSAMT data inversion results are shown in Figures 6–9. From Table 1, we know that the weathered granite has a low resistivity. By contrast, the limestone and dolomites are a bit higher than granite in resistivity. In Figure 6, the resistivity of the stations on the right side of 1000 station describe the third stratum (T_2g^3) of the Gejiu Formation. The low-resistivity anomaly at the depth of 400–500 m between the station of 1200 and 1700 is inferred to be due to the mineral deposits.

In Figure 7, the fault F2 is still Triassic. The fault zone is expressed at the surface between stations 1400 m and 1500 m along profile A. It extends to around 300 m below the surface, and the interface between the shallow layer and the granite argillaceous limestone overlaps. There is a low resistivity anomaly band between the stations 1600 and 2000 at a depth of 600 m. The high resistivity anomalies in Figures 7 and 8 are due to the dolomites in this area.

The RMS misfit of the inversion is shown in Figure 10. All RMS inversion residual reduce fast before the third iteration. After that, four curves of the profiles decrease small. When the percent of residual change is smaller than 1%, the inversion is stopped.

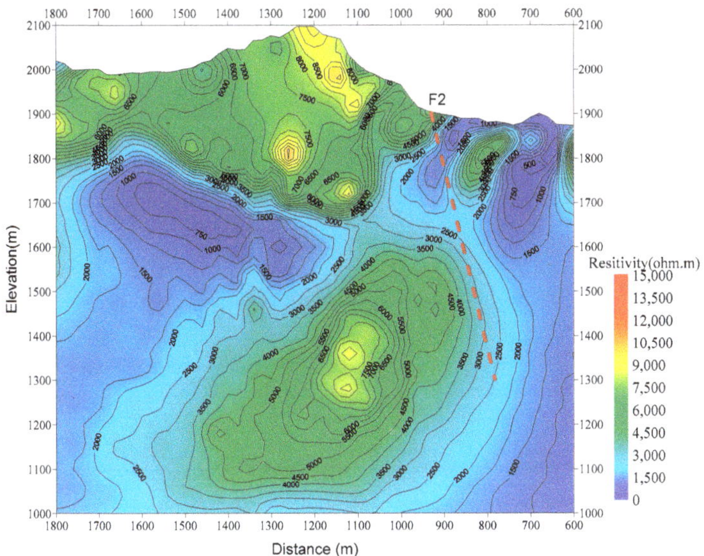

Figure 6. CSAMT data inversion result of A profile, where distance equals to station number. The red dash line is fault F2.

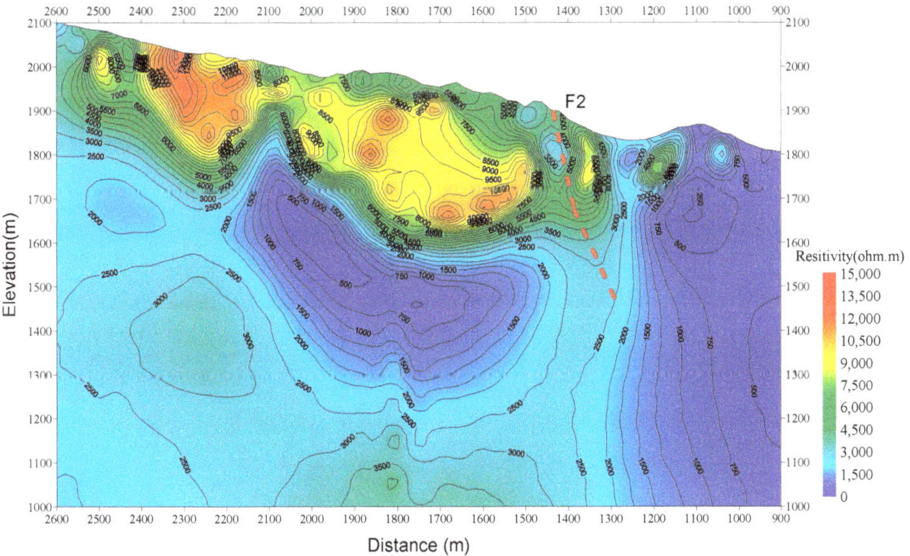

Figure 7. CSAMT data inversion result of B profile, where distance equals station number. The red dash line is fault F2.

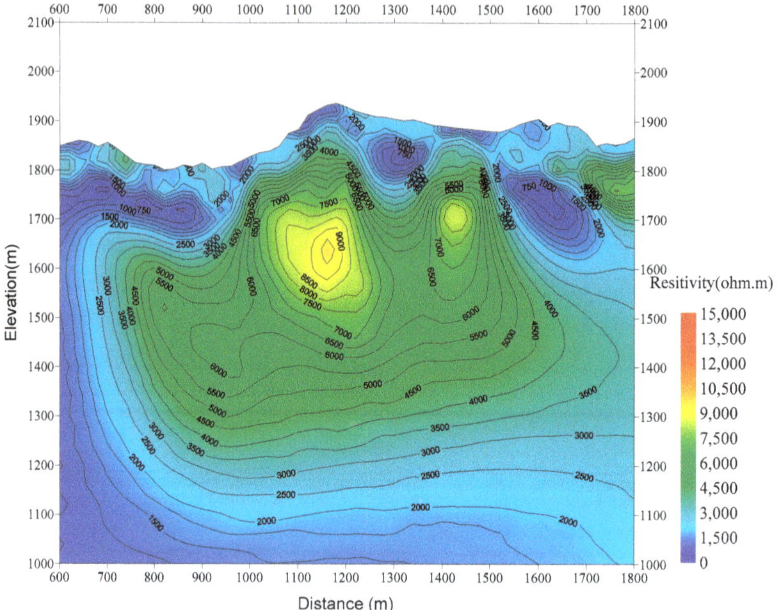

Figure 8. CSAMT data inversion result of C profile, where distance equals station number.

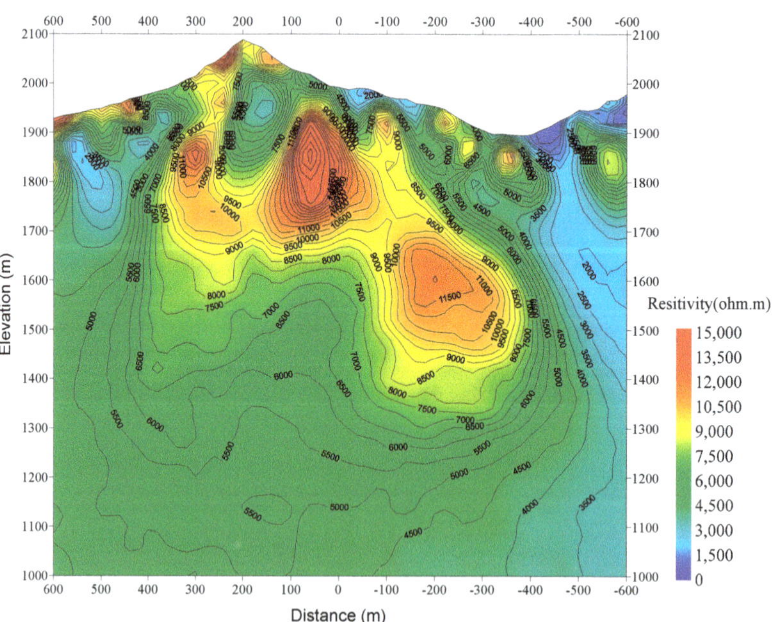

Figure 9. CSAMT data inversion result of D profile, where distance equals station number.

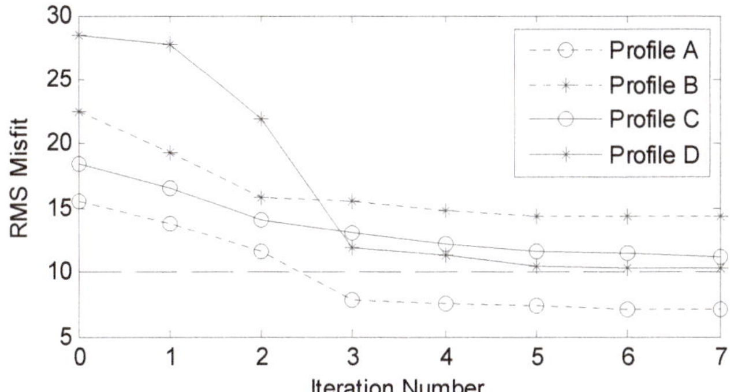

Figure 10. The RMS misfit of CSAMT data inversion.

4.3. CSAMT Results and Interpretation

In order to describe the resistivity distribution in three-dimensions, we imaged the resistivity slices as shown in Figure 11. As is illustrated in Figure 11, the low-resistivity anomaly shown in profile A continued to profile B and disappeared in profile C. Moreover, this anomaly is related to the IP anomaly shown in Figure 3. The high amplitude of induced polarization anomaly has the same location as the low-resistivity anomaly. Based on this information, we inferred that the anomaly was due to the mineral deposits.

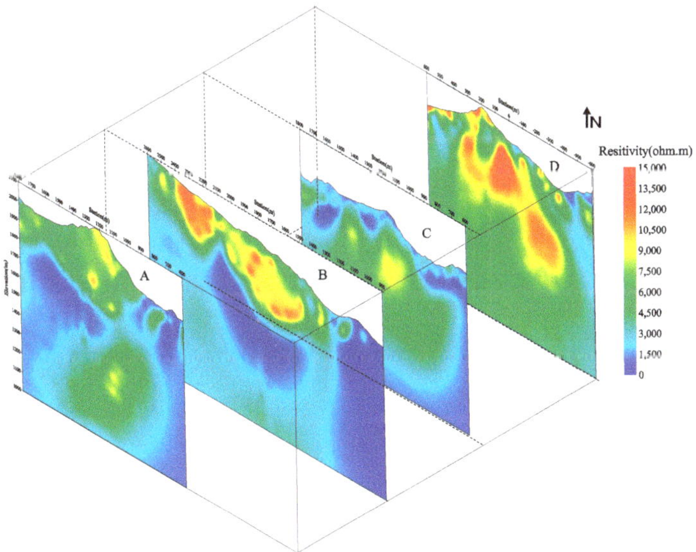

Figure 11. Resistivity profiles of CSAMT data inversion.

By combining the amplitude and resistivity characteristics of the survey area (Figures 3 and 4) with the given rock properties of Table 1, we defined four IP anomaly areas marked as IP1, IP2, IP3 and IP4 (Figure 12). The interface between the granite and the second stratum (T_2g^2) is deep and steep. The ore-preserving structure is caused by the fault F2 in small scale. The larger deposit is located in the north-western part of the mining area (IP1), which is produced at the contact zone between the granite and the argillaceous limestone.

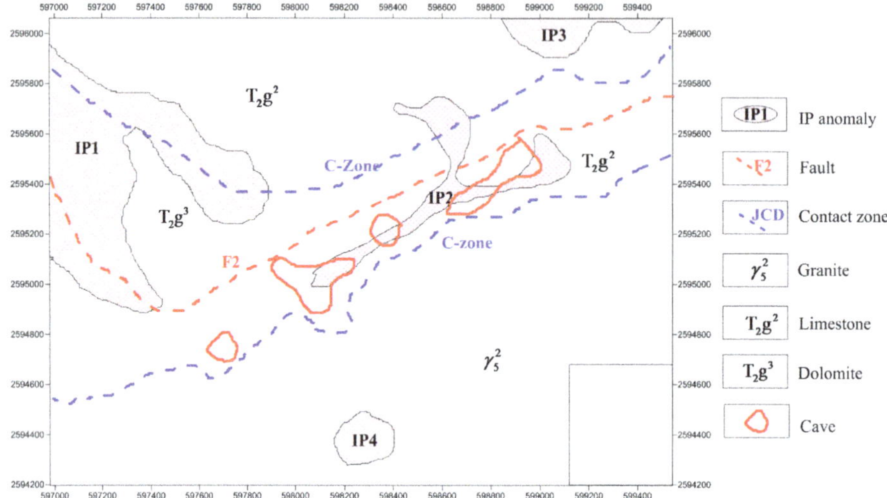

Figure 12. The interpretation map of the study area.

The IP1 anomaly zone is the largest anomaly in this region, and the area is around 0.6 km^2. IP2 zone is located at the center of the survey area. The anomaly zone looks like a "Y", and the mineralized rocks were visible in some caves in the area. In earlier exploration, the orebody deposits were discovered at three different locations. IP surveying in this study confirmed that all three orebodies belong to the same anomaly IP2 zone. Another strong anomaly IP3 is located north of the area. We infer this target is shallow buried at the anomaly IP3 zone. Anomaly IP4 circle zone is located south of the area with a strong amplitude of induced polarization.

In this study area, the IP value of granite and dolomites are less than 2%, so we consider them as background. The limestone has a higher amplitude than granite, but it is not as high as the Pb–Zn deposits. Anomaly IP4 zone has a value less than 3%, similar to the limestone. However, the IP1, IP2 and IP3 describe strong anomaly like effects of the Pb–Zn deposits.

Figure 13 shows the interpretation of the CSAMT data inversion result of profile A. The red anomaly represents the Pb–Zn orebody in Figure 13b, which is proved by drilling well. The Pb–Zn ore body is found between 373.70 m and 407.35 m in the well.

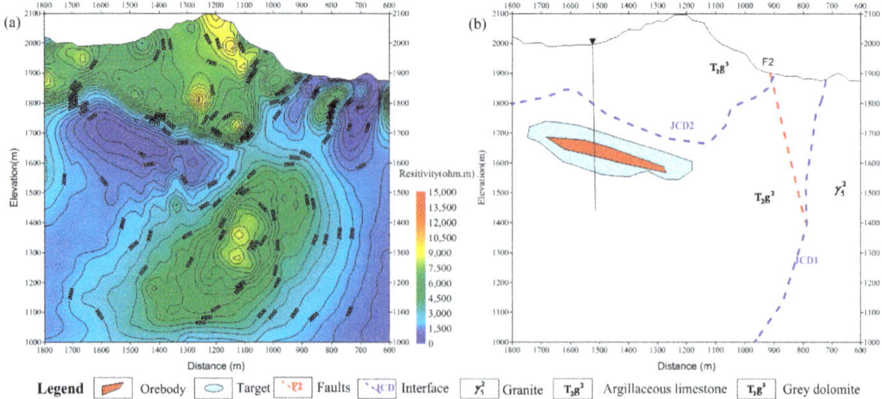

Figure 13. CSAMT data inversion result (**a**) and interpretation (**b**) of A profile, where distance equals to station number.

5. Discussion and Conclusions

In this paper, IP and CSAMT methods were used to test their combined effectiveness for use in exploration of Pb–Zn deposits in the Eagles-Nest region. The IP method provides high sensitivity to the metal deposits [14]. So it is necessary to discover the IP anomaly from a geophysical map scanning. However, the CSAMT method has the deep resolution advantages compared to the ERT or TEM method [9]. Combining the advantages of IP high sensitivity and CSAMT high resolution, the study performs well.

The conclusions of geophysical investigations in Jianshui, Southwest of China are the following:

Based on results of IP and CSAMT surveys (in Figures 11 and 12), we can infer the Pb–Zn deposits correlated with the anomaly of high induced polarization and low resistivity, which are shown in the geophysical properties in Table 1. The applied methods (CSAMT and IP) are effective for Pb–Zn deposits investigation in this area. The predicted deposits zones are generally consistent with the IP anomaly zones. The IP method allows easy selection of the interesting zones for the orebody targets. The inversion results from CSAMT method can predict the depth of the orebody targets.

In summary, this case study shows a successful geophysical survey to detect Eagles-Nest Pb–Zn deposits in the Jianshui area. Based on this experimental study, these methods may be an effective exploration strategy and geophysical model to detect similar potential deposits in the surrounding areas.

Author Contributions: Conceptualization, Z.G. and L.H.; methodology, C.C. and C.L.; software, C.C.; validation, C.L., R.L. and J.L.; formal analysis, Z.G.; investigation, Z.G.; resources, C.C.; data curation, R.L.; writing—original draft preparation, Z.G. and L.H.; writing—review and editing, J.L.; visualization, R.L.; supervision, J.L.; project administration, J.L.; funding acquisition, C.C.

Funding: This research is supported by the Natural Science Foundation of China (NSFC) [41804073; 41674080]. The research also funded by Open Research Fund Program of Key Laboratory of Metallogenic Prediction of Nonferrous Metals and Geological Environment Monitoring (Central South University), Ministry of Education [Grant: 2019YSJS10]. Furthermore, the research got the Grant from the Open Foundation of Key Laboratory of Submarine Geosciences, SOA [KLSG1905].

Acknowledgments: We thank the Geology Survey Institute of Hunan province for allowing us to publish the results. We appreciate the jobs by the colleagues on the data acquisition.

Conflicts of Interest: The authors declare no conflict of interest.

References

1. Cagniard, L. Basic theory of the magnetotelluric method of geophysical prospecting. *Geophysics* **1953**, *18*, 605–635. [CrossRef]
2. Goldstein, M.A. Magnetotelluric Experiments Employing an Artificial Dipole Source. Ph.D. Thesis, University of Toronto, Toronto, ON, Canada, 1971; pp. 56–69.
3. Sandberg, S.K.; Hohmann, G.W. Controlled-source audiomagnetotellurics in geothermal exploration. *Geophysics* **1982**, *47*, 100–116. [CrossRef]
4. Mustopa, E.J. 2D Interpretation of Controlled Source Audio Magnetotelluric (CSAMT) data integrated with borehole data in Kamojang geothermal field west Java, Indonesia. *J. Phys. Conf. Ser.* **2019**, *1127*, 012021.
5. Wannamaker, P. Tensor CSAMT survey over the Sulphur Springs thermal area, Valles Calder, New Mexico, USA, Part 1: Implications for structure of the western Caldera. *Geophysics* **1997**, *62*, 451–465. [CrossRef]
6. Fu, C.; Di, Q.; An, Z. Application of the CSAMT method to groundwater exploration in a metropolitan environment. *Geophysics* **2013**, *78*, B201–B209. [CrossRef]
7. Chen, W.; Liu, H.; Liu, J.; Sun, X.; Zeng, Q. Integrated geophysical exploration for the Longtoushan Ag-Pb-Zn deposit in the southeast of the Da Xing'an Ling mountains, Inner Mongolia, northern China. *Explor. Geophys.* **2010**, *41*, 279–288. [CrossRef]
8. Hu, X.; Peng, R.; Wu, G.; Wang, W.; Huo, G.; Han, B. Mineral Exploration using CSAMT data: Application to Longmen region metallogenic belt, Guangdong Province, China. *Geophysics* **2013**, *78*, B111–B119. [CrossRef]
9. An, Z.; Di, Q. Investigation of geological structures with a view to HLRW disposal, as revealed through 3D inversion of aeromagnetic and gravity data and the results of CSAMT exploration. *J. Appl. Geophys.* **2016**, *135*, 204–211. [CrossRef]

10. Schlumberger, C. *Study of Underground Electrical Prospecting*; University of California Libraries LLC.: Sacramento, CA, USA, 2017; p. 99.
11. Revil, A.; Murugesu, M.; Prasad, M.; Le Breton, M. Alteration of volcanic rocks: A new non-intrusive indicator based on induced polarization measurements. *J. Volcanol. Geotherm. Res.* **2017**, *341*, 351–362. [CrossRef]
12. Ghorbani, A.; Revil, A.; Coperey, A.; Ahmed, A.S.; Roque, S.; Heap, M.J.; Grandis, H.; Viveiros, F. Viveiros Complex conductivity of volcanic rocks and the geophysical mapping of alteration in volcanoes. *J. Volcanol. Geotherm. Res.* **2018**, *357*, 106–127. [CrossRef]
13. Basokur, A.T.; Rasmussen, T.M.; Kaya, C.; Altun, Y.; Aktas, K. Comparison of induced polarization and controlled-source audio-magnetotellurics methods for massive chalcopyrite exploration in a volcanic area. *Geophysics* **1997**, *62*, 1087–1096. [CrossRef]
14. Bao, G.; He, J. Dual-frequency and multi-parameter IP instrument and its application research. *J. Cent. South Univ. Technol.* **1996**, *3*, 12–16. [CrossRef]
15. Constable, S.C.; Parker, R.L.; Constable, C.G. Occam's inversion—A practical algorithm for generating smooth models from electromagnetic sounding data. *Geophysics* **1987**, *52*, 289–300. [CrossRef]

Article

Joint Inversion of 2D Gravity Gradiometry and Magnetotelluric Data in Mineral Exploration

Rongzhe Zhang * and Tonglin Li

College of Geo-Exploration Sciences and Technology, Jilin University, Changchun 130026, China
* Correspondence: zhangrongzhe_jlu@163.com

Received: 23 July 2019; Accepted: 4 September 2019; Published: 7 September 2019

Abstract: We have developed a mineral exploration method for the joint inversion of 2D gravity gradiometry and magnetotelluric (MT) data based on data-space and normalized cross-gradient constraints. To accurately explore the underground structure of complex mineral deposits and solve the problems such as the non-uniqueness and inconsistency of the single parameter inversion model, it is now common practice to perform collocated MT and gravity surveys that complement each other in the search. Although conventional joint inversion of MT and gravity using model-space can be diagnostic, we posit that better results can be derived from the joint inversion of the MT and gravity gradiometry data using data-space. Gravity gradiometry data contains more abundant component information than traditional gravity data and can be used to classify the spatial structure and location of underground structures and field sources more accurately and finely, and the data-space method consumes less memory and has a shorter computation time for our particular inversion iteration algorithm. We verify our proposed method with synthetic models. The experimental results prove that our proposed method leads to models with remarkable structural resemblance and improved estimates of electrical resistivity and density and requires shorter computation time and less memory. We also apply the method to field data to test its potential use for subsurface lithofacies discrimination or structural classification. Our results suggest that the imaging method leads to improved characterization of geological targets, which is more conducive to geological interpretation and the exploration of mineral resources.

Keywords: gravity gradiometry; magnetotelluric; model-space; data-space; joint inversion

---

1. Introduction

The necessity for the accurate exploration of subsurface structures and conditions are driving the development of various strategies for the joint use of information from multiple geophysical data [1–9]. Different geophysical data can provide different information on the distribution of subsurface properties, while different geophysical exploration methods have different detection capabilities for underground targets. However, the uncertainty and non-uniqueness of the inversion interpretation is a common problem in geophysics [10]. To suppress the inversion results' non-uniqueness and obtain more accurate information for underground media, comprehensive geophysical methods that study the same geological target from different aspects have led to very effective research methods and development trends [11].

Joint inversion is a combination of multiple geophysical data to invert the same subsurface target through the correlation of the petrophysical and geometric parameters of the geological body. Joint inversion can improve the non-uniqueness of geophysical inversion. After more than a decade of development, geophysical workers have proposed many different joint inversion methods. The study of joint inversion for geophysical exploration can be classified into two categories. The first category relies on the coupling of petrophysical properties. By determining the petrophysical relationship functions of different physical properties (e.g., how petrophysical characteristics relate resistivity and seismic velocity in porous media), the coupling inversion of different physical properties was realized [12,13]. However, this method is limited by its difficulty in finding the accurate physical relationships of rock in complex underground areas. Therefore, a drawback of joint inversion based on the empirical relationship of the rock's physical properties restricts the development of joint inversion. The second category is based on the similar structural distribution of subsurface parameters. Joint inversion is realized by minimizing structural differences [14–16]. Among these differences, the cross-gradient inversion algorithm proposed by Gallardo and Meju [16] is a joint inversion method that has received extensive attention. This method assumes that the boundaries of the anomalous bodies are identical or partially identical in different geophysical fields and have been widely used in the comprehensive interpretation of geophysics [17–24].

Gravity and magnetotelluric explorations have the characteristics of wide coverage, high efficiency, and low cost, so they have been widely used in resource exploration and are especially popular in the comprehensive interpretation of joint inversion [8,21–24]. In the above study, joint inversion based on cross-gradient constraints is mainly applied to gravity data and other geophysical data and is rarely applied to the gravity gradiometry data and other geophysical data. With the commercialization of gravity gradiometry measurement and the rapid development of computer technology, gravity gradiometry inversion has been widely applied in oil, gas, and mineral exploration [25–27]. Gravity gradiometry data can provide more abundant multi-component information, as each gradient component contains different information content. Compared with the gravity data, this data has a higher resolution in reflecting the details of the anomalous body [28–30]. If gravity gradiometry data are added to the joint inversion, they will help to better describe the spatial structure and location of the field source and improve the level of geological interpretation. Moreover, the above joint inversion methods are performed in model-space, which require extremely long computing times and high levels of memory usage if the number of meshes is large. If the inversion calculation process is transformed from the model-space into data-space, it may effectively solve the problem of long calculation times and large memory usage [31,32].

In this paper, we present an approach for the joint inversion of magnetotelluric (MT) and gravity gradiometry data based on normalized cross-gradient constraints and data-space. Considering that the gravity gradiometry data contains more information than traditional gravity data, and due to the lack of vertical resolution of the gravity gradiometry method, we constructed a new joint inversion objective functional, including a multi-component gravity gradiometry and MT data misfit terms, as well as their model constraints and normalized cross-gradient constraints. The joint inversion of the gravity gradiometry and MT data is superior to the separate inversion or the joint inversion of gravity and MT data, which could reduce the non-uniqueness and improve the resolution of the inversion results. Moreover, we applied the data-space method to the joint inversion calculation, the data-space joint inversion consumes less memory and has a shorter computation time. We used synthetic models of the subsurface to test the resolution, computational efficiency, and memory usage of the algorithm. Then, we applied this approach to the interpretation of geophysical field data collected in a mining study area from Jilin Province, China.

2. Forward Problem

2.1. Gravity and Gravity Gradiometry Forward Problem

We begin with a brief description of the gravity and gravity gradiometry fields. The gravity field, which is the first derivatives of the gravity potential $V(\mathbf{r})$, is defined as

$$\mathbf{g} = \nabla V = \begin{bmatrix} \frac{\partial V}{\partial x} \\ \frac{\partial V}{\partial y} \\ \frac{\partial V}{\partial z} \end{bmatrix} = \begin{bmatrix} g_x \\ g_y \\ g_z \end{bmatrix} \tag{1}$$

The gravity gradiometry field, which is the second derivatives of the gravity potential $V(\mathbf{r})$, is defined as

$$\hat{\mathbf{g}} = \nabla\nabla V = \begin{bmatrix} \frac{\partial^2 V}{\partial x^2} & \frac{\partial^2 V}{\partial xy} & \frac{\partial^2 V}{\partial xz} \\ \frac{\partial^2 V}{\partial yx} & \frac{\partial^2 V}{\partial yy} & \frac{\partial^2 V}{\partial yz} \\ \frac{\partial^2 V}{\partial zx} & \frac{\partial^2 V}{\partial zy} & \frac{\partial^2 V}{\partial zz} \end{bmatrix} = \begin{bmatrix} g_{xx} & g_{xy} & g_{xz} \\ g_{yx} & g_{yy} & g_{yz} \\ g_{zx} & g_{zy} & g_{zz} \end{bmatrix} \quad (2)$$

where

$$V(\mathbf{r}) = 2G\rho \iint_s \ln \frac{1}{\left((x-\xi)^2 + (z-\zeta)^2\right)^{1/2}} d\xi d\zeta \quad (3)$$

here, $\mathbf{r}$ is the space vector of the observation point $p(x,z)$ to the anomalous point $Q(\xi,\zeta)$, and G is the gravitational constant, while ρ is the anomalous density distribution with a domain s.

The most common method to calculate the gravity and gravity gradiometry fields from the subsurface density is to divide the 2D domain into geometrically simple bodies with constant densities. In this case, the domain studied is divided into a finite number of rectangular units of uniform densities (the analytical formulas for the fast computation of gravity fields caused by a polygon are given by Singh [33], and gravity gradiometry fields caused by a polygon are given by Won [34]). Considering that there are N_d observations and N_m rectangular units, the discrete forward modeling operators for gravity or gravity gradiometry can be expressed in a matrix form:

$$\mathbf{d} = \mathbf{G} \cdot \mathbf{m} \quad (4)$$

where, $\mathbf{d}$ is a vector of the observed data (gravity or gravity gradient) of the order N_d, $\mathbf{m}$ is a vector of the remaining densities of the order N_m, and $\mathbf{G}$ is expressed as a forward problem kernel functional, a rectangular matrix of a size $N_\mathrm{d} \times N_\mathrm{m}$. According to Equation (4), the observed data are linearly related to the residual density, and it can be observed that the Jacobian matrix of the observed data is independent of the residual density and only depends on the position of the rectangular unit relative to the observation point.

2.2. MT Forward Problem

In a non-uniform electrical structure, the MT field can be decomposed into primary and secondary fields. The primary field is the field generated by the background, and the secondary field is the field generated by the anomalous body. In the two-dimensional case, take the y-axis as strike direction. From the Maxwell equation of secondary field, two types of polarization modes (transverse electric (TE) and transverse magnetic (TM)) of the expression can be obtained. Two models of the secondary field Helmholtz equations can be obtained through the derivation of the Maxwell equation [35].

TE-mode expression:

$$\frac{\partial}{\partial x}\left(\frac{1}{\hat{z}}\frac{\partial E_{ys}}{\partial x}\right) + \frac{\partial}{\partial z}\left(\frac{1}{\hat{z}}\frac{\partial E_{ys}}{\partial z}\right) - \hat{x}E_{ys} = \Delta\hat{x}E_{yp} \quad (5)$$

TM-mode expression:

$$\frac{\partial}{\partial x}\left(\frac{1}{\hat{x}}\frac{\partial H_{ys}}{\partial x}\right) + \frac{\partial}{\partial z}\left(\frac{1}{\hat{x}}\frac{\partial H_{ys}}{\partial z}\right) - \hat{z}H_{ys} = -\frac{\Delta k^2}{\hat{x}}H_{yp} + \frac{\partial}{\partial z}\left(\frac{\Delta\hat{x}}{\hat{x}}\right)E_{rp} \quad (6)$$

where, $\hat{x} = \sigma + i\omega\varepsilon$ is the admittivity, $\hat{z} = i\omega\mu_0$ is the impedance rate, $\Delta\hat{x}$ is the difference of the admittivity between two-dimensional inhomogeneous medium and a one-dimensional homogeneous medium, and the subscript s and p, respectively, represent the secondary field and the primary field, $\Delta k^2 = -\Delta\hat{x}\hat{z}$. The primary field, E_{xp}, E_{yp}, E_{zp}, H_{xp}, H_{yp}, H_{zp}, can be obtained via the analytic solution of a one-dimensional homogeneous layered medium. Wanamaker [35] used triangular meshes to solve the partial differential Equations (5) and (6) by the finite element method, In this way, the secondary field E_{ys} and H_{ys} can be calculated, and the reciprocity principle can be used to calculate the Jacobian matrix of the MT sounding response.

3. Joint Inversion Methodology

The traditional gravity and MT joint inversion objective functional contains the data misfit terms of gravity and MT method, and the model smoothing constraint terms [8,21–24]. In the proposed method of gravity gradiometry and MT joint inversion, the objective functional contains three components of the gravity gradiometry data misfit term and the MT data misfit term, along with added model smoothing constraints and normalized cross-gradient constraints. The expressions are as follows:

$$\Phi = (\mathbf{d} - \mathbf{f}(\mathbf{m}))^{\mathrm{T}} \cdot \mathbf{C}_d^{-1}(\mathbf{d} - \mathbf{f}(\mathbf{m})) + \boldsymbol{\alpha} \cdot (\mathbf{m} - \mathbf{m}_0)^{\mathrm{T}} \cdot \mathbf{C}_m^{-1} \cdot (\mathbf{m} - \mathbf{m}_0) \tag{7}$$

$$\text{Constraint conditions: } \boldsymbol{\tau}(\mathbf{m}_1, \mathbf{m}_2) = \nabla(\mathbf{m}_1/\chi_1) \times \nabla(\mathbf{m}_2/\chi_2) = 0$$

where $\mathbf{m} = [\mathbf{m}_1^{\mathrm{T}}, \mathbf{m}_2^{\mathrm{T}}]^{\mathrm{T}}$, $\mathbf{m}_0 = [\mathbf{m}_{01}^{\mathrm{T}}, \mathbf{m}_{02}^{\mathrm{T}}]^{\mathrm{T}}$, $\mathbf{d} = [\mathbf{d}_1^{\mathrm{T}}, \mathbf{d}_2^{\mathrm{T}}, \mathbf{d}_3^{\mathrm{T}}, \mathbf{d}_4^{\mathrm{T}}]^{\mathrm{T}}$, $\mathbf{f}(\mathbf{m}) = [\mathbf{f}_1^{\mathrm{T}}(\mathbf{m}), \mathbf{f}_2^{\mathrm{T}}(\mathbf{m}), \mathbf{f}_3^{\mathrm{T}}(\mathbf{m}), \mathbf{f}_4^{\mathrm{T}}(\mathbf{m})]^{\mathrm{T}}$, $\mathbf{C}_d = diag[\mathbf{C}_{d1}, \mathbf{C}_{d2}, \mathbf{C}_{d3}, \mathbf{C}_{d4}]$, $\mathbf{C}_m = diag[\mathbf{C}_{m1}, \mathbf{C}_{m2}]$, $\boldsymbol{\alpha} = [\alpha_1, \alpha_2]$.

Here, $\mathbf{C}_{d1}$, $\mathbf{C}_{d2}$, and $\mathbf{C}_{d3}$ are respectively the data covariance matrix of the gravity gradiometry observation data $\mathbf{d}_1$, $\mathbf{d}_2$, and $\mathbf{d}_3$. $\mathbf{C}_{d4}$ is the data covariance matrix of the MT observation data, $\mathbf{C}_{m1}$ and $\mathbf{C}_{m2}$ are the model covariance matrix of density $\mathbf{m}_1$ and resistivity $\mathbf{m}_2$, respectively, $\mathbf{m}_{01}$ and $\mathbf{m}_{02}$ are the prior model parameters, $\boldsymbol{\alpha}_1$ and $\boldsymbol{\alpha}_2$ are damping parameters, $\mathbf{f}_1(\mathbf{m})$, $\mathbf{f}_2(\mathbf{m})$ and $\mathbf{f}_3(\mathbf{m})$ are forward responses of the three components of the gravity gradiometry, $\mathbf{f}_4(\mathbf{m})$ is an MT forward response, ∇ is a gradient, $\boldsymbol{\tau}$ is a normalized cross-gradient functional as a parameter gradient of the density and resistivity model, and χ_1 and χ_2 are the normalized operators of resistivity and density, respectively. For our particular test cases, the normalized operators are determined by $\chi_1 = \mathrm{Max}(\mathbf{m}_{1\_sep}) - \mathrm{Min}(\mathbf{m}_{1\_sep})$, and $\chi_2 = \mathrm{Max}(\mathbf{m}_{2\_sep}) - \mathrm{Min}(\mathbf{m}_{2\_sep})$. Max and Min are the maximum and minimum values, respectively. $\mathbf{m}_{1\_sep}$ and $\mathbf{m}_{2\_sep}$ are represented as a separate inversion density and resistivity model, respectively.

In this paper, the Gaussian Newton algorithm (GN) is used to solve the joint inversion objective functional (7) in model-space [19–24]. Firstly, the nonlinear objective functional (7) and the constraint condition are transformed into linear equations by the Taylor expansion. Then, the Lagrangian operator method is used [36,37], which adds a constraint condition to the objective functional (7):

$$\begin{aligned}\Psi = \; & (\hat{\mathbf{d}} - \mathbf{A} \cdot (\mathbf{m} - \mathbf{m}_0))^{\mathrm{T}} \cdot \mathbf{C}_d^{-1} \cdot (\hat{\mathbf{d}} - \mathbf{A} \cdot (\mathbf{m} - \mathbf{m}_0)) + \boldsymbol{\alpha} \cdot (\mathbf{m} - \mathbf{m}_0)^{\mathrm{T}} \cdot \mathbf{C}_m^{-1} \cdot (\mathbf{m} - \mathbf{m}_0) \\ & + 2\boldsymbol{\Lambda} \cdot (\hat{\boldsymbol{\tau}} + \mathbf{B} \cdot (\mathbf{m} - \mathbf{m}_0))\end{aligned} \tag{8}$$

where $\hat{\mathbf{d}} = \mathbf{d} - \mathbf{f}(\mathbf{m}_0)$, $\hat{\boldsymbol{\tau}} = \boldsymbol{\tau}(\mathbf{m}_0)$

$$\mathbf{A} = \frac{\partial \mathbf{f}(\mathbf{m})}{\partial \mathbf{m}} = \begin{bmatrix} \cdots & \cdots & \cdots \\ \cdots & A_{ij} & \cdots \\ \cdots & \cdots & \cdots \end{bmatrix}, \quad A_{ij} = \frac{\partial \mathbf{f}_i(\mathbf{m})}{\partial \mathbf{m}_j}, \quad i = 1,2,3,4; \; j = 1,2; \tag{9}$$

$$\mathbf{A} = \begin{bmatrix} \mathbf{A}_{11} & \\ \mathbf{A}_{21} & \\ \mathbf{A}_{31} & \\ & \mathbf{A}_{42} \end{bmatrix}, \quad \mathbf{B} = [\frac{\partial \boldsymbol{\tau}}{\partial \mathbf{m}_1} \frac{\partial \boldsymbol{\tau}}{\partial \mathbf{m}_2}] \tag{10}$$

where $\mathbf{A}$ and $\mathbf{B}$ represent the Jacobian matrix of the forward response $\mathbf{f}(\mathbf{m})$ and the normalized cross-gradient functional $\boldsymbol{\tau}$, and $\boldsymbol{\Lambda}$ is a column vector of the Lagrange multipliers.

The solution of Equation (8) is determined using iterative model updates from a starting model by solving the equation $\partial\Psi/\partial\mathbf{m} = 0$. The model update is expressed as:

$$\Delta\mathbf{m} = \mathbf{N}^{-1}\cdot\mathbf{n} - \mathbf{N}^{-1}\cdot\mathbf{B}^{\mathrm{T}}\cdot\boldsymbol{\Lambda} \tag{11}$$

$$\mathbf{N} = \mathbf{A}^{\mathrm{T}}\cdot\mathbf{C}_d^{-1}\cdot\mathbf{A} + \alpha\cdot\mathbf{C}_m^{-1},\ \mathbf{n} = \mathbf{A}^{\mathrm{T}}\cdot\mathbf{C}_d^{-1}\cdot\hat{\mathbf{d}}. \tag{12}$$

Substituting Equation (11) into the constraint condition, the expression of the Lagrange multiplier $\boldsymbol{\Lambda}$ is given as

$$\boldsymbol{\Lambda} = \left(\mathbf{B}\cdot\mathbf{N}^{-1}\cdot\mathbf{B}^{\mathrm{T}}\right)^{-1}\left(\hat{\boldsymbol{\tau}} + \mathbf{B}\cdot\mathbf{N}^{-1}\cdot\mathbf{n}\right) \tag{13}$$

Finally, we obtained the model expression of the model-space as follows:

$$\mathbf{m} = \mathbf{m}_0 + \left(\mathbf{A}^{\mathrm{T}}\cdot\mathbf{C}_d^{-1}\cdot\mathbf{A} + \alpha\cdot\mathbf{C}_m^{-1}\right)^{-1}\cdot\mathbf{A}^{\mathrm{T}}\cdot\mathbf{C}_d^{-1}\cdot\hat{\mathbf{d}} - \left(\mathbf{A}^{\mathrm{T}}\cdot\mathbf{C}_d^{-1}\cdot\mathbf{A} + \alpha\cdot\mathbf{C}_m^{-1}\right)^{-1}\cdot\mathbf{B}^{\mathrm{T}}\cdot\boldsymbol{\Lambda} \tag{14}$$

In this paper, we transformed the inversion computing space from a model-space to a data-space. The formula for the joint inversion in the data-space can be directly derived from those in the model-space using the matrix identity that changes the size of the inverse matrix. We obtained the model expression of the data-space as follows:

$$\begin{aligned}\mathbf{m} = \mathbf{m}_0 + \mathbf{C}_m\cdot\mathbf{A}^{\mathrm{T}}\cdot\left(\mathbf{A}\cdot\mathbf{C}_m\cdot\mathbf{A}^{\mathrm{T}} + \alpha\cdot\mathbf{C}_d\right)^{-1}\cdot\hat{\mathbf{d}} - \\ \left(\mathbf{I} - \mathbf{C}_m\cdot\mathbf{A}^{\mathrm{T}}\cdot\left(\mathbf{A}\cdot\mathbf{C}_m\cdot\mathbf{A}^{\mathrm{T}} + \alpha\cdot\mathbf{C}_d\right)^{-1}\cdot\mathbf{A}\right)\cdot\mathbf{C}_m\cdot\mathbf{B}^{\mathrm{T}}\cdot\boldsymbol{\Lambda}\end{aligned} \tag{15}$$

$$\mathbf{N}^{-1}\cdot\mathbf{n} = \mathbf{C}_m\cdot\mathbf{A}^{\mathrm{T}}\cdot\left(\mathbf{A}\cdot\mathbf{C}_m\cdot\mathbf{A}^{\mathrm{T}} + \alpha\cdot\mathbf{C}_d\right)^{-1}\cdot\hat{\mathbf{d}} \tag{16}$$

$$\mathbf{N}^{-1} = \left(\mathbf{I} - \mathbf{C}_m\cdot\mathbf{A}^{\mathrm{T}}\cdot\left(\mathbf{A}\cdot\mathbf{C}_m\cdot\mathbf{A}^{\mathrm{T}} + \alpha\cdot\mathbf{C}_d\right)^{-1}\cdot\mathbf{A}\right)\cdot\mathbf{C}_m \tag{17}$$

Comparing the model expressions Equations (14) and (15) in different spaces, $\mathbf{A}^{\mathrm{T}}\cdot\mathbf{C}_d^{-1}\cdot\mathbf{A} + \alpha\cdot\mathbf{C}_m^{-1}$ is an $\mathrm{M}\times\mathrm{M}$ dimensions matrix, and $\mathbf{A}\cdot\mathbf{C}_m\cdot\mathbf{A}^{\mathrm{T}} + \alpha\cdot\mathbf{C}_d$ is an $\mathrm{N}\times\mathrm{N}$ dimensions matrix. M is the number of model parameters and N is the number of data. In general, the number of model parameters will be much larger than the number of data, and the Equation (14) needs to solve the inverse of the M $\times$ M dimension matrix during the joint inversion process with the model-space method, which results in increased storage requirements and computational time. Therefore, cross-gradient joint inversion can be more effectively implemented in data-space than model-space.

In this paper, the model covariance matrices of the two methods are constructed differently. The model covariance matrix ($\mathbf{C}_m$) is directly required for the data-space method, while its inverse is required in the model-space method. In the model-space method, the inverse of the model covariance is usually implemented as a sparse model roughness operator [38–40]. The exact inverse of a specific roughness operator cannot be determined in practice because of the size of this matrix, which, in general, will be full. In the data-space method, the model covariance matrix is not directly constructed, and the products of the model covariance matrix with any model vector are computed by solving a diffusion equation [31,32]. Thus, the diffusion equation has to be repeatedly solved in QMR (quasi-minimal residual), unlike in the model space method, but using the operator splitting solution [32] makes it less computationally expensive. In summary, we use different model covariance matrices in model-space and data-space, so we will get two different inversion results.

4. Synthetic Example

4.1. Compare Gravity and Gravity Gradiometry Data

To verify the validity and feasibility of our proposed data-space joint inversion of the gravity gradiometry and MT data, we tested the separate inversion, the joint inversion of gravity and MT data, and the joint inversion of the gravity gradiometry and MT data in the data-space using a complex combined model that is more similar to the real underground structure, including some typical geological elements and fault structures, as shown in Figure 1. Four anomalous bodies are embedded in the uniform half-space E. Anomalous bodies A, B, and C are isolated crustal rocks of different sizes and different buried depths. Anomalous body D is a low-resistivity and high-density body in the deep region of the model. It is buried under the high-resistivity and low-density anomaly bodies B and C, with a ladder fracture structure on its left. The density model and the resistivity model adopt the same mesh (140 × 60) in the joint inversion mesh region, and the resistivity model requires the extension of the mesh (156 × 76) outside the joint inversion mesh region. Gravity and gravity gradiometry methods use the same observation point. There are 30 observation points with a point spacing of 0.2 km from a horizontal distance from 0 to 6 km. The MT data contain the apparent phase and apparent resistivity for 10 frequencies in the range of 1 to 1000 Hz for nine stations spaced 0.6 km from each other. To simulate the noisy data, a 5% random Gaussian noise was added to the MT, gravity, and gravity gradiometry forward response data.

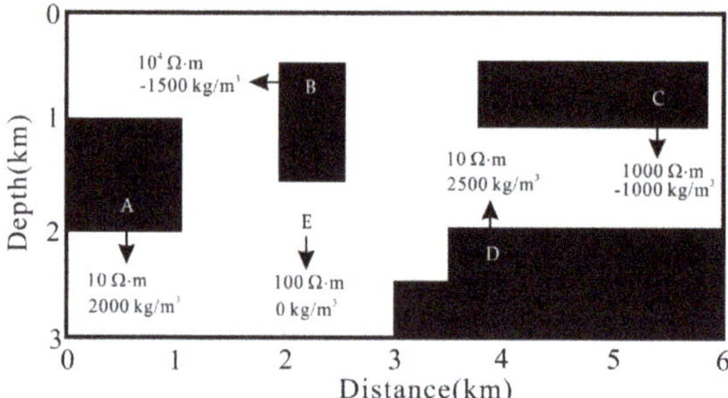

Figure 1. Theoretical model 1.

The initial model of all inversions uses a uniform half-space with a density of 0 kg/m^3 and a resistivity of 100 Ω·m. The reference model is set to the initial model. Figures 2 and 3 show the forward response results of resistivity and density models of different methods. The forward response calculated by either the separate inversion or joint inversion results is basically consistent with theoretical model responses. Figure 4 shows the separate inversion results of the MT, gravity and gravity gradiometry for the complex combined model. The data misfit of MT, gravity and gravity gradiometry inversion are $RMS_{MT} = 0.95$, $RMS_{Grv} = 0.71$, and $RMS_{Grad} = 0.80$, respectively (Figure 5a). Notably, the gravity inversion (Figure 4b) cannot recover the true space geometry and position of the anomalous bodies, and the vertical resolution of the inversion results is very poor, and it is difficult to accurately identify low-density anomalous bodies. However, the inversion result of the gravity gradiometry (Figure 4c) can roughly reflect the spatial position and geometry of the true anomalous bodies. Compared with the results of the gravity inversion, the distribution of the recovered underground anomaly can be improved, but the gravity gradiometry inversion results still have problems, such as the upset of the center of the anomalous bodies, the blurring of the boundary between the anomalous bodies and the surrounding rock, and the inability to recover the exact physical parameter value. The resistivity model (Figure 4a) recovers the gross structure of the synthetic model better than the density model (Figure 4b,c), as the MT data cover an adequate frequency range for distinguishing deep anomalous bodies. The results of the joint inversion are shown in Figure 6. For the joint inversion of gravity and MT (Figure 6a,b), the final models of resistivity and density are obtained with the data misfit (RMS_{MT} = 0.66, RMS_{Grv} = 1.0), as shown in Figure 5b. The structural features of the anomalous bodies are very similar in the resistivity and density models as required by our joint inversion algorithm. The deep anomalous body of the density model can be identified as an indirect contribution propagated from the resistivity model by the cross-gradient constraints. For the joint inversion of the gravity gradiometry and MT (Figure 6c,d), the final models of resistivity and density are obtained with data misfit (RMS_{MT} = 0.83, RMS_{Grad} = 0.80), as shown in Figure 5c. The resistivity and density models show improved features when compared to the above separate and joint inversion results. The density models' anomalous bodies boundaries become clearer, and the geometrical and physical values of the anomaly more closely reflect the true model, as an indirect result of the resistivity model using the cross-gradient constraints. In particular, the vertical resolution has been significantly improved. The resistivity model also improves the horizontal resolution due to the structural similarity of the density. The results of the gravity gradiometry and MT joint inversion are better than the results of the separate inversion and gravity and MT joint inversion, both in space form and the numerical recovery of physical properties.

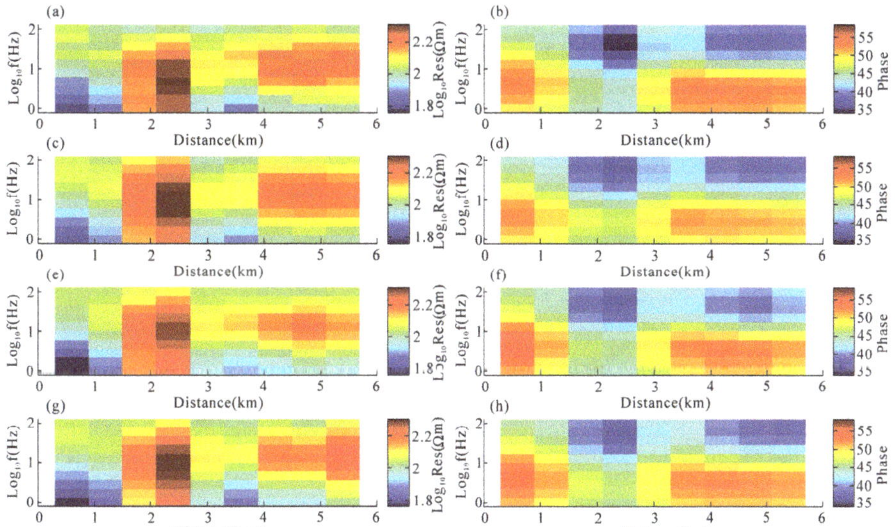

Figure 2. Pseudo-sections illustrating the forward responses of apparent resistivity (**a**,**c**,**e**,**g**) and apparent phase (**b**,**d**,**f**,**h**) by the transverse magnetic mode. Forward response of theoretical resistivity model (**a**,**b**), forward response of separate inversion of magnetotelluric (MT) (**c**,**d**), forward response of joint inversion of gravity and MT (**e**,**f**), forward response of joint inversion of gravity gradiometry and MT (**g**,**h**).

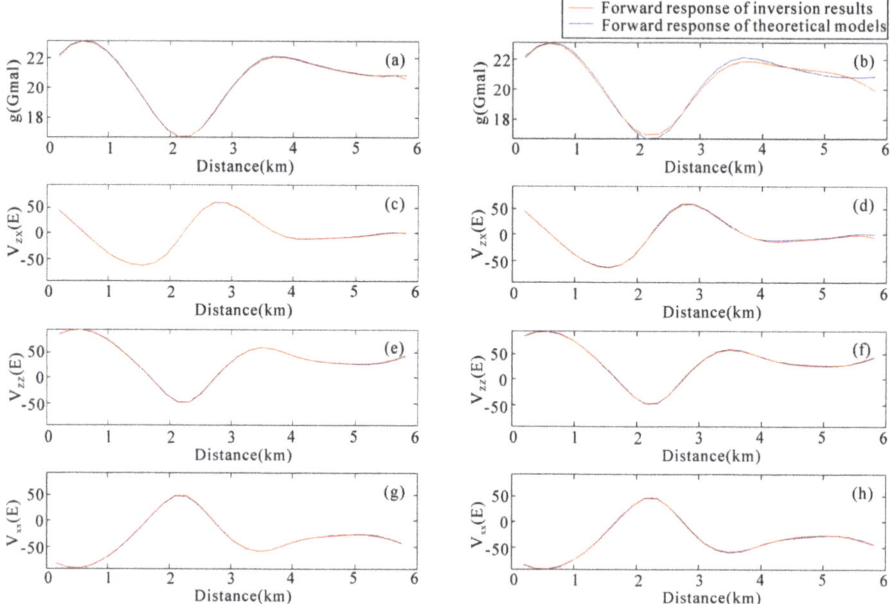

Figure 3. Forward responses curves of gravity (**a**,**b**) and gravity gradiometry (**c**,**d**,**e**,**f**,**g**,**h**). Forward response of the separate inversion of the gravity (**a**), forward response of the separate inversion of the gravity gradiometry (**c**,**e**,**g**), forward response of the joint inversion of the gravity and MT (**b**), forward response of the joint inversion of the gravity gradiometry and MT (**d**,**f**,**h**). The blue line represents the forward response of the theoretical models. The red line represents the forward response of the inversion results.

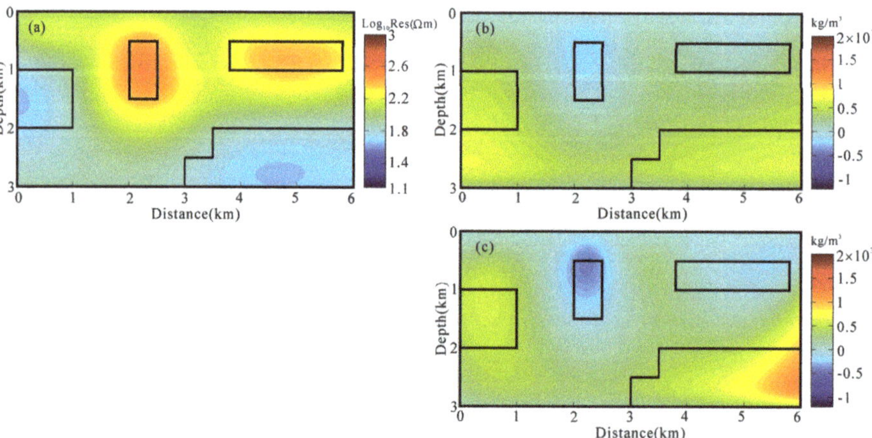

Figure 4. The data-space separate inversion results of model 1. Separate inversion results of MT (**a**), gravity (**b**) and gravity gradiometry (**c**), resistivity model (**a**), density model (**b**,**c**).

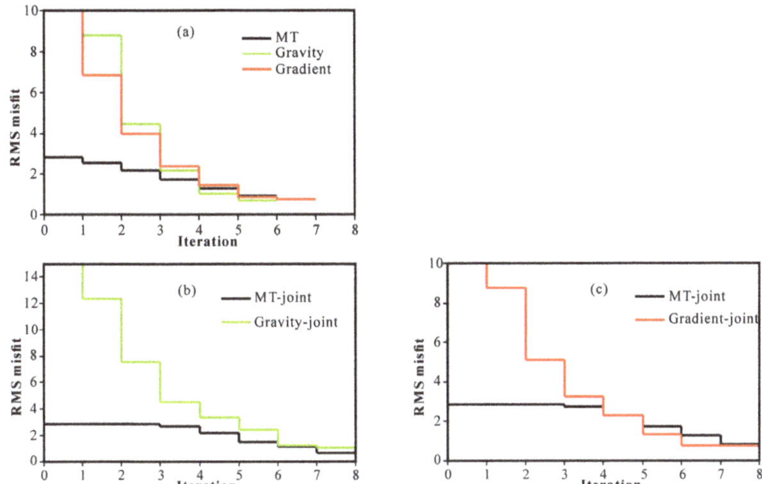

Figure 5. The iterative curve of the root mean square (RMS) misfit of the separate inversion (**a**), joint inversion of gravity and MT data (**b**), and joint inversion of gravity gradiometry and MT data (**c**).

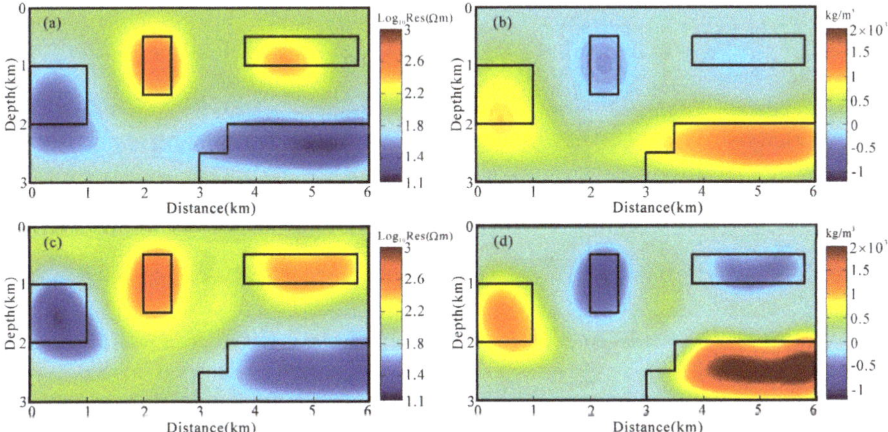

Figure 6. The data-space joint inversion results of model 1. Joint inversion results of MT and gravity (**a**,**b**), MT and gravity gradiometry (**c**,**d**), resistivity model (**a**,**c**), density model (**b**,**d**).

We also calculate the cross-gradient values of every pair of models from the data-space separate inversion of MT and gravity data (Figure 4a,b), the data-space separate inversion of MT and gravity gradiometry data (Figure 4a,c), the data-space joint inversion of MT and gravity data (Figure 6a,b), and the data-space joint inversion of MT and gravity gradiometry data (Figure 6c,d). The final cross-gradient values obtained for every model combination after applying separate and joint inversions are illustrated in the detailed maps of the cross-gradient values, as shown in Figure 7. Note that the joint inversion yields much smaller cross-gradient values for every pair of models than those of separate inversions. Meanwhile, the cross-gradient values of joint inversion base on MT and gravity gradiometry data are much smaller than those of joint inversion based on MT and gravity. This result serves as quantitative evidence that the proposed method ensures a higher level of structural conformity.

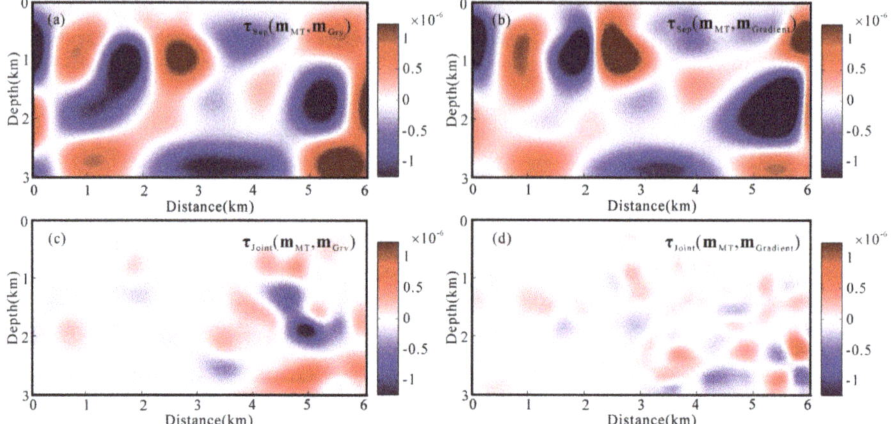

Figure 7. Cross-gradient values attained for every pair of models for the data-space separate inversion of gravity and MT data (**a**), gravity gradient and MT data (**b**), and for the data-space joint inversion of gravity and MT data (**c**), gravity gradiometry and MT data (**d**).

4.2. Test the Partially Structurally Consistent and Inconsistent Model

To test the adaptability and accuracy of our proposed data-space joint inversion of the gravity gradiometry and MT data in various scenarios, we tested the models that were partially structurally consistent and inconsistent. The mesh division and distribution of the observation points of the model are the same as those of theoretical model 1. We also added 5% Gaussian noises to the resistivity and density model responses.

The initial model of all inversions uses a uniform half-space with a density of 0 kg/m^3 and a resistivity of 100 Ω·m. The reference model is set to the initial model. In the partially structurally consistent model (Figure 8a,b), the joint inversion results are shown in Figure 9a,b. The data misfit of the MT and gravity gradiometry inversion are $RMS_{MT} = 0.75$ and $RMS_{Grad} = 0.95$, respectively. In the structurally inconsistent model (Figure 10a,b), the joint inversion results are shown in Figure 11a,b. The data misfit of MT and gravity gradiometry inversion are $RMS_{MT} = 0.97$ and $RMS_{Grad} = 0.73$, respectively. All joint inversions converge to the RMS misfit threshold of one or less. We find that the proposed method can recover the geometry and physical parameter values of the underground anomaly sources in the partially structurally consistent and inconsistent models. The results show that the joint inversion of MT and gravity gradiometry is suitable for various scenarios (structural consistent and structural inconsistent models), which can recover the true model. Different geophysical methods are mutually constrained by the structural similarities at the same model boundaries, while the structural similarity constraints are not effective at different model boundaries, and joint inversion only performs smoothing model constraints.

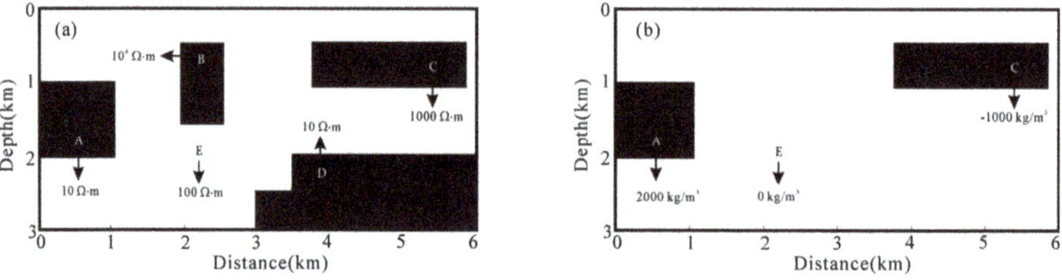

Figure 8. The partially structurally consistent model. Resistivity model (**a**), density model (**b**).

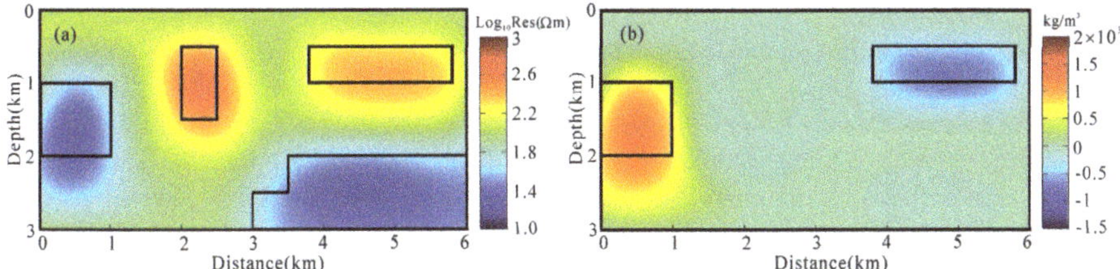

Figure 9. Joint inversion results of partially structurally consistent model. Resistivity model (**a**), density model (**b**).

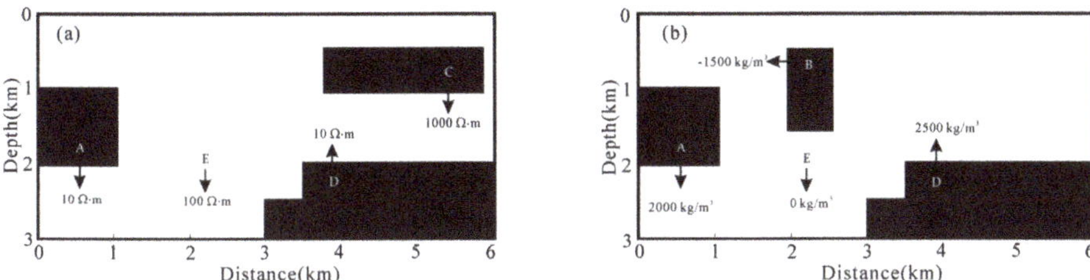

Figure 10. The structurally inconsistent model. Resistivity model (**a**), density model (**b**).

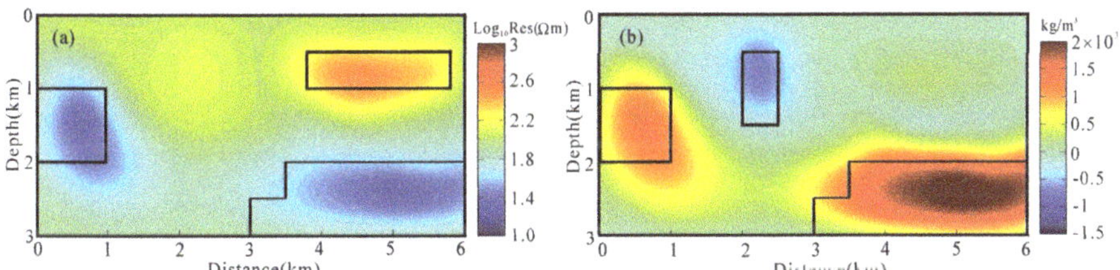

Figure 11. Joint inversion results of structurally inconsistent model. Resistivity model (**a**), density model (**b**).

4.3. Compare Joint Inversion of Model-Space and Data-Space

In this section, we will test the advantages of the joint inversion of gravity gradiometry and MT data in the data-space compared to the model space in terms of the computational time and memory storage. We designed a density and resistivity model that increases the number of meshes and observation points, as shown in Figure 12. The density model and the resistivity model adopt the same mesh (341 × 121) in the joint inversion mesh region, and the resistivity model requires an extension of the mesh (357 × 135) outside the joint inversion mesh region. There are two rectangular bodies embedded in the uniform half-space. The sizes of the two rectangular bodies are 0.6 × 1.0 km^2. The gravity gradiometry method has 79 observation points, with a point spacing of 0.1 km from a horizontal distance of −1 to 7 km. The MT data contain the apparent phase and apparent resistivity for 10 frequencies in the range of 1 to 1000 Hz for 39 stations spaced 0.2 km from each other.

Regardless of the data-space or the model-space method, the initial model uses a uniform half-space, with a density of 0 kg/m^3 and a resistivity of 100 Ω·m. The reference model is set to the initial model. For the model-space joint inversion of gravity gradiometry and MT, the sixth iteration model (Figure 13a,b) attained data misfits of 1.05 and 1.14 for gravity gradiometry and MT, respectively. For the data-space joint inversion of gravity gradiometry and MT, the sixth iteration model (Figure 13c,d) attained data misfits of 1.05 and 1.00 for gravity gradiometry and MT. We can find that the inversion results of the two methods can recover the geometry of the double anomaly.

However, the resistivity and density models (Figure 13c,d) show improved features when compared to the model-space joint inversion results (Figure 6b,e). The density and resistivity models' anomalous body boundaries become clearer, and the geometrical and physical values of the anomaly more closely reflect those of the true model. The inconsistencies of the inversion results are mainly due to the different model covariance matrices used by the two methods. In terms of inversion calculation time, the model-space method consumes approximately 13.15 h, whereas the data-space method consumes approximately 3.92 h for our particular test example. In addition, the maximum memory requirements are approximately 4.1 and 0.62 GB, respectively, as shown in Table 1. The data-space method can be applied to the joint inversion MT and gravity gradiometry, which can greatly reduce memory consumption and effectively improve the calculation speed. This method allows us to invert the multi-parameter joint inversion of large areas of large data volume with a personal computer (PC) in a relatively short period of time.

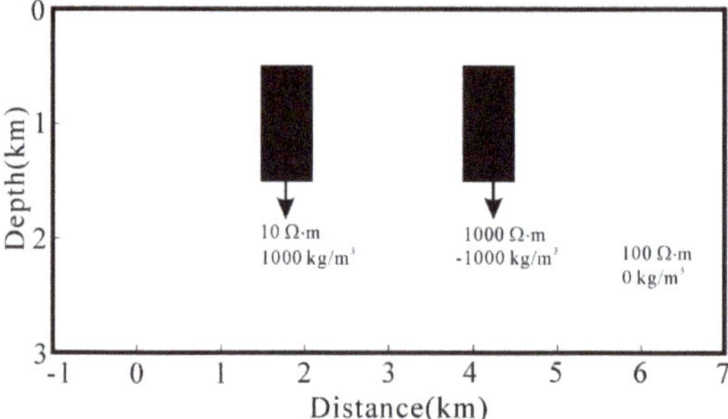

Figure 12. Theoretical model 3.

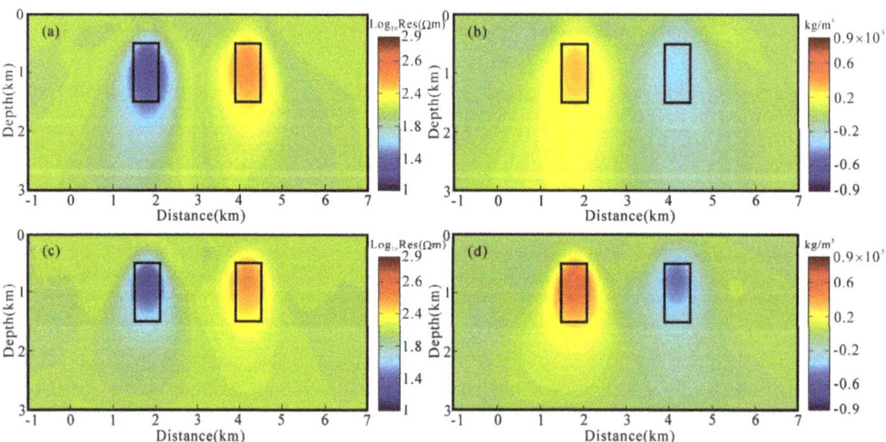

Figure 13. The data-space and model-space joint inversion results of MT and gravity gradiometry data. the model-space joint inversion results (**a**,**b**), the data-space joint inversion results (**c**,**d**), resistivity model (**a**,**c**), density model (**b**,**d**).

Table 1. Comparison of computational time and memory storage in the data-space and model-space joint inversion.

	Model-Space	Data-Space
Computational Time	13.15 h	3.92 h
Memory Storage	4.1 GB	0.62 GB

5. Field Example

5.1. Geologic Background of the Study Area

The study area is located in Liangjiang Town, Antu County, Jilin Province (Figure 14). The area belongs to the eastern part of the northern margin of the North China plate in the geotectonic position, and its northeast side is adjacent to the Xingmeng-Jihei orogenic belt. The oldest formations exposed in the study area are the Paleoproterozoic formations, followed by the Middle Proterozoic, Mesozoic, and Cenozoic formations. However, Paleozoic formations are generally missing. Due to the special geotectonic environment in the study area, the magmatic action in the area is strong, and intrusive rocks of different ages are widely developed. According to the characteristics of the times, these rocks can be divided into two parts: Late Archean and Mesozoic granitic rocks. The area is located in the eastern part of the Jihei trough fold system, and the Jiamusi block and the north edge of the North China plate collide with the wrinkled orogenic belt and are in the superposition of the western Pacific tectonic domain (NE-trending structure) and the ancient Asian domain (EW-trending structure). The fold structure and fault structure are very developed. The mineral resources in the area are also very rich, with many types of deposits, many mineralization periods, large scales, as well as high gold grade and concentrated distribution. The main minerals include gold, molybdenum, iron, copper, and nickel. The representative deposit in the area is the Haigou gold mine. The formation and distribution of the Haigou gold deposits are closely controlled by the structure of the area. The deposit is a large-scale sulphide-like quartz vein- type gold deposit, belonging to the orogenic gold deposit. Other mineral resources in the area include coal mines, peat, and oil shale deposits, which are important metallogenic prospects for endogenous metals and non-metals.

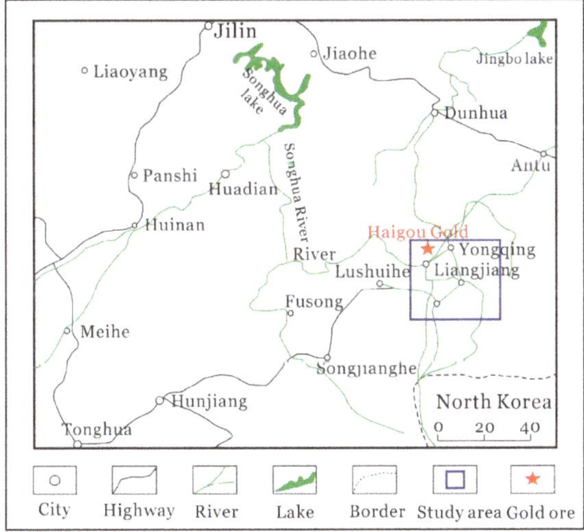

Figure 14. Location of the study area.

5.2. Data Acquisition and Inversion

We conducted a geophysical survey along with one profile in the study area. Figure 15 shows the geological map of the study area. The black solid line indicates the position of the survey line. The survey line runs in a north-east direction and measures 18 km in length. A total of 181 controlled-source audio-frequency magnetotelluric (CSAMT) points are collected, and the distance between the observation points is 100 m. Each observation point acquired the apparent resistivity of 10 frequencies (16–8192 Hz) in TM mode. Using the full regional apparent resistivity method [41], we convert the CSAMT apparent resistivity data to the near-source corrected apparent resistivity data, as shown in Figure 16a. Meanwhile, the gravity Bouguer data were extracted with 40 m spacing along our study profile to furnish 451 gravity data for inversion. We transform the gravity data (Figure 16b) into gravity gradiometry data (Figure 16c–e) by Fourier transform, i.e., the space domain and frequency domain integration and derivation [42].

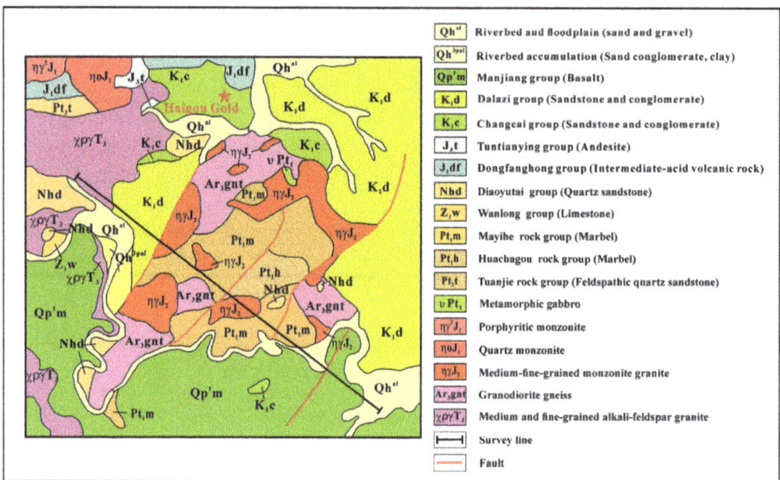

Figure 15. The geological map of the study area.

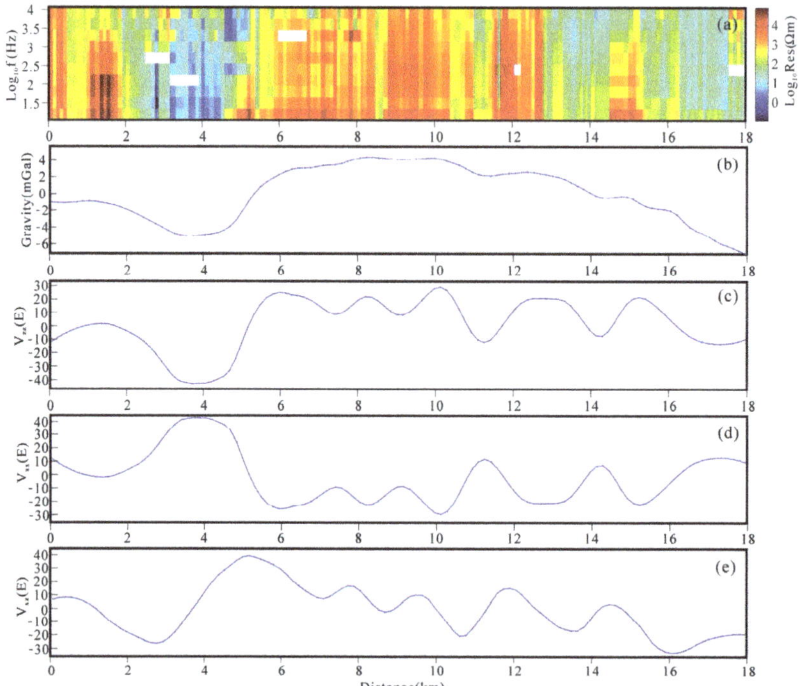

Figure 16. Field observation data. The apparent resistivity in the TM mode after conversion (**a**), the gravity data (**b**), the gravity gradiometry data V_{zz} (**c**), the gravity gradiometry data V_{xx} (**d**), the gravity gradiometry data V_{xz} (**e**).

The CSAMT, gravity, and gravity gradiometry data were inverted to resistivity and residual density models using the separate inversion, the joint inversion of CSAMT and gravity data using the model-space method, and the joint inversion of CSAMT and gravity gradiometry data using the data-space method. The density model and the resistivity model use the same mesh (327 × 61) in the joint inversion region. Outside the joint inversion region, the resistivity model also needs to expand the mesh size (343 × 73). The initial model is a uniform half-space with a density of 0 kg/m^3 and a resistivity of $10^{2.5}$ Ω·m. The results of the separate inversion are shown in Figures 17 and 18. In the separate inversion using the model-space method, the data misfit of the CSAMT, gravity, and gravity gradiometry are RMS_{CSAMT} = 1.08, RMS_{Grv} = 0.48, and RMS_{Grad} = 0.97, respectively. In the separate inversion using the data-space method, the data misfit of the CSAMT, gravity, and gravity gradiometry are RMS_{CSAMT} = 1.08, RMS_{Grv} = 0.93 and RMS_{Grad} = 0.99, respectively. The separate inversion results obtained by the two methods are, generally, features of the same trend. The inconsistencies of the inversion results are mainly due to the different covariances of the models used by the two methods. The structural similarities of the separate inversion results of the three geophysical exploration methods are very different in both the model-space and data-space methods. These differences create great difficulties for geological interpretation. Compared with the gravity inversion results, the gravity gradiometry inversion results identify greater underground structural distribution and can obtain more underground information and help geological interpretation.

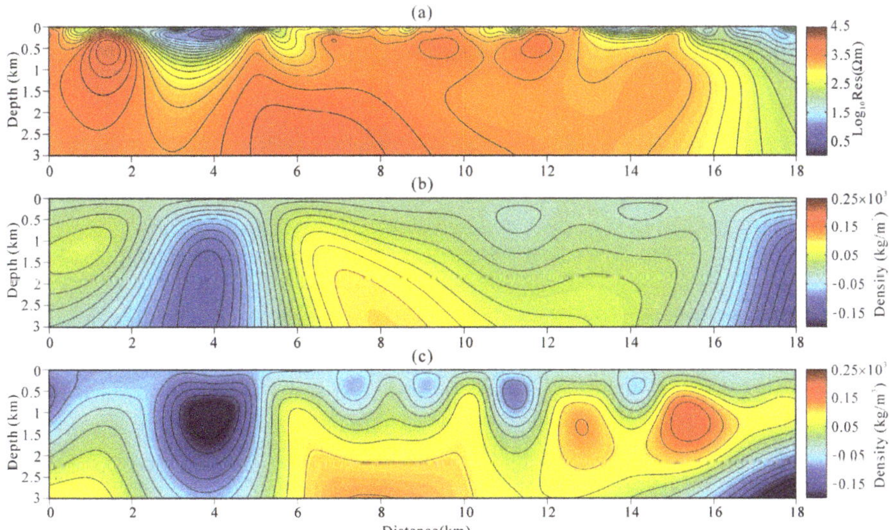

Figure 17. Separate inversion results of the field data using the model-space. The controlled-source audio-frequency magnetotelluric (CSAMT) separate inversion (**a**), gravity separate inversion (**b**), gravity gradiometry separate inversion (**c**).

The results of the joint inversion are shown in Figures 19–21. For the joint inversion of gravity and CSAMT data using the model-space method (Figure 19a,b), the final models of resistivity and density are obtained with the data misfit (RMS_{CSAMT} = 1.08, RMS_{Grv} = 0.64). For the joint inversion of gravity gradiometry and CSAMT data using the model-space method (Figure 20a,b), the final models of resistivity and density are obtained with data misfit (RMS_{CSAMT} = 1.00, RMS_{Grad} = 0.99). For the joint inversion of gravity gradiometry and CSAMT data using the data-space method (Figure 21a,b), the final models of resistivity and density are obtained with data misfit (RMS_{CSAMT} = 1.64, RMS_{Grad} = 1.29). Note that the structural similarity between the different physical properties in the results of joint inversion, which is due to the contribution of cross-gradient constraints in the objective functional, and the result of the joint inversion is more conducive to geological interpretation. Comparing the joint inversion results of the two different data sets, we find that the results of the joint inversion of gravity and MT can roughly divide the underground structure. However, the results of the joint inversion of the gravity gradiometry and MT can clearly and meticulously identify the underground structural information, which facilitates more accurate geological interpretation and detection of geological resources. Meanwhile, the joint inversion based on data-space (Figure 21a,b) has more advantages in memory storage and computing time than model-space (Figure 20a,b).

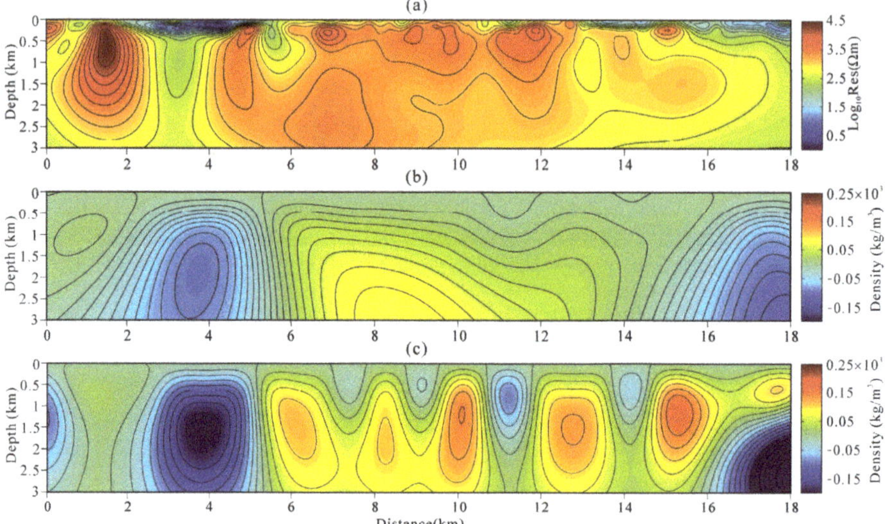

Figure 18. Separate inversion results of field data using the data-space. CSAMT separate inversion (**a**), gravity separate inversion (**b**), gravity gradiometry separate inversion (**c**).

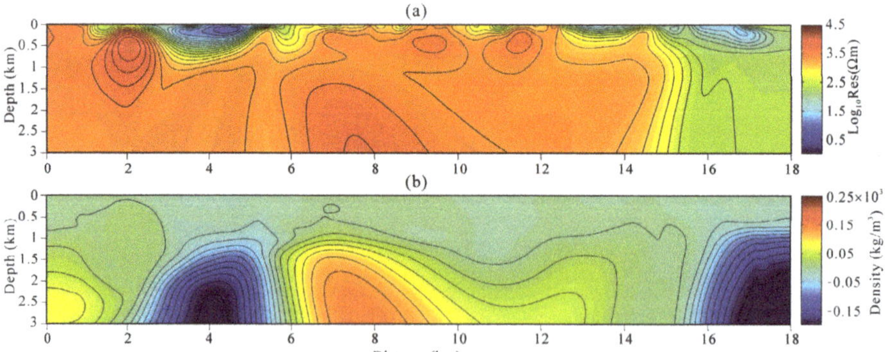

Figure 19. Joint inversion results of CSAMT and gravity data using the model-space method. Resistivity model (**a**), density model (**b**).

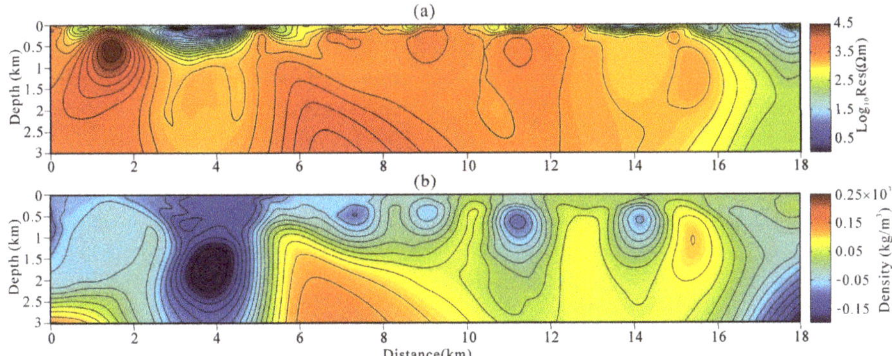

Figure 20. Joint inversion results of CSAMT and gravity gradiometry data using the model-space method. Resistivity model (**a**), density model (**b**).

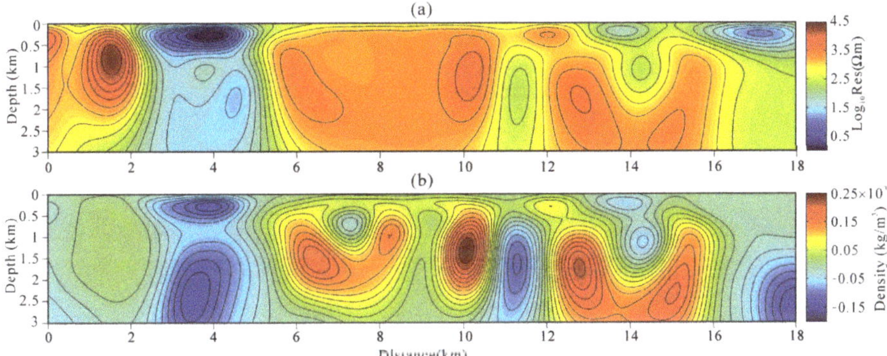

Figure 21. Joint inversion results of CSAMT and gravity gradiometry data using the data-space method. Resistivity model (**a**), density model (**b**).

5.3. Geological Interpretation

The geophysical models selected for the final interpretation are represented in the composite form of an RGB (red-green-blue) image, which facilitates an integrated analysis of the multiple models and a visualization of the significant units. Figure 22 shows the RGB images composed of separately and jointly estimated models using the data-space method. By comparing the composite image of the separate inversion and joint inversion, it can be found that the composite images obtained by joint inversion are superior to those obtained by separate inversion, and a clearer anomalous source boundary can be obtained, mainly due to the inconsistency of the separate inversion results of the different methods. The composite image color contains the resistivity and residual density information, from which the inversion results can be analyzed intuitively to accurately identify, and possibly classify, the underground geological structures. We predict that the combined analysis of the constituent parameter values in the RGB image will be of immense value to the search for multiple parameter cross-correlations that can help in lithology classification and the understanding of complex subsurface processes.

We produced a geological interpretation for this area using the RGB composite image (Figure 22c) and a geological map (Figure 15). First, we divided the survey line into four segments (I–IV) according to the distribution of gravity anomalies from the northwest side to the southeast side of the profile from high to low to high and then low. Segment I is located at the northwest end of the study area, and the surface reveals the medium and fine-grained alkali-feldspar granite with low density and high resistivity. The gravity anomaly of this segment is characterized by a mid–high anomaly with a gentle change, and it can be inferred that it is a comprehensive reflection of the background value increase caused by the relatively mid–high density Qingbaikouan period Diaoyutai formation near the northwest end of the survey line. The surface lithology of segment II is the sandstone and conglomerate of the Cretaceous Dalizi formation. The gravity anomaly value of this segment is the smallest, and the resistivity is low. This resistivity is presumed to be caused by the Cretaceous Dalizi formation and the Changcai formation, with large sediment thickness caused by the Yalu River fault. The surface of segment III reveals the Diaoyutai formation, Mayihe rock formation, and Archean gneiss. The gravity anomaly value of this segment is the highest, and the resistivity is high. This resistivity is mainly caused by the Archean granodiorite gneiss and Early Proterozoic Mayihe rock formation marble. The local peak depression of the gravity gradiometry anomaly is caused by the Jurassic medium-fine-grained monzonite granite. Segment IV is located at the southeast end of the survey line, is a low-gravity anomaly area. Here, there are Manjiang formations and Cretaceous Dalazi formations on the surface. It was speculated that this segment was caused by the basalt of the Manjiang formation and the sandstone of the Cretaceous Dalazi formation under a cover layer, the Jurassic Tuntianying formation andesite, and the Jurassic medium-fine-grained monzonite granite with relatively low-density and low-resistivity. Moreover, through the geological map and the RGB composite map, we can infer the distribution of the fault structure (F_1–F_4) below the survey line. Based on the above inference results, we finally obtained a comprehensive interpretation profile, as shown in Figure 23. The corresponding lithostratigraphic units in Figure 23 are shown in Table 2.

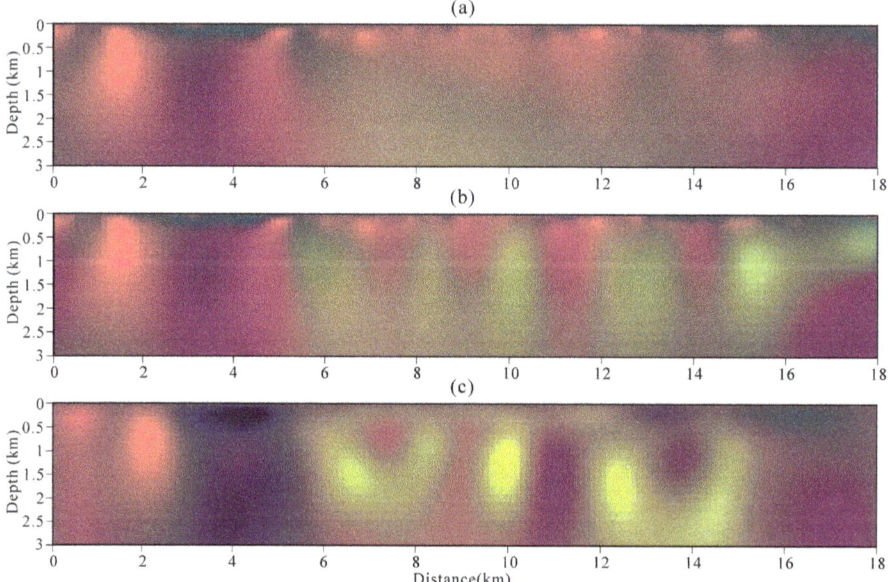

Figure 22. The RGB composite image of the separate inversion of gravity and CSAMT data using the data-space method (**a**), the separate inversion of gravity gradiometry and CSAMT data using the data-space (**b**), and the joint inversion of gravity gradiometry and CSAMT data using the data-space (**c**).

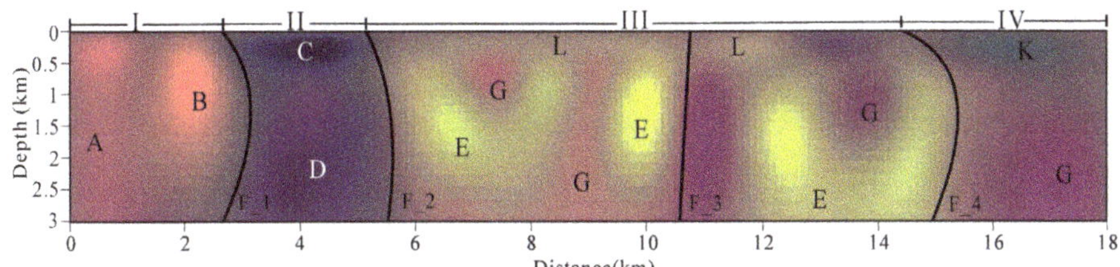

Figure 23. Comprehensive geological interpretation profile.

Table 2. Stratigraphic lithology.

Geological Time		Lithostratigraphic Units	Lithology	Unit
Era	Period	Formation		
Cenozoic	Neogene	Manjiang (Qp^1m)	Basalt	K
	Triassic	($\chi\rho\gamma T_3$)	Medium and fine-grained alkali-feldspar granite	A
Mesozoic		Tuntianying (J_3t)	Andesite	D
	Jurassic	Dongfanghong (J_1df)	Intermediate-acid volcanic rock	D
		($\eta\gamma J_2$)	Medium-fine-grained monzonite granite	G
	Cretaceous	Dalazi (K_1d)	Sandstone	C
		Changcai (K_1c)	Sandstone	C
Proterozoic	Qingbaikouan	Diaoyutai (Nhd)	quartz sandstone	B
Paleoproterozoic		Mayihe (Pt_1m)	Marble	L
Archeozoic		(Ar_3gnt)	Granodiorite gneiss	E

The Haigou gold deposit discovered in the study area is mainly controlled by the fault structure. The vein is mainly distributed in the ore-controlling fault zone. The main mineralized surround rock of the gold vein is monzonitic granite. According to the comprehensive interpretation profile results (Figure 23) and the deposit formation of the known mining area, it can be helpful to determine the ore-forming target area of the profile. We preliminarily predicted that the ore-forming target area would be mainly concentrated near the fault structure F_3. The interlaminar fault zone is the center of the regional tectonic, magmatic activity, and later ore-forming hydrothermal activities. Prospecting work should pay attention to the mineralization clues of the contact sites between the Mesozoic intrusive rocks and Paleoproterozoic Mayihe rock formation, particularly the Jurassic medium-fine-grained monzonite granite. In this paper, the data-space joint inversion of MT and gravity gradiometry data could accurately divide the stratum structure and fault zone and have the ability to find the prospecting target area, thereby providing a powerful basis for the mining of the deposit.

6. Conclusions

We developed an approach to the joint inversion of MT and gravity gradiometry data based on normalized cross-gradient constraints and data-space. This study shows that the normalized cross-gradient and data-space techniques can be successfully adapted to the problem of 2D joint inversion of collocated MT and gravity gradiometry data from measurements at different spatial scales. Numerical modeling results showed that joint inversion of MT and gravity gradiometry data lead to models that are in superior structural accord and closer to the ground-truth than those derived from separate methods and conventional joint inversion of MT and gravity data. The joint inversion models recovered the salient features of the range of resistivity and density amplitude, low-density zones and steep features that posed significant imaging problems to the separate methods. Moreover, the data-space method can be applied to the joint inversion MT and gravity gradiometry, which can greatly reduce memory consumption and effectively improve calculation speed. This method allows us to invert the joint inversion of a large data volume with a personal computer (PC) in a relatively short period of time. Note that the resistivity and density estimates provided improved information that can facilitate the lithotype and structural classification of the subsurface. Application of this approach to field data sets from a mining study area are available yielded CSAMT and gravity gradiometry models that are in excellent geometrical accord and remarkable consistency. We suggest that joint inversion of collocated MT and gravity gradiometry profiles and the use of the RGB composite image results in structural or lithological classification will lead to improved subsurface characterization in complicated geological terrains and should be seen as the way forward in deep subsurface minerals and petroleum resources exploration studies.

Author Contributions: Conceptualization, R.Z.; methodology, R.Z. and T.L.; software, R.Z.; validation, T.L.; formal analysis, R.Z. and T.L.; investigation, R.Z. and T.L.; resources, R.Z. and T.L.; data curation, R.Z. and T.L.; writing—original draft preparation, R.Z.; writing—review and editing, R.Z. and T.L.; supervision, T.L.; project administration, T.L.; funding acquisition, T.L.

Funding: This research was funded by the Nation Key R&D Program of China (2017YFC0601606) and Jilin University PhD Graduate Interdisciplinary Research Funding Project (10183201838).

Conflicts of Interest: The authors declare no conflict of interest.

References

1. Auken, E.; Pellerin, L.; Christensen, N.B. A survey of current trends in near-surface electrical and electromagnetic methods. *Geophysics* **2006**, *71*, G249–G260. [CrossRef]
2. Kowalsky, M.B.; Chen, J.; Hubbard, S.S. Joint inversion of geophysical and hydrological data for improved subsurface characterization. *Lead. Edge* **2006**, *25*, 730–731. [CrossRef]
3. Colombo, D.; De Stefano, M. Geophysical modeling via simultaneous joint inversion of seismic, gravity, and electromagnetic data: Application to prestack depth imaging. *Lead. Edge* **2007**, *26*, 326–331. [CrossRef]
4. Alpak, F.O.; Torres-Verdin, C.; Habashy, T.M. Estimation of in situ petrophysical properties from wireline formation tester and induction logging measurements: A joint inversion approach. *J. Pet. Sci. Eng.* **2008**, *63*, 1–17. [CrossRef]
5. Fregoso, E.; Gallardo, L.A. Cross-gradients joint 3D inversion with applications to gravity and magnetic data. *Geophysics* **2009**, *74*, L31–L42. [CrossRef]
6. Linde, N.; Tryggvason, A.; Peterson, J.E. Joint inversion of cross hole radar and seismic traveltimes acquired at the South Oyster Bacterial Transport Site. *Geophysics* **2008**, *73*, G29–G37. [CrossRef]
7. Moorkamp, M.; Lelièvre, P.G.; Linde, N.; Khan, A. *Integrated Imaging of the Earth: Theory and Applications*; John Wiley & Sons: New Jersey, NJ, USA, 2016.
8. Moorkamp, M.; Heincke, B.; Jegen, M.; Roberts, A.W.; Hobbs, R.W. A framework for 3D joint inversion of MT, gravity and seismic refraction data. *Geophys. J. Int.* **2011**, *184*, 477–493. [CrossRef]
9. Gallardo, L.A.; Meju, M.A. Structure-coupled multiphysics imaging in geophysical sciences. *Rev. Geophys.* **2011**, *49*, RG1003. [CrossRef]
10. Tarantola, A. *Inversion Problem Theory and Methods for Model Parameter Estimation*; SIAM: Philadelphia, PA, USA, 2005.
11. Yang, W.C. *Theory and Methods of Geophysical Inversion*; Geological Publishing House: Beijing, China, 1997. (In Chinese)
12. Heincke, B.; Jegen, M.; Hobbs, R. Joint inversion of MT, gravity and seismic data applied to sub-basalt imaging. *SEG Expand Abstr.* **2006**, *2006*, 784–789.
13. Gao, G.; Abubakar, A.; Habashy, T.M. Joint petrophysical inversion of electromagnetic and full-waveform seismic data. *Geophysics* **2012**, *77*, WA3–WA18. [CrossRef]
14. Haber, E.; Oldenburg, D. Joint inversion: A structural approach. *Inverse Probl.* **1997**, *13*, 63–77. [CrossRef]
15. Zhang, J.; Morgan, F.D. Joint seismic and electrical tomography. In *Symposium on the Application of Geophysics to Engineering and Environmental Problems*; SEG: Denver, CO, USA, 1997; pp. 391–396.
16. Gallardo, L.A.; Meju, M.A. Characterization of heterogeneous near-surface materials by joint 2d inversion of dc resistivity and seismic data. *Geophys. Res. Lett.* **2003**, *30*, 1658. [CrossRef]

17. Abubakar, A.; Gao, G.; Havashy, T.M.; Liu, J. Joint inversion approaches for geophysical electromagnetic and elastic full-waveform data. *Inverse Probl.* **2012**, *28*, 055016. [CrossRef]
18. Hamdan, H.A.; Vafidis, A. Joint inversion of 2D resistivity and seismic travel time data to image saltwater intrusion over karstic areas. *Environ. Earth Sci.* **2013**, *68*, 1877–1885. [CrossRef]
19. Li, T.L.; Zhang, R.Z.; Pak, Y.C. Joint Inversion of magnetotelluric and first-arrival wave seismic traveltime with cross-gradient constraints. *J. Jilin Univ. Earth Sci. Ed.* **2015**, *45*, 952–961.
20. Gallardo, L.A.; Meju, M.A. Joint two-dimensional DC resistivity and seismic travel time inversion with cross-gradients constraints. *J. Geophys Res. Solid Earth* **2004**, *109*, B03311. [CrossRef]
21. Gallardo, L.A.; Fontes, S.L.; Meju, M.A.; Buonora, M.P.; De Lugao, P.P. Robust geophysical integration through structure-coupled joint inversion and multispectral fusion of seismic reflection, magnetotelluric, magnetic, and gravity images: Example from santos basin, offshore brazil. *Geophysics* **2012**, *77*, B237–B251. [CrossRef]
22. Li, T.L.; Zhang, R.Z.; Pak, Y.C. Multiple joint inversion of geophysical data with sub-region cross-gradient constraints. *Chin. J. Geophys.* **2016**, *59*, 2979–2988.
23. Zhang, R.Z.; Li, T.L.; Deng, H. 2D joint inversion of MT, gravity, magnetic and seismic first-arrival traveltime with cross-gradient constraints. *Chin. J. Geophys.* **2019**, *62*, 2139–2149.
24. Zhang, R.Z.; Li, T.L.; Zhou, S.; Deng, X.H. Joint MT and gravity inversion using structural constraints: A case study from the Linjiang copper mining area, Jilin, China. *Minerals* **2019**, *9*, 407. [CrossRef]
25. Zhdanov, M.S.; Ellis, R.; Mukherjee, S. Three-dimensional regularized focusing inversion of gravity gradient tensor component data. *Geophysics* **2004**, *69*, 1–4. [CrossRef]
26. Martinez, C.; Li, Y.; Krahenbuhl, R. 3D inversion of airborne gravity gradiometry data in mineral exploration: A case study in the Quadrilátero Ferrífero, Brazil. *Geophysics* **2013**, *78*, B1–B11. [CrossRef]
27. Geng, M.; Huang, D.; Yang, Q. 3D inversion of airborne gravity-gradiometry data using cokriging. *Geophysics* **2014**, *79*, G37–G47. [CrossRef]
28. Pilkington, M. Analysis of gravity gradiometer inverse problems using optimal design measures. *Geophysics* **2012**, *77*, G25–G31. [CrossRef]
29. Pilkington, M. Evaluating the utility of gravity gradient tensor components. *Geophysics* **2014**, *79*, G1–G14. [CrossRef]
30. Kivior, I.; Markham, S.; Hagos, F. Improved imaging of the subsurface geology in the Mowla Terrace, Canning Basin using gravity gradiometry data. *ASEG Ext. Abstr.* **2018**, *2018*, 1–10. [CrossRef]
31. Siripunvaraporn, W.; Egbert, G. An efficient data-subspace inversion method for 2-D magnetotelluric data. *Geophysics* **2000**, *65*, 791–803. [CrossRef]
32. Siripunvaraporn, W.; Egbert, G.; Lenbury, Y.; Uyeshima, M. Three–dimensional magnetotelluric inversion: Data-space method. *Phys. Earth Planet. Inter.* **2005**, *150*, 3–14. [CrossRef]
33. Singh, B. Simultaneous computation of gravity and magnetic anomalies resulting from a 2D object. *Geophysics* **2002**, *67*, 801–806. [CrossRef]
34. Won, I.J. Computing the gravitational and magnetic anomalies due to a polygon: Algorithms and Fortran subroutines. *Geophysics* **1987**, *52*, 202–205. [CrossRef]
35. Wannamaker, P.E.; Stodt, J.A.; Rijo, L. A stable finite element solution for two-dimensional magnetotelluric modeling. *Geophys. J. Int.* **1987**, *88*, 277–296. [CrossRef]
36. Menke, W. *Geophysical Data Analysis: Discrete Inverse Theory, Revised Version: International Geophysics*; Academic Press: San Diego, CA, USA, 1989.
37. Tarantola, A. *Inverse Problem Theory: Methods for Data Fitting and Model Parameter Estimation*; Elsevier: New York, NY, USA, 1987.
38. Constable, S.C.; Parker, R.L.; Constable, C.G. Occam's inversion: A practical algorithm for generating smooth models from electromagnetic sounding data. *Geophysics* **1987**, *52*, 289–300. [CrossRef]
39. deGroot-Hedlin, C. Removal of static shift in two dimensions by regularized inversion. *Geophysics* **1991**, *56*, 2102–2106. [CrossRef]
40. deGroot-Hedlin, C.; Constable, S. Inversion of magnetotelluric data for 2D structure with sharp resistivity contrasts. *Geophysics* **2004**, *69*, 78–86. [CrossRef]

41. Li, H.; Li, T.; Wu, L. Transformation of all-time apparent resistivity of CSAMT and analysis of its effect. *Prog. Geophys. (In Chinese)* **2015**, *30*, 0889–0893.
42. ÖzgüArisoy, M.; Dikmen, Ü. Potensoft: MATLAB-based software for potential field data processing, modeling and mapping. *Comput. Geosci.* **2011**, *37*, 935–942.

Article

Joint MT and Gravity Inversion Using Structural Constraints: A Case Study from the Linjiang Copper Mining Area, Jilin, China

Rongzhe Zhang [1,*], Tonglin Li [1,*], Shuai Zhou [1,*] and Xinhui Deng [1,2]

1 College of Geo-Exploration Sciences and Technology, Jilin University, Changchun 130026, China
2 Changchun Institute of Technology, Changchun 130021, China
* Correspondence: zhangrongzhe_jlu@163.com (R.Z.); litl@jlu.edu.cn (T.L.); zhoushuai@jlu.edu.cn (S.Z.)

Received: 30 May 2019; Accepted: 29 June 2019; Published: 2 July 2019

Abstract: We present a joint 2D inversion approach for magnetotelluric (MT) and gravity data with elastic-net regularization and cross-gradient constraints. We describe the main features of the approach and verify the inversion results against a synthetic model. The results indicate that the best fit solution using the L2 is overly smooth, while the best fit solution for the L1 norm is too sparse. However, the elastic-net regularization method, a convex combination term of L2 norm and L1 norm, can not only enforce the stability to preserve local smoothness, but can also enforce the sparsity to preserve sharp boundaries. Cross-gradient constraints lead to models with close structural resemblance and improve the estimates of the resistivity and density of the synthetic dataset. We apply the novel approach to field datasets from a copper mining area in the northeast of China. Our results show that the method can generate much more detail and a sharper boundary as well as better depth resolution. Relative to the existing solution, the large area divergence phenomenon under the anomalous bodies is eliminated, and the fine anomalous bodies boundary appeared in the smooth region. This method can provide important technical support for detecting deep concealed deposits.

Keywords: MT; gravity; elastic-net regularization; cross-gradients constraints; joint inversion

---

1. Introduction

Many inverse problems in geophysics are ill-posed and the solutions may be non-unique due to errors inherent in the observational data, limitations in the range of observations, the discretization of the observation data, and the simplification of the underground model. Additionally, constraints are often incorporated into the inversion to seek stable solutions. The penalty function regularization technique is one of the most commonly used, where the fundamental idea is that some prior constraints are added to the least-squares misfit term to preserve the stability of the inversion solution [1–3].

Different types of penalty functions can be employed in geophysical inversion. The L2 norm regularization imposing smoothness priors is one of the most successful methods [4–8]. This regularization improves the stability of the inversion at the expense of its resolution: the inversion results overestimate the size of the source and blur the boundaries of buried geological contacts as well as their associated physical properties. The total variation (TV) regularization method is a filter to decrease the noise in the data [9–12] given that a dataset denoised by TV-denoising tends to contain blocks with a constant gray level, separated by intensity gaps, even in what should be smooth areas (the so-called "staircase effect") [13]. The focusing regularization can obtain focused reconstruction images and even sharper boundaries than the conventional smooth stabilizer. However, if the inversion is over-focused, it sometimes distorts the structural form and makes the inversion result inaccurate. In recent years, with the continuous development of compressive sensing theory and related algorithms, the sparse regularization method has been widely used in many fields, such as signal processing, image denoising, magnetic resonance imaging, and geophysical inversion [14–17]. This regularization mainly obtains the sub-surface non-zero element, and the response produced by the non-zero element must fit the observed data. The minimization of the L0 norm is a non-deterministic polynomial-time hard problem (NP-hard), which indicates that the optimization algorithms used to solve the problem cannot be completed in polynomial terms [18]. Thus, the L1 norm is usually used to approximate the L0 norm to solve the NP-hard problem. It can be seen that L1 norm regularization has received considerable interest in solving inversion problems [19–22] and can efficiently capture the small-scale details of the inversion solution. Although L1 norm regularization has shown success in some situation, it has some limitations. When the correlation of a set of variables is very high, L1 norm regularization often only selects one variable and does not care whether another variable is selected. That is, group selection is not possible [23]. Zou and Hastie [24] proposed a new regularization method named the elastic-net regularization method. Elastic-net regularization not only has the advantage of L1 norm regularization, but also takes into account the ability to select groups of related variables and can retain the characteristic variables of the data as much as possible. Elastic-net regularization has been extensively applied in statistics, computer science, and medical imaging, but has not been applied in geophysical inversion [25–27].

Each geophysical method has a different resolution and only reflects one physical property of the underground; thus, separate inversion methods have some limitations. Methods to overcome such problems in geophysical inversion are the focus of much research in the inversion community [28]. Currently, the most effective comprehensive interpretation method is joint inversion, and joint inversion can reduce the multi-solution of inversion. The study of integrating multiple data for geophysical exploration can be classified into two categories. The first category is based on coupling the empirical relationships between different physical properties, e.g., fluid saturation and porosity [29–32]. Unfortunately, this method has some limitations: the lithology of different zones differs, and it is difficult to find the exact relationship between the various physical properties to produce a single relationship.

The second category utilizes structural coupling, which is not dependent on the petrophysics and depends on the similarity of the spatial structure of different physical models to the coupling data as has been reported [33–40]. Gallardo and Meju [33] first published the cross-gradient function, which is used to identify the structure boundary as the cross product between different physical gradients. They reported on a two-dimensional cross-gradient joint inversion study of the seismic time and the direct current (DC) resistivity that achieved good results. Moorkamp [41] compared the results of a petrophysics constrained inversion and a cross-gradient joint inversion of gravity, magnetotelluric (MT), and seismic data. Gallardo et al. [42] first applied the cross-gradients joint inversion algorithm of four methods—namely, gravity, magnetic, MT, and seismic reflection data. Other applications of the cross-gradient approach include integrating ground-penetrating radar travel-time data, cross-hole electrical resistance data, and seismic travel-time data for the better determination of lithologic boundaries in hydrogeologic studies [43,44]. Zhdanov et al. [45] developed a generalized approach to joint inversion based on Gramian constraints by using the formalism of Gramian determinants, which when applied to the parameters' gradients reduces to a constraint similar to cross-gradients. Haber and Gazit [46] proposed a novel approach based on a joint total variation function. The joint total variation is a convex function for joint inversion and can improve the inversion of models with a sharp interface. The above structural coupling joint inversion mainly uses a separate smooth or focusing regularization method and rarely uses a hybrid regularization method. The obtained inversion results may be too smooth or too sparse, affecting the accuracy of the inversion results. Furthermore, the cross-gradient constraint proposed above is directly structurally coupled through model parameters. Since there are different units and magnitudes between different physical parameters, direct coupling may affect the inversion results. To address these issues, we first investigated an elastic-net regularization method as a convex combination term of the L2 norm and L1 norm can not only enforce the stability to preserve large-scale smoothness, but can also efficiently capture the small-scale details of the inversion solution and enforce the sparsity to preserve sharp boundaries. Next, we considered whether the normalization method could effectively eliminate large differences in magnitude between the model parameters.

In this paper, we present an original approach to the joint inversion of MT and gravity data based on elastic-net regularization and cross-gradient constraints. Considering the fact that the L2 norm penalty always makes the inversion results overly smooth and the L1 norm penalty always makes the inversion results too sparse, we investigated an elastic-net regularization method with a stronger convex combination of the L1 norm and the L2 norm for an ill-posed joint inversion problem. We constructed a new joint inversion penalty function, that is, based on the data misfit terms, we added L1 norm and L2 norm regularization to form an elastic-net regularization. The elastic-net regularization can effectively improve the blur degree of the L2 inversion result and the over-sparse degree of the L1 norm inversion result, so that the inversion is not only stable, but can also enforce the sparsity to preserve sharp boundaries. We used synthetic models of the subsurface to gauge the performance of the algorithm and test its accuracy. Then, we applied this approach to the interpretation of geophysical field data collected in the Linjiang (Jilin Province) copper mining area in Northeast China.

2. Joint Inversion Methodology

To avoid the instability and multi-solution problems caused by solving ill-posed inverse problems, the Tikhonov regularization method is used to solve linear equations. We constructed the joint inversion objective function of MT and gravity data, which is expressed as follows:

$$\begin{aligned} \Phi(m_1,m_2) = \; & \Phi_d(m_1,m_2) + \alpha\cdot\Phi_m(m_1,m_2) \\ & = \|W_{d1}(d_1 - f_1(m_1))\|^2 + \|W_{d2}(d_2 - f_2(m_2))\|^2 + \\ & \alpha_1\cdot\|W_{m1}(m_1 - m_{1\_ref})\|^2 + \alpha_2\cdot\|W_{m2}(m_2 - m_{2\_ref})\|^2\Phi(m_1,m_2) \\ & \text{Subject to } \tau(m_1,m_2) = 0 \end{aligned} \tag{1}$$

where, $\Phi_d(m_1,m_2)$ is represented as a data misfit term; $\Phi_m(m_1,m_2)$ is represented as a model constraint term; and α is a regularization factor. d_1(apparent resistivity and phase) and d_2 (gravity Bouguer anomaly) are the observed data. are the computed apparent resistivities and phases and is the computed gravity Bouguer anomaly. m_1 and m_2 are the vectors of resistivity and density, respectively. m_{1_ref} and m_{2_ref} are the prior model of resistivity and density, respectively. α_1 and α_2 are represented as the MT and gravity regularization factor, respectively. W_{d1} and W_{d2} are the noise error diagonal inverse matrix of the MT and gravity data, respectively. W_{m1} and W_{m2} are the MT and gravity model weighting matrix where the smooth roughness model matrix is usually adopted, respectively. τ is the cross-gradient constraint term. A convenient un-normalized way to measure the geometrical similarity of the models can be obtained through the use of the cross-gradient function [33,34] given by

$$\tau(m_1,m_2) = \nabla m_1(x,y,z)\times\nabla m_2(x,y,z) = 0 \tag{2}$$

where ∇m_1 and ∇m_2 are the resistivity and density property gradients, respectively. Please note that resistivity and density have different orders of magnitude and units. If the physical parameter gradients are directly coupled, it may produce inaccurate inversion results. In this paper, we added a normalized factor to the cross-gradient constraints term to overcome the effects of different physical parameter differences. The cross-gradient constraints term expression that is added to the normalized factor is as follows:

$$\tau(m_1,m_2) = \nabla(\kappa_1^{-1}m_1)\times\nabla(\kappa_2^{-1}m_2) = 0 \tag{3}$$

where κ_1 and κ_2 are the normalized operators of resistivity and density, respectively. The normalized operator is determined by the amount of change in the model parameters of the separate inversion results.

Our objective function (1) is nonlinear, since the MT forward problem and the cross-gradient constraints are nonlinear. Our solution to Equation (1) is thus achieved by linearization. For the

MT forward problem, we computed forward responses using the OCCAM2DMT code developed by deGroot-Hedlin and Constable [47], which uses the FE forward modeling code developed by Wannamaker [48]. For the gravity forward problem, we used analytical solutions for polygonal prisms [49]. The nonlinear problem was transformed into a linear problem by using a first-order Taylor expansion (neglecting higher orders), namely,

$$f_1(m_1) \cong f_1(m_{1\_0}) + A_1 \cdot (m_1 - m_{1\_0}) \tag{4}$$

$$f_2(m_2) \cong f_2(m_{2\_0}) + A_2 \cdot (m_2 - m_{2\_0}) \tag{5}$$

where A_1 defines the resistivity Jacobian matrix, which calculates the Jacobian matrix using reciprocity. A_2 defines the density Jacobian matrix. m_{1_0} and m_{2_0} are the initial models of resistivity and density, respectively.

Similarly, the linearization of the cross-gradient constraints in Equation (3) can be accomplished by using a first-order Taylor expansion (neglecting higher orders), namely,

$$\tau(m_1, m_2) \cong \tau(m_{1\_0}, m_{2\_0}) + B \cdot \begin{pmatrix} m_1 - m_{1\_0} \\ m_2 - m_{2\_0} \end{pmatrix} \tag{6}$$

where the derivative of τ for the model parameters is given by B.

Using Equations (4)–(6), the linearized equivalent of Equation (1) is stated as

$$\begin{aligned} \Phi(m_1, m_2) \quad &= \|W_{d1}(d_1 - f_1(m_{1\_0}) - A_1 \cdot (m_1 - m_{1\_0}))\|^2 + \\ &\|W_{d2}(d_2 - f_2(m_{2\_0}) - A_2 \cdot (m_2 - m_{2\_0}))\|^2 + \\ &\alpha_1 \cdot \|W_{m1}(m_1 - m_{1\_ref})\|^2 + \alpha_2 \cdot \|W_{m2}(m_2 - m_{2\_ref})\|^2 \\ &\text{Subject to } \tau(m_{1\_0}, m_{2\_0}) + B \cdot \begin{pmatrix} m_1 - m_{1\_0} \\ m_2 - m_{2\_0} \end{pmatrix} = 0 \end{aligned} \tag{7}$$

For convenience, we introduced an integrated model parameter $m = [m_1^T, m_2^T]^T$ and an integrated data parameter $d = [d_1^T, d_2^T]^T$. Then, the objective function in Equation (7) can be rewritten as

$$\begin{aligned} \Phi(m) = \quad &\Phi_d(m) + \Phi_m(m) \\ &= \|W_d(d - f(m_0) - A \cdot (m - m_0))\|^2 + \alpha \cdot \|W_m(m - m_{ref})\|^2 \\ &\text{Subject to } \tau(m_0) + B \cdot (m - m_0) = 0 \end{aligned} \tag{8}$$

where $m_0 = [m_{1\_0}^T, m_{2\_0}^T]^T$; ; $f(m_0) = [f_1^T(m_{1\_0}), f_2^T(m_{2\_0})]^T$; $W_d = \text{diag}[W_{d1}, W_{d2}]$; $W_m = \text{diag}[W_{m1}, W_{m2}]$; $\alpha = [\alpha_1, \alpha_2]$; and $A = \text{diag}[A_1, A_2]$.

The solution of Equation (8) can be determined using Lagrange multipliers [50], by solving the system of equations:

$$\frac{\partial}{\partial m}\left\{ \begin{array}{l} \|W_d(d - f(m_0) - A \cdot (m - m_0))\|^2 + \alpha \cdot \|W_m(m - m_{ref})\|^2 + \\ 2\Gamma \cdot (\tau(m_0) + B \cdot (m - m_0)) \end{array} \right\} = 0 \tag{9}$$

$$\tau(m_0) + B \cdot (m - m_0) = 0 \tag{10}$$

The solutions to Equations (9) and (10), after some algebra, are

$$\Gamma = (B \cdot X^{-1} \cdot B^T)^{-1} (\tau(m_0) + B \cdot X^{-1} \cdot Y) \tag{11}$$

$$m = X^{-1} \cdot Y - X^{-1} \cdot B^T \cdot \Gamma + m_0 \tag{12}$$

where $X = A^T \cdot W_d^T W_d \cdot A + \alpha \cdot W_m^T W_m$; $Y = A^T \cdot W_d^T W_d \cdot (d - f(m_0))$.

Equation (12) is used as a model update expression, where the first term on the right side of Equation (12) corresponds to the regularized least squares solution (no cross-gradient constraints), whereas the second term corresponds to the structural constraints solution.

3. Regularization Inversion Methodology

To reconstruct the stable inversion model, some sort of regularization model constraints are required. Regularization model constraints come in numerous forms, that is, $\Phi_m(m)$ in Equation (8) can choose different forms. Different inversion methods can be developed by changing the model constraints such as smooth inversion based on the L2 norm, sparse inversion based on the L1 norm as well as focusing inversion and total variation inversion, and so on. Below, we discuss the principles and characteristics of these inversion methods.

3.1. L2 Norm Regularization

The L2 norm defined by $||m||^2 = \left(\sum_{i=1}^{n} m_i^2\right)^{1/2}$ refers to the square root of the squares of the elements in the vector. The self-multiplication of each element in the L2 norm means that the element with a large value contributes a large amount to the whole, thus minimizing the L2 norm of a model produces a relatively smoothly changing model. Occam inversion is also essentially an inversion based on the L2 norm inversion. Its model constraint represents the difference of adjacent element model parameters. Minimizing the model constraint causes the difference in the adjacent element model parameters to be minimized, so the inversion result is smooth. In this paper, the L2 norm inversion used the same model constraints as the Occam inversion. The expression of the L2 norm regularization term for a 2D model is as follows

$$\Phi_{m\_L2}(m) = ||W_m m||^2 = ||\partial_x m||^2 + ||\partial_z m||^2 \tag{13}$$

$$W_m = [\partial_x, \partial_z]^T = \nabla \tag{14}$$

where ∂_x is a roughening matrix that differences the model parameters of laterally adjacent elements, and ∂_z is a roughening matrix that differences the model parameters of vertically adjacent elements. ∇ is a gradient operator. The details of the roughening matrix mathematical expression can be found in deGroot-Hedin [47]. Equation (14) is brought into Equation (12) to obtain a model update based on the L2 norm.

3.2. L1 Norm Regularization

The L1 norm defined by $||m||^1 = \sum_{i=1}^{n} |m_i|$ refers to the sum of the absolute values of the elements in the vector and is also known as the sparse operator. The introduction of the sparse operator can remove the features without information in the inversion, that is, the weight corresponding to these features is set to zero. The feature with information corresponds to the anomalous area of the underground medium, while the feature without corresponds to the background area of the underground medium. Therefore, sparseness highlights the distribution features of subsurface anomalies. The model solution of the sharp boundary can be obtained by inversion based on the L1 norm. The expression of the L1 norm regularization term is as follows

$$\Phi_{m\_L1}(m) - ||m||^1. \tag{15}$$

To avoid singularity in the case of m = 0, the L1 norm regularization of the model is generally approximated by

$$\Phi_{m\_L1\_app}(m) = (m^2 + \varepsilon^2)^{1/2} \tag{16}$$

where ε is a small number, so we set $\varepsilon = 10^{-8}$. A possible method for solving Equation (16) is to approximate the L1 norm minimization as an iteratively reweighted L2 norm minimization problem [51–53].

$$\Phi_{m\_L1\_app}(m) = ||W_m m||^2 \tag{17}$$

$$W_m = \left(\frac{\partial \Phi_{m\_L1\_app}(m)}{\partial m} / m\right)^{1/2} = \left(m^2 + \varepsilon^2\right)^{(-1/4)}. \tag{18}$$

Equation (18) is brought into Equation (12) to obtain a model update based on the L1 norm.

3.3. Focusing Regularization

Last and Kubic [54] proposed a minimum support function (MS) as a model constraint, which can minimize the section area of the anomalous body. The expression of the minimum support function can be expressed as:

$$\Phi_{m\_Ms}(m) = \frac{m^2}{m^2 + e^2} \tag{19}$$

$$\Phi_{m\_Ms}(m) = ||W_m m||^2 \tag{20}$$

$$W_m = \left(\frac{\partial \Phi_{m\_MS}(m)}{\partial m} / m\right)^{1/2} = \left(\frac{2e^2}{\left(m^2 + e^2\right)^2}\right)^{(1/2)} \tag{21}$$

where e is a focusing parameter determining the sharpness of the produced image [55]. Inversion based on focusing regularization can recover a model with clearer boundaries and contrasts than the traditional smooth inversion. However, it tends to produce the minimum volume of a model, which usually underestimates the distribution space of the recovered model. Equation (20) is brought into Equation (12) to obtain a model update based on the focusing regularization.

3.4. Total Variation Regularization

The total variation regularization method was first proposed as an image denoising technique for image processing [56]. It uses a variation function as a penalty function and belongs to a class of bounded variation function, which allows discontinuous points. The total variation method can effectively reconstruct non-smooth images with "corner points", that is, the case of a discontinuous solution, and has better edge preservation. However, the main disadvantage is that the reconstruction results often show a blocking phenomenon, which is prone to the "staircase effect". The total variation function is defined as

$$\Phi_{m\_TV}(m) = \sqrt{|\nabla m|^2}. \tag{22}$$

Various approaches can be used to approximate the original TV term to make it differentiable at the origin. One common approach to make the original TV term differentiable is to introduce a small smoothing parameter ω such that

$$\Phi_{m\_TV}(m) = \sqrt{|\nabla m|^2 + \omega^2} \tag{23}$$

$$\Phi_{m\_TV}(m) = ||W_m m||^2 \tag{24}$$

$$W_m = \left(\frac{\partial \Phi_{m\_TV}(m)/\partial m}{\nabla m} / m\right)^{1/2} = \left(\frac{\nabla \cdot \nabla}{\sqrt{m^2 + \omega^2}}\right)^{(1/2)}. \tag{25}$$

Equation (25) is brought into Equation (12) to obtain a model update based on the TV regularization.

3.5. Elastic-Net Regularization

Zou and Hastie [24] proposed a new regularization method named the elastic-net regularization method. The elastic-net regularization method, a convex combination term of the L2 norm and L1 norms, can not only enforce the stability to preserve large-scale smoothness, but can also efficiently capture the small-scale details of the inversion solution and enforce the sparsity to preserve sharp boundaries. It is similar to a stretchable fishing net that retains "all the big fish". This method has been extensively applied in statistics, computer science, and medical imaging, but is rarely used in geophysical inversion. Elastic-net regularization is defined as

$$\Phi_{m\_En}(m) = \lambda_1 \Phi_{m\_L1}(m) + \lambda_2 \Phi_{m\_L2}(m) \tag{26}$$

where λ_1 and λ_2 are the non-negative parameters used to control the relative weighting of the L1 norm and the L2 norm. Let $\beta = \lambda_2/(\lambda_1+\lambda_2)$ be a convex combination regularization term equal to Equation (26) can be expressed as

$$\Phi_{m\_En}(m) = (1-\beta)\Phi_{m\_L1}(m) + \beta\Phi_{m\_L2}(m) \tag{27}$$

$$\Phi_{m\_En}(m) = ||W_m m||^2 \tag{28}$$

$$W_m = \left(\frac{\partial \Phi_{m\_En}(m)}{\partial m}/m\right)^{1/2} = \left[(1-\beta)(m^2+\varepsilon^2)^{(-1/2)} + \beta\nabla\cdot\nabla\right]^{1/2}. \tag{29}$$

The $\beta = 1$ regularization term (Equation (27)) reduces to L2 norm regularization, while the $\beta = 0$ regularization term (Equation (27)) reduces to L1 norm regularization. In this paper, we assigned a non-zero value for β (i.e., $1 > \beta > 0$), since we wanted to emphasize the structural features of sparsity for all iterations. This will typically reduce overly smooth behavior while well maintaining the sharp edges. Equation (29) is brought into Equation (12) to obtain a model update based on the elastic-net regularization. Please note that gravity inversion requires model weighting matrices. In this paper, the model weighting functions were based on the depth-weighting function used to counteract the depth-dependent decay of the sensitivity matrix. When denoting the model weighting function as Z, W_m is replaced by diag(Z)$\cdot W_m$ in all of the above equations.

4. Synthetic Example

4.1. Comparison Regularization Methods

To illustrate the feasibility and efficiency of the elastic-net regularization method, we inverted the MT and gravity data using different regularization methods by different typical resistivity and density synthetic models, as shown in Figures 1a and 2a. The physical property values of the resistivity and residual density models are listed in Table 1. The resistivity model features a conductive wedge underlying a highly resistive surface layer and overlying a basement of intermediate resistivity, representative of sediments below basalts. The residual density model consists of adjacent low and high-density blocks in the homogeneous subsurface surrounding rock. For the resistivity model, apparent resistivity and phase data for both the transverse electric (TE) and transverse magnetic (TM) modes were generated at 14 stations, spaced at 3 km intervals. The frequency ranges were chosen so that the penetration depth corresponded to the top unit at the highest frequency and the bottom unit at the lowest frequency (at 10 frequencies from 25 to 0.05 Hz). For the density model, the gravity data were Bouguer corrected at 41 stations spaced uniformly from a horizontal distance −20 km to 20 km. To simulate the noisy data, 5% random Gaussian noise was added to the MT and gravity data prior to inversion. The density model grid was set to 125 × 41 in the horizontal and vertical directions, respectively. However, the resistivity model grid needed to be extended from the density model grid, and the grid size was 141 × 55. The starting model was a homogeneous medium with a resistivity of 100 Ω·m and a residual density of 0 kg/m^3.

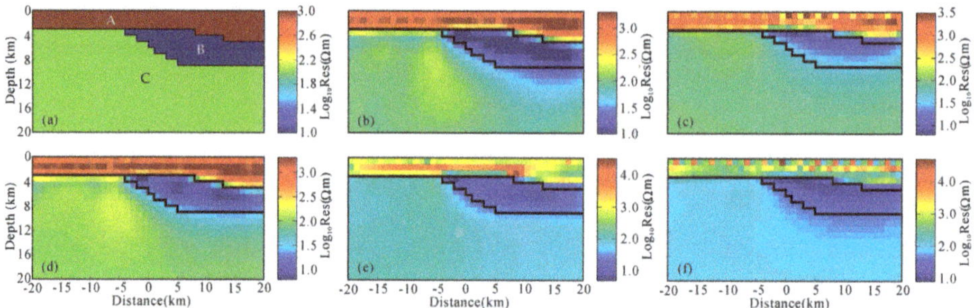

Figure 1. MT separate inversion results obtained using different regularization methods. (**a**) Resistivity true model; (**b**) Resistivity inversion model using L2 regularization; (**c**) Resistivity inversion model using L1 regularization; (**d**) Resistivity inversion model using elastic-net regularization; (**e**) Resistivity inversion model using TV regularization; (**f**) Resistivity inversion model using focusing regularization. The black box indicates the boundary of the anomalous body.

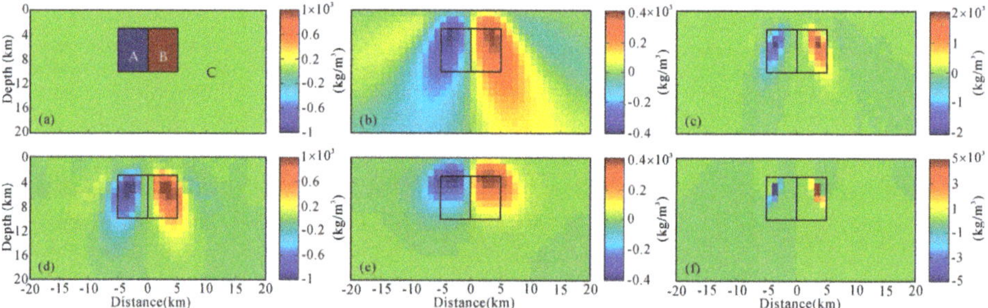

Figure 2. Gravity separate inversion results obtained using different regularization methods. (**a**) Density true model; (**b**) Density inversion model using L2 regularization; (**c**) Density inversion model using L1 regularization; (**d**) Density inversion model using elastic-net regularization; (**e**) Density inversion model using TV regularization; (**f**) Density inversion model using focusing regularization. The black box indicates the boundary of the anomalous body.

Table 1. The resistivities and residual densities of synthetic model 1.

Unit	Resistivity Model	Residual Density Model
A	1000 Ω·m	−1000 kg/m^3
B	10 Ω·m	1000 kg/m^3
C	100 Ω·m	0 kg/m^3

For comparison, we conducted these tests on four other regularization methods: the L2 norm, L1 norm, TV, and focusing, respectively. The best-fitting resistivity models were achieved by five different regularization inversions (Figure 1b–f), the evolution of the RMS misfit at every iteration of the five different regularization inversion processes is shown in Figure 3a. Inversion using the L2 norm regularization yielded a relatively stable and smooth result, as shown in Figure 1b, but the results blurred the wedge sharp boundaries of the subsurface geological contact. The L1 norm regularization method (Figure 1c) exhibited sparse characteristics; the wedge sharp boundaries were inverted, but the inversion convergence was poor, showing some instability, and the anomalous amplitude of recovery was larger than the true model. The TV regularization method could reconstruct non-smooth images, as shown in Figure 1e. However, the inversion result differed from the true model, and the convergence was poor, showing instability. The focusing regularization method seeks the source distribution with the minimum volume. As a result, the inversion result (Figure 1f) was too focused and even provided the wrong wedge boundaries; moreover, the value of resistivity was greater than the amplitude of the true model. A comparison of the solution with the true wedge sharp shown superimposed on the model indicated that both the unit conductivities and boundary locations had been well imaged (Figure 1d). In particular, the shape of both the top and low wedge boundaries were well recovered within the limits afforded by the stairstep approximation. The elastic-net regularization method could not only enforce the stability to preserve large-scale smoothness, but also efficiently enforced the sparsity to preserve sharp boundaries.

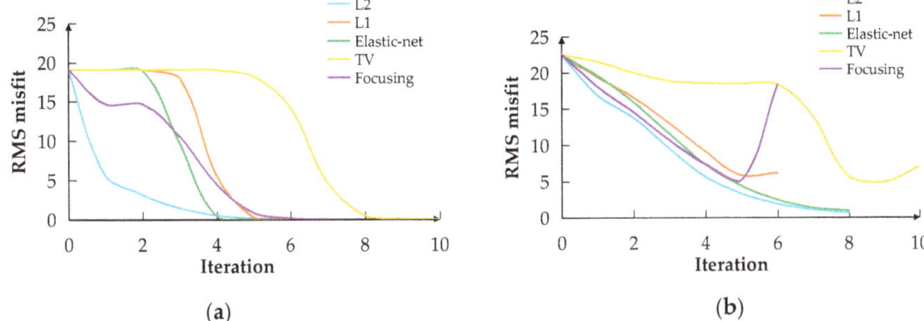

Figure 3. RMS misfit evolution of the MT separate inversion (**a**) and gravity separate inversion (**b**).

The best-fitting density models were achieved by five different regularization inversions (Figure 2b–f); the evolution of the RMS misfit at every iteration of the five different regularization inversion processes is shown in Figure 3b. Inversion using the L2 norm regularization yielded a relatively stable and smooth result, which revealed a large area of ambiguous divergence, and it was difficult to identify the deep boundary of anomalous bodies, as shown in Figure 2b. The L1 norm regularization method inversion results underestimated the size of anomalous bodies, and the gravity inversion obtained a physical property value of anomalous bodies that was larger than the true model, as illustrated in Figure 2c. The TV regularization method inversion result could reconstruct block images, but the center of the anomalous body moved up, as shown in Figure 2e. The anomalous body recovered from the focusing regularization was too focused and it severely underestimated the size of the anomaly, and even provided the wrong model boundaries, as shown in Figure 2f. The elastic-net regularization inversion (Figure 2d) could effectively improve the blur degree of the L2 inversion result and the over-sparse degree of the L1 norm inversion result, so the inversion was not only stable, but also could enforce the sparsity to preserve sharp boundaries.

4.2. Comparison Separate and Joint Inversion

In this section, we explore the advantages of the elastic-net regularization joint inversion compared to separate inversion and traditional joint inversion using an upper and lower anomalous bodies model, as shown in Figures 4a and 5a. The physical property values of the resistivity and residual density models are listed in Table 2. For the resistivity model, apparent resistivity and phase data for both the transverse electric (TE) and transverse magnetic (TM) modes were generated at nine stations, spaced at 0.6 km intervals. The frequency ranges were chosen so that the penetration depth corresponded to the top unit at the highest frequency and the bottom unit at the lowest frequency (at 10 frequencies from 1000 to 1 Hz). For the density model, the gravity data were Bouguer corrected at 30 stations spaced uniformly from a horizontal distance 0 km to 6 km. To simulate noisy data, 5% random Gaussian noise was added to the MT and gravity data before inversion. The density model grid was set to 141 × 81 in the horizontal and vertical directions, respectively. However, the resistivity model grid needed to be extended from the density model grid, and the grid size was 157 × 95. The starting model was a homogeneous medium with a resistivity of 100 Ω·m and a residual density of 0 kg/m^3.

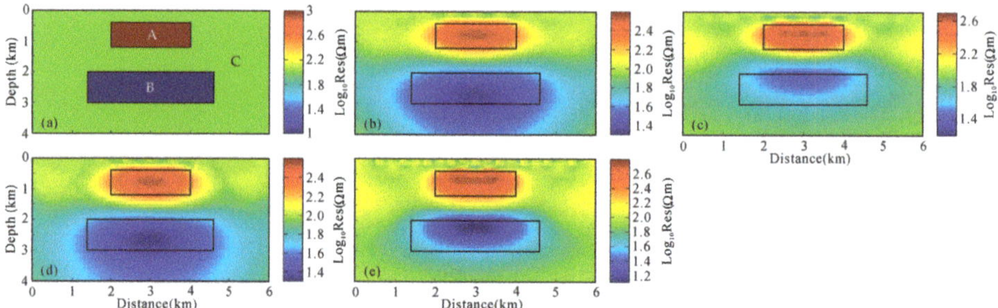

Figure 4. The MT inversion results obtained using separate and joint inversion. (**a**) Resistivity true model; (**b**) Resistivity separate inversion model using L2 regularization; (**c**) Resistivity separate inversion model using elastic-net regularization; (**d**) Resistivity joint inversion model using L2 regularization; (**e**) Resistivity joint inversion model using elastic-net regularization. The black box indicates the boundary of the anomalous body.

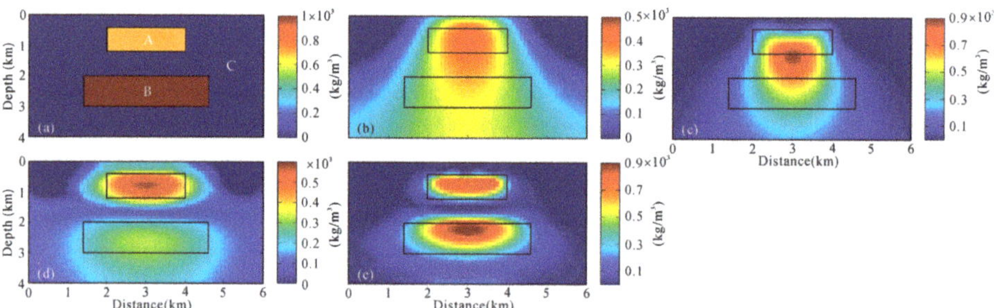

Figure 5. The gravity inversion results obtained using separate and joint inversion. (**a**) Density true model; (**b**) Density separate inversion model using L2 regularization; (**c**) Density separate inversion model using elastic-net regularization; (**d**) Density joint inversion model using L2 regularization; (**e**) Density joint inversion model using elastic-net regularization. The black box indicates the boundary of the anomalous body.

Table 2. The resistivities and residual densities of synthetic model 2.

Unit	Resistivity Model	Residual Density Model
A	1000 Ω·m	700 kg/m^3
B	10 Ω·m	1000 kg/m^3
C	100 Ω·m	0 kg/m^3

For comparison, we performed these tests on the separate and joint inversion using the L2 and elastic-net regularization methods. The best-fitting resistivity models were achieved by separate and joint inversion using the L2 and elastic-net regularization methods (Figure 4b–e), the evolution of the RMS misfit at every iteration of the inversion process is shown in Figure 6a. Separate and joint inversion of the MT method using the L2 norm regularization yielded a relatively stable and smooth result, as shown in Figure 4b,d, but the results blurred the boundaries of the underground block, in particular, the deep low-resistance block exhibited a serious divergence. Although the MT method in the joint inversion incorporated the structural similarity constraint of the gravity method, the vertical resolution of the gravity was relatively poor, so joint inversion also found it difficult to improve the vertical resolution of the MT method. The elastic-net regularization separate and joint inversion (Figure 4c,e) could effectively improve the blur degree of the deep low-resistance block and more clearly identify the lower boundary of the deep block. Please note that the physical parameters of the block resistivity obtained from the joint inversion of the elastic-net regularization were closer to the true model, which was due to the contribution of the gravity inversion structural similarity constraints.

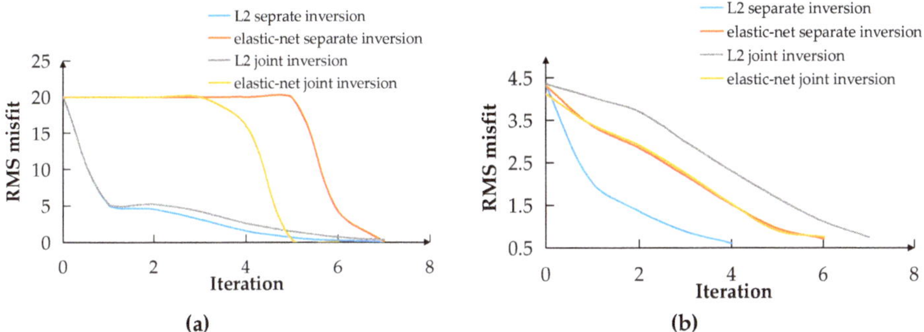

Figure 6. The RMS misfit evolution of the MT inversion (**a**) and gravity inversion (**b**).

The best-fitting density models were achieved by separate and joint inversion using the L2 and elastic-net regularization methods (Figure 5b–e), the evolution of the RMS misfit at every iteration of the inversion process is shown in Figure 6b. Separate inversion density models of the L2 and elastic-net regularization methods could not distinguish the upper and lower blocks, showing an anomalous region with high density, as shown in Figure 5b,c. Although the elastic-net regularization was added to the inversion objective function of the gravity, we found that the inversion of the gravity still did not solve the problem of vertical low resolution. Joint inversion density models of the L2 and elastic-net regularization methods (Figure 5d,e) could distinguish the upper and lower blocks as an indirect contribution propagated from the resistivity model by the cross-gradient constraints, showing improved features when compared to the above separate inversion results. Although the density model of the L2 regularization joint inversion (Figure 5d) could identify the upper and lower anomalous blocks, the inversion showed that the high-density anomalous block was above, and the low-density anomalous block was below, which was quite different from the true model. However, in the joint inversion of the elastic-net regularization method (Figure 5e), we found that the boundaries of the anomalous blocks became clearer, and the geometrical and physical values of the anomaly more closely reflected the true model. In particular, the vertical resolution of the density models was significantly improved, and the upper and lower anomalies could be identified as an indirect contribution propagated from the resistivity model by the cross-gradient constraints.

Compared with all of the above inversion methods, the method based on elastic-net joint inversion has its advantages, and not only can it enforce the stability to preserve large-scale smoothness, but it can also efficiently enforce the sparsity to preserve sharp boundaries.

4.3. Noise Effect and Sensitivity Analysis of Elastic-Net Joint Inversion

In this section, we tested the impact of noise on the inversion results and analyzed the sensitivity of the inversion method. We designed three models for MT and gravity methods, each with a different source size, as shown in Figure 7a–f. All models included two rectangular boxes of the same size embedded in a half-space. The distribution of the observation points and mesh sizes of all of the models were the same. The physical property values of the resistivity and residual density models are listed in Table 3. For the resistivity model, apparent resistivity for transverse magnetic (TM) mode were generated at 29 stations, spaced at 0.2 km intervals (at 10 frequencies from 1000 to 1 Hz). For the density model, the gravity data were Bouguer corrected at 29 stations spaced uniformly from a horizontal distance 0 km to 6 km. The density model grid was set to 141 × 61 in the horizontal and vertical directions, respectively. The resistivity model grid needed to be extended from the density model grid, and the grid size was 157 × 75. We added 5%, 10%, and 20% random Gaussian noise to the resistivity and density model responses, respectively. The model response results after adding noise are shown in Figure 8. We found that useful signals would be covered by the increase of noise, especially when using the MT method. At the same time, when the anomalous sources became small, the noise may completely cover the useful signals.

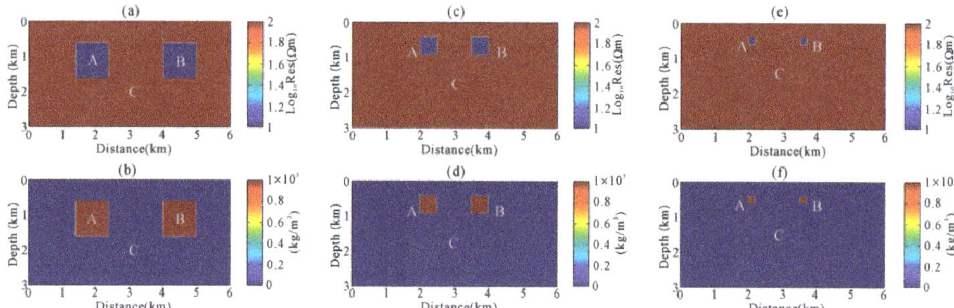

Figure 7. The true model of MT and gravity methods. Model 1 (**a**,**b**); Model 2 (**c**,**d**); Model 3 (**e**,**f**); Resistivity models (**a**,**c**,**e**) ; Density models (**b**,**d**,**f**).

Table 3. The resistivities and residual densities of the synthetic model.

Unit	Resistivity Model	Residual Density Model	Size	
A	10 Ω·m	1000 kg/m^3	1 × 1 km^2	
B	10 Ω·m	1000 kg/m^3	1 × 1 km^2	Model 1
C	100 Ω·m	0 kg/m^3	-	
A	10 Ω·m	1000 kg/m^3	0.5 × 0.5 km^2	
B	10 Ω·m	1000 kg/m^3	0.5 × 0.5 km^2	Model 2
C	100 Ω·m	0 kg/m^3	-	
A	10 Ω·m	1000 kg/m^3	0.2 × 0.2 km^2	
B	10 Ω·m	1000 kg/m^3	0.2 × 0.2 km^2	Model 3
C	100 Ω·m	0 kg/m^3	-	

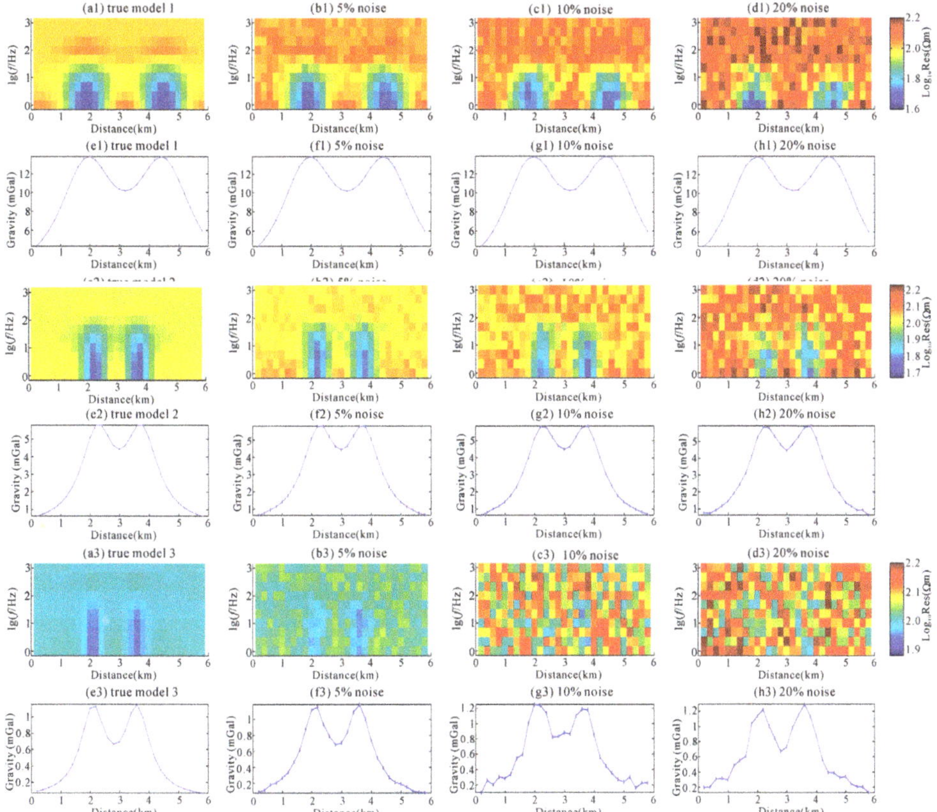

Figure 8. The model response results for three models of resistivity and density. The theoretical model responses of resistivity models obtained by TM mode (**a1–a3**); The theoretical model responses of density models (**e1–e3**); The resistivity model responses with 5% noise obtained by TM mode (**b1–b3**); The density model responses with 5% noise (**f1–f3**); The resistivity model responses with 10% noise obtained by TM mode (**c1–c3**); The density model responses with 5% noise. (**g1–g3**); The resistivity model responses with 20% noise obtained by TM mode (**d1–d3**); The density model responses with 20% noise (**h1–h3**).

We first performed inversion tests on model response datasets with 5%, 10%, and 20% random Gaussian noise, respectively (Figure 9a1,e1). All the starting models were homogeneous medium with a resistivity of 100 Ω·m and a residual density of 0 kg/m^3. It was found that the model response datasets after adding noise in model 1 could still recover inversion results similar to the true model, as shown in Figure 9a1–h2. This shows that the inversion method has certain anti-noise ability and is suitable for field measurement data processing with high noise. Next, we performed an inversion sensitivity analysis. We reduced the size of the anomalous sources of the model (Figure 9a2,e2) and added 5%, 10%, and 20% random Gaussian nosie to the theoretical model response. The inversion results showed that when the anomalous sources shrank to 0.5 × 0.5 km^2, the inversion method could still recover similar results to the true model. However, when the anomalous sources shrank to 0.2 × 0.2 km^2 (Figure 9a3,e3), we found that accurate inversion results might not be obtained with the increase of noise, mainly because the response amplitude of the model was smaller than the noise magnitude, especially the MT method. At this time, our inversion method may not have identified less than 0.2 × 0.2 km^2 anomalous sources, so when using this inversion for field data inversion, an anomalous source smaller than 0.2 × 0.2 km^2 needs to be interpreted with caution, as it may be a false anomaly. For anomalous sources above 0.2 × 0.2 km^2, the inversion could be accurately identified and identified as a true anomalous source.

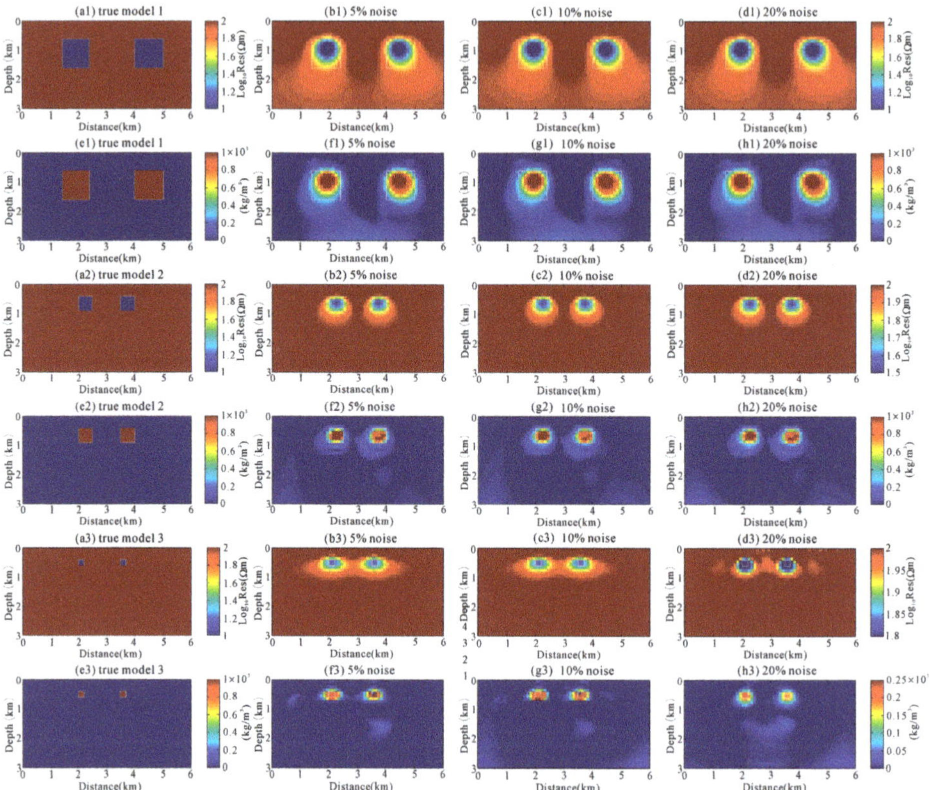

Figure 9. The MT and gravity inversion results obtained using elastic-net joint inversion. Resistivity true model (**a1–a3**); Density true model (**e1–e3**); The MT inversion results with 5% noise (**b1–b3**); The gravity inversion results with 5% noise (**f1–f3**); The MT inversion results with 10% noise (**c1–c3**); The gravity inversion results with 10% noise (**g1–g3**); The MT inversion results with 20% noise (**d1–d3**); The gravity inversion results with 20% noise (**h1–h3**).

5. Field Example

5.1. Geologic Background of the Survey Area

The Linjiang area of Jilin Province is located in the proterozoic Liaoji Rift Valley. The north and south sides are the Wolf Frost block and Longgang block, respectively (Figure 10). The Linjiang copper mine is located in Liudaogou Town, Linjiang City. The exposed strata in the mining area are mainly Laoling group metamorphic rocks, Mesozoic effusion rocks, and Tertiary basalts. The Laoling group stratum is affected by multi-stage tectonic formation, mainly composed of marble and hornstone, and local schist and phyllite. The Mesozoic intrusive rocks are mainly composed of a set of medium-acid volcanic rocks. The formation and distribution of the Linjiang copper deposit are closely controlled by the structure of the area. The ore deposit is located on the north-east side of the deep fault of the Yalu River. The northeastward fault of the main ore body controls the magmatism and mineralization in the mining area. The junction of the north-east fault and the east-west fault intersects and produces the granodiorite body, which controls the distribution of the copper deposit and is the most important ore-conducting structure in the mining area. The magmatic rocks in the mining area are widely distributed and mainly include the Yanshanian hornblende gabbro, the Yanshanian early acid volcanic complex, the Yanshanian mid-granitic granodiorite associated with mineralization, and the re-invasive quartz diorite. Therefore, the Linjiang copper mining area has good metallogenic conditions, large ore-forming intensity, various types of deposits, and potential for finding skarn-type copper deposits in the periphery and deep mining area.

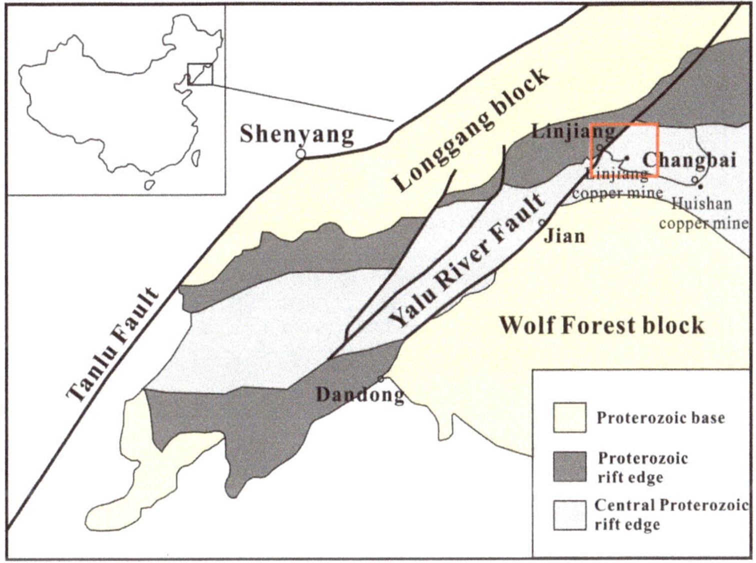

Figure 10. Location (red rectangle) and tectonic background map of the study area.

5.2. Data Acquisition

The Linjiang copper mining area is mainly covered by basalts of varying thicknesses and is a semi-concealed deposit. We conducted a geophysical survey along one profile in the study area. Figure 11 shows the geological map of the study area. The black solid line indicates the position of the survey line. The survey line runs in a north-south direction and measures 20.7 km in length. Multiple geophysical surveys, which included CSAMT (controlled-source audio-frequency magnetotelluric) and gravity exploration, were carried out along the survey line. Gravity and CSAMT data were collected along the survey line using a high-precision metal spring gravity meter (Burris, USA) and a GDP-32II (Zonge Company, USA), respectively. A total of 207 CSAMT points were collected at an interval of 100 m and 17 frequencies were collected at each point. The observed data were the apparent resistivity in TM mode (Figure 12a), which were obtained by the full-region apparent resistivity conversion. Furthermore, 526 gravity observation points (Figure 12b) were collected at a spacing of 40 m.

collected at a spacing of 40 m.

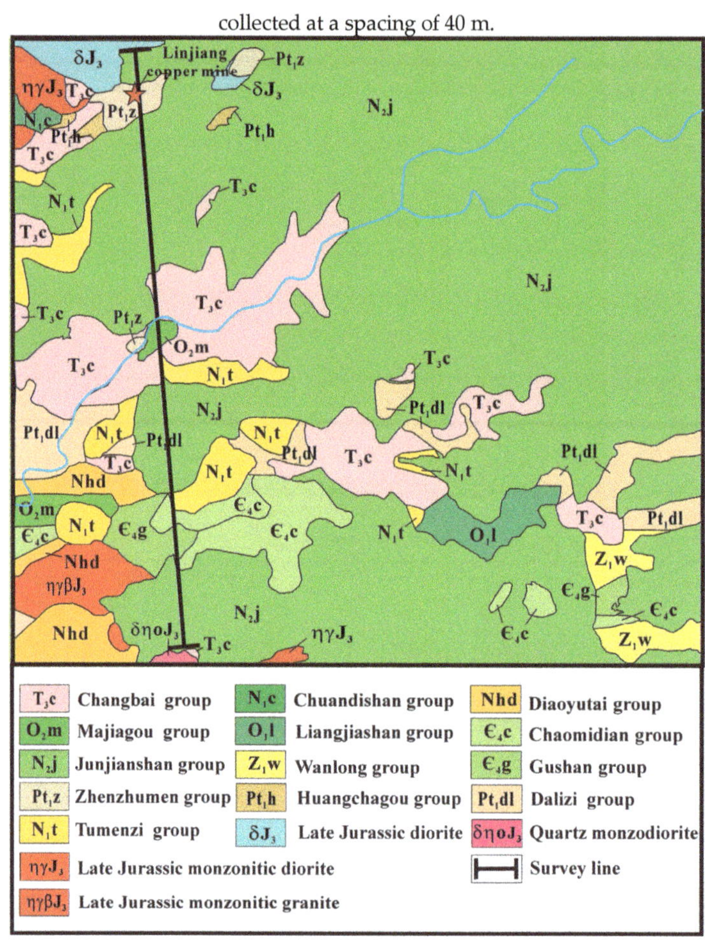

Figure 11. The geological map of the Linjiang copper mining area.

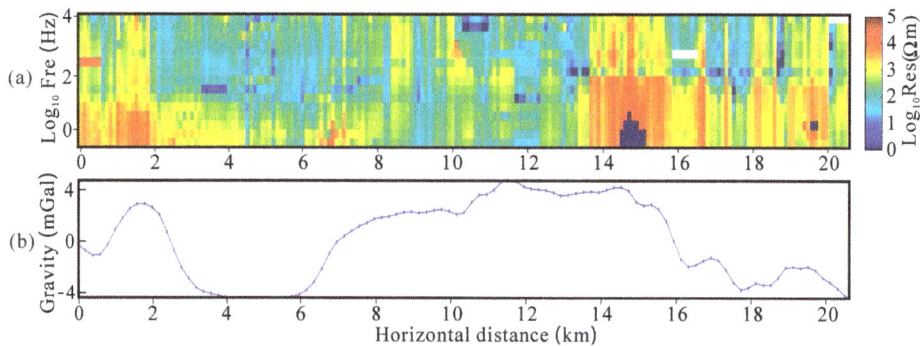

Figure 12. Observation data of the survey line: apparent resistivity in TM mode (**a**) and residual anomaly (**b**).

5.3. Inversion Models of CSAMT and Gravity Data

The CSAMT and gravity data were inverted to 2D resistivity and residual density models using separate inversion, joint inversion using smooth regularization method, and joint inversion using the elastic-net regularization method. For our inversion, our base model of the subsurface was discretized on rectangular cells 100 m wide and 10 to 100 m thick. This base model was padded by logarithmically spaced cells beyond its four edges to allow for the decay of the electromagnetic fields. These models were initially assigned homogeneous property values of $10^{2.5}$ Ω·m, 0.0 kg/m^3 in rock formation. The models obtained from the separate inversion are shown in Figure 13. The RMS misfit values found after separate inversion were as follows: RMS_{CSAMT} = 1.39 and RMS_{GRV} = 0.45. The separate inversion models showed major structural differences, and an interpreter will be faced with difficulties in interpreting them.

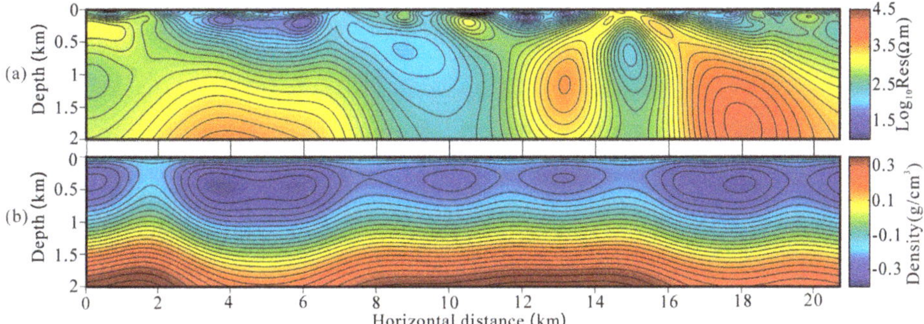

Figure 13. The results of separate inversion: resistivity model (**a**), residual density model (**b**).

For the joint inversion experiment, the process started from the same homogeneous model as that of the separate inversion and the process converged consistently in terms of data fit until the misfits were reduced to values similar to those obtained for the corresponding separate inversion experiments. The RMS misfit values found after the smooth joint inversion were as follows: RMS_{CSAMT} = 1.39 and RMS_{GRV} = 0.55. Figure 14 shows the joint inversion results obtained by the smooth regularization method. We found a structural similarity between the density and resistivity models in the results of the smooth joint inversion, which was due to the contribution of the cross-gradient constraints in the objective function. However, the inversion results were relatively smooth, and a large area of divergence was exposed below the anomalous bodies. It was difficult to depict the sharp boundaries of the subsurface geological contact. The corresponding RMS misfit for each data set after the elastic-net joint inversion did not increase significantly and the values attained were: RMS_{CSAMT} = 1.42 and RMS_{GRV} = 0.24. Figure 15 shows the joint inversion results obtained by the elastic-net regularization method. We found that the elastic-net joint inversion method could generate much greater detail and a sharper boundary as well as better depth resolution. Compared with the smooth joint inversion results, the large area divergence phenomenon under the anomalous bodies was eliminated, and the fine anomalous bodies boundary appeared in the smooth region. The elastic-net joint inversion method provided more detailed information about the structure of the sources for further analysis than the separate inversion and smooth joint inversion methods.

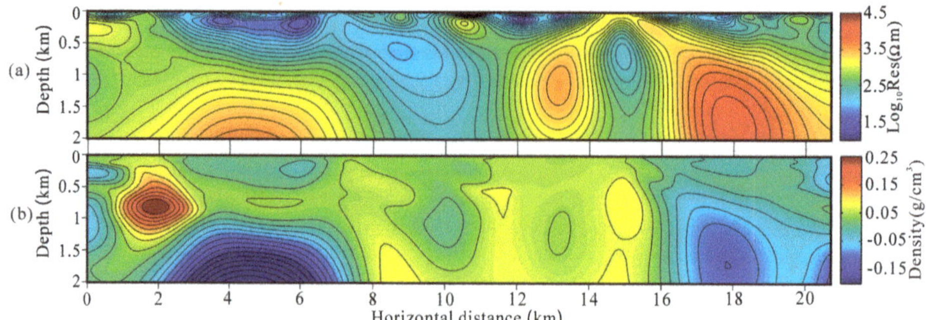

Figure 14. The results of joint inversion using smooth regularization: resistivity model (**a**), residual density model (**b**).

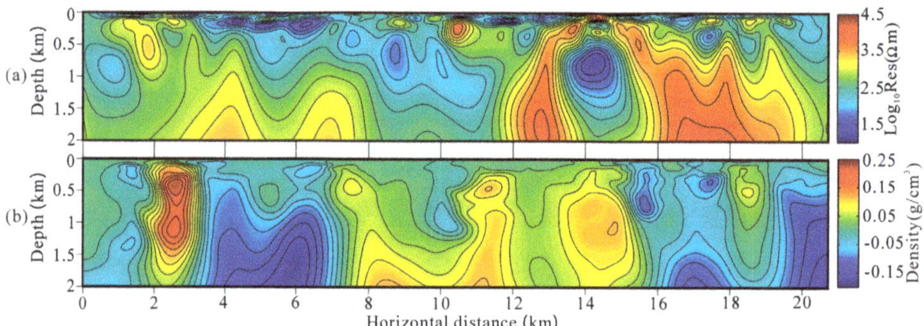

Figure 15. The results of joint inversion using elastic-net regularization: resistivity model (**a**), residual density model (**b**).

5.4. Geological Interpretation of the Mining Area

We conducted a geological interpretation of the higher resolution elastic-net joint inversion results. To facilitate geological interpretation, we combined the resistivity and residual density models obtained from the inversion into an RGB composite map of the (red–green–blue) mode. The RGB color space model is an additive color mixing model where red, green, and blue colors are superimposed on each other to achieve color mixing. The resistivity is represented in red, where a redder hue indicates the greater the resistivity; the residual density is in green, where a greener hue is a greater residual density; the blue was set to a value that had no change. The RGB composite map of the joint inversion model was generated by recombining the elastic-net joint inversion results (Figure 16). The composite image color contained the resistivity and residual density information, from which the inversion results could be analyzed intuitively to accurately identify, and possibly classify, the underground geological structures.

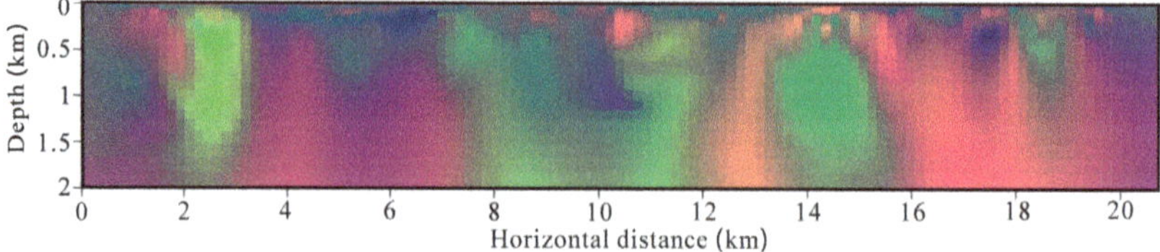

Figure 16. The RGB composite image of the elastic-net joint inversion.

We carried out a geological interpretation for this area using the RGB composite image (Figure 16), the geological map (Figure 11), and a statistical table of stratigraphic lithology (Table 4). First, we divided the survey line into four segments (I–IV) according to the distribution of gravity anomalies. Segment I is located at the northern end of the study area and has the characteristics of a high-gravity anomaly. The surface reveals the Palaeozoic Majiagou formation and the Paleoproterozoic Zhenzumen formation, basalt is directly covered on the Paleoproterozoic, and it can be inferred that this was caused by the Paleoproterozoic basement uplift, as shown in unit A in Figure 17. Segment II appears as a low-gravity anomaly. Except for a large number of the Triassic Changbai formation and Tertiary Tumenzi formation, the others are covered by a large number of basalt of the Neogene Chuandishan formation. The resistivity of different depths of the Segment II showed obvious differences and could be roughly divided into two layers, unit B, which is a low resistive layer less than 1 km deep and unit C, a highly resistive layer greater than 1 km in depth. We concluded that unit B was caused by the volcanic subsidence basins of the Mesozoic and Cenozoic and unit C was caused by the intrusion of medium-acid volcanic rocks with low density and high resistivity. Combined with the Late Jurassic diorite exposed on the surface, it can be inferred that unit C is the intrusive Late Jurassic diorite (Figure 17). Segment III is a high-gravity anomaly zone. The surface of the Paleoproterozoic Dalizi formation, the Diaoyutai formation of the Qingbaikouan Period, and the Chaomidian and Gushan formations of the Late Cambrian are exposed. The rock density values of formation above-mentioned are relatively high, and the density of the overlying Mesozoic and Cenozoic caprocks are quite different. Therefore, the Paleozoic basin caused high gravity anomalies. The base of this segment is the Paleoproterozoic Dalizi formation (unit E), which is a favorable area for finding precious metals such as gold. Unit G has a high-resistance medium density characteristic, which is presumed to be the Diaoyutai formation quartz sandstone of the Qingbaikouan Period. Segment IV is located at the southern end of the study area and is a low-gravity anomaly area spread near the east and west. There are intrusive rocks such as Late Jurassic monzonitic granites and quartz monzodiorite on the surface. It was speculated that this segment was caused by the intrusion of volcanic rocks and medium-acid rocks with relatively low-density and high resistivity. Unit K is composed of Late Jurassic monzonitic granites and unit L is quartz monzodiorite. Moreover, through the geological map and the RGB composite map, we could infer the distribution of the fault structure (F_1–F_4) below the survey line. Based on the above inference results, we finally obtained a comprehensive interpretation profile, as shown in Figure 17. The corresponding lithostratigraphic units in Figure 17 are shown in Table 4.

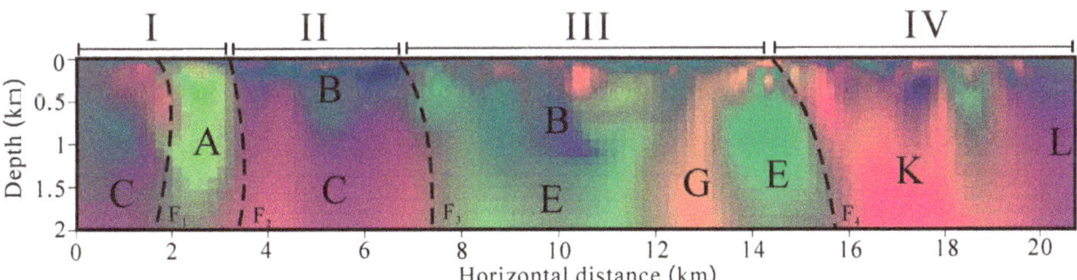

Figure 17. Comprehensive geological interpretation profile.

Table 4. Statistical table of stratigraphic lithology.

Geological Time		Lithostratigraphic Units	Lithology	Density	Resistivity	Unit
Era	Period	Formation		Average (g/cm^3)	Average (Ω·m)	
Cenozoic	Neogene	Junjianshan (N_2j)	Basalt	2.55	500	B
		Chuandishan (N_1c)	Basalt	2.55	506	B
		Tumenzi (N_1t)	Basalt	2.57	404	B
Mesozoic	Triassic	Changbai (T_3c)	Tuff	2.6	274	B
Paleozoic	Ordovician	Majiagou (Q_2m)	Limestone	2.75	3386	
		Liangjiashan (Q_1l)	Micrite	2.65	524	
	Cambrian	Caomidian ($Є_3$c)	Limestone	2.70	2975	
		Gushan ($Є_3$g)	Schist	2.69	2666	
Proterozoic	Sinian	Wanlong (Z_1w)	Limestone	2.72	6244	
	Qingbaikouan	Diaoyutai (Nhd)	Feldspathic quartz sandstone	2.62	3227	G
Paleoproterozoic		Dalizi (Pt_1dl)	Eryun schist with marble	2.75	1140	E
		Huashan (Pt_1h)	Cloud schist	2.80	1957	A
		Zhenzhumen (Pt_1z)	Dolomitic marble	2.78	2615	A

At present, Segment I has been mined. The copper deposit is produced in the inter-layer fault zone of the Paleoproterozoic aging group. The ore-bearing horizon coincides with the intrusion space of the parent rock, that is, along the inter-laminar fault zone of the thick layer of dolomite in the Zhenzhumen formation. The output law of the ore body group is mainly based on the inner contact belt ore body, followed by the outer contact part. The cause of the Segment I deposit is due to the magma carrying mineral components and forming skarns along the contact zone at the edge of the rock mass, and the copper and sulfur being combined to precipitate and form the ore. The deposit belongs to the hydrothermal skarn type, and the skarn type ore is mainly composed of chalcopyrite, bornite, and molybdenite.

Compared to the above actual situation, we found that the comprehensive geological interpretation profile provided an accurate formation (Unit A) and fault structure position (F_1–F_2), which verified the accuracy of the proposed algorithm. According to the comprehensive interpretation profile results and the deposit formation of the known mining area, we have speculated on the ore-forming target area of other segments. It was preliminarily predicted that the ore-forming target area would be mainly concentrated near the fault structure (F_3–F_4). The interlaminar fault zone is the center of the regional tectonic, magmatic activity, and later ore-forming hydrothermal activities. Prospecting work should pay attention to the mineralization clues of the contact sites between the Mesozoic intrusive rocks and the Precambrian shallow metamorphic strata, particularly the Yanshanian intermediate-acid granites. Mineralization and alteration should also be given sufficient attention, and silicidation and muddy mixtures that are easily overlooked may also have mineralization indication significance. In this paper, the elastic-net joint inversion method could accurately divide the stratum structure and fault zone, and had the ability to find the prospecting target area, which provides a powerful basis for the mining of the deposit.

6. Conclusions

We presented a novel 2D joint inversion approach for imaging collocated magnetotelluric (MT) and gravity data with elastic-net regularization and cross-gradient constraints. We described the main features of the novel approach and first applied it to the synthetic model. Numerical modeling results indicated that the L2 norm penalty always performed the solution overly smooth and the L1 norm penalty always presented the solution too sparse. However, the elastic-net regularization method, a convex combination term of the L2 norm and L1 norm, could not only enforce the stability to preserve local smoothness, but could also enforce the sparsity to preserve sharp boundaries. Meanwhile, we found that the cross-gradient constraints could effectively complement the defects of different geophysical methods, improve the horizontal resolution of MT and the vertical resolution of gravity methods, and obtained a model with a remarkable structural resemblance. Moreover, this paper provided a field example in a regional exploration context of the elastic-net joint inversion of gravity and MT data. This illustrated that structurally reconciled models of a 20.7 km profile in the Linjiang copper mining area in Jilin Province, northeast China, fused into an RGB composite image, yielded a more detailed and meaningful integrated interpretation of the subsurface. This case history highlights the advantages of using structurally reconciled models for integrated geophysical interpretations and showcases the power of RGB imaging in facilitating interpretation. Elastic-net joint inversion leads to more detailed and sharp boundary models that depict more property zones and may leads to the less ambiguous distinction of geologic materials. The RGB composite image adequately characterized the stratigraphic sequences and fault structure, and showed compatibility with the underground distribution structure of the identified deposits. This provides a favorable basis for unexplored areas, both in the formation sequence and fault location.

Whereas the proposed method proved its suitability for improved and integrated geophysical interpretation, it also illustrated that an improved resolution was largely driven by the accuracy of the individual geophysical modeling strategies and geophysical data. It is evident that strategies such as the incorporation of magnetic data, high precision seismic reflection data, or cross-hole data should lead to a higher resolution in imaging. Furthermore, these particular strategies can be adopted within the elastic-net joint inversion scheme.

Author Contributions: Conceptualization, R.Z.; methodology, R.Z., T.L.; software, R.Z.; validation, T.L.; formal analysis, R.Z., T.L.; investigation, R.Z., T.L.; resources, R.Z., T.L.; data curation, R.Z., T.L.; writing—original draft preparation, R.Z.; writing—review and editing, R.Z., T.L., S.Z., X.D.; supervision, T.L.; project administration, T.L.; funding acquisition, T.L., S.Z.

Funding: This research was funded by the Nation Key R&D program of China (2017YFC0601606), Jilin University PhD Graduate Interdisciplinary Research Funding Project (10183201838), the National Postdoctoral Program for Innovative Talents (B 2017-014), and China Postdoctoral Science Foundation funded project (2018M630323).

Conflicts of Interest: The authors declare no conflict of interest.

References

1. Engl, H.W.; Ramlau, R. *Regularization of Inverse Problems*; Springer: Berlin, Germany, 2015.
2. Honerkamp, J.; Weese, J. Tikhonovs regularization method for ill-posed problems. A comparison of different methods for the determination of the regularization parameter. *Contin. Mech. Thermodyn.* **1990**, *2*, 17–30. [CrossRef]
3. Tikhonov, A.N.; Dsh, B.; Griogolava, I.V. Interaction of ubisemiquinone with the high-potential iron-sulfur center of submitochondrial particle succinate dehydrogenase. EPR study at 240 and 12 degrees K. *Biofizika* **1977**, *22*, 734. [PubMed]
4. Li, Y.G.; Oldenburg, D.W. 3D inversion of magnetic data. *Geophysics* **1996**, *61*, 394–408. [CrossRef]
5. Li, Y.G.; Oldenburg, D.W. Fast inversion of large-scale magnetic data using wavelet transforms and a logarithmic barrier method. *Geophys. J. Int.* **2003**, *152*, 251–265. [CrossRef]
6. Pilkington, M. 3D magnetic imaging using conjugate gradients. *Geophysics* **1997**, *62*, 1132–1142. [CrossRef]
7. Cella, F.; Fedi, M. Inversion of potential field data using the structural index as weighting function rate decay. *Geophys. Prospect.* **2012**, *60*, 313–336. [CrossRef]
8. Paoletti, V.; Ialongo, S.; Florio, G.; Fedi, M.; Cella, F. Self-constrained inversion of potential fields. *Geophys. J. Int.* **2013**, *195*, 854–869. [CrossRef]
9. Chopra, A.; Lian, H. Total variation, adaptive total variation and nonconvex smoothly clipped absolute deviation penalty for denoising blocky images. *Pattern Recognit.* **2010**, *43*, 2609–2619. [CrossRef]
10. Figueiredo, M.A.T.; Dias, J.B.; Oliveira, J.P.; Nowak, R.D. On total variation denoising: A new majorization-minimization algorithm and an experimental comparison with wavalet denoising. In Proceedings of the IEEE International Conference on Image Processing (ICIP 2006), Atlanta, GA, USA, 8–11 October 2006; pp. 2633–2636.
11. Li, Q.; Micchelli, C.A.; Shen, L. A proximity algorithm accelerated by Gauss-Seidel iterations for L1/TV denoising models. *Inverse Probl.* **2012**, *28*, 095003. [CrossRef]

12. Guitton, A. Blocky regularization schemes for full waveform inversion. *Geophys. Prospect.* **2012**, *60*, 870–884. [CrossRef]
13. Strong, D.; Chan, T. Edge-preserving and scale-dependent properties of total variation regularization. *Inverse Probl.* **2003**, *19*, 165–187. [CrossRef]
14. Daubechies, I.; DeVore, R.; Fornasier, M. Iteratively reweighted least squares minimization for sparse recovery. *Commun. Pure Appl. Math.* **2010**, *63*, 1–38. [CrossRef]
15. Donoho, D.L. Compressed sensing. *IEEE Trans. Inf. Theory* **2006**, *52*, 1289–1306. [CrossRef]
16. Jin, B.T.; Maass, P. Sparsity Regularization for Parameter Identification Problems. *Inverse Probl.* **2012**, *28*, 123001. [CrossRef]
17. Loris, I.; Nolet, G.; Daubechies, I.; Dahlen, F.A. Tomographic inversion using ℓ_1-norm regularization of wavelet coefficients. *Geophys. J. Int.* **2007**, *170*, 359–370. [CrossRef]
18. Wang, Y.; Yang, C.; Cao, J. On Tikhonov regularization and compressive sensing for seismic signal processing. *Math. Models Meth. Appl. Sci.* **2012**, *22*, 1150008. [CrossRef]
19. Farquharson, C.G.; Oldenburg, D.W.; Routh, P.S. Simultaneous 1D inversion of loop-loop electromagnetic data for magnetic susceptibility and electrical conductivity. *Geophysics* **2003**, *68*, 1857–1869. [CrossRef]
20. Farquharson, C.G. Constructing piecewise-constant models in multidimensional minimum-structure inversions. *Geophysics* **2008**, *73*, K1–K9. [CrossRef]
21. Loke, M.H.; Acworth, I.; Dahlin, T. A comparison of smooth and blocky inversion methods in 2D electrical imaging surveys. *Explor. Geophys.* **2003**, *34*, 182–187. [CrossRef]
22. Van Zon, T.; Roy-Chowdhury, K. Structural inversion of gravity data using linear programming. *Geophysics* **2006**, *71*, J41–J50. [CrossRef]
23. Anbari, M.E.; Mkhadri, A. Penalized regression combining the L1 norm and a correlation based penalty. *Sankhya B* **2014**, *76*, 82–102. [CrossRef]
24. Zou, H.; Hastie, T. Regularization and variable selection via the elastic net. *J. R. Stat. Soc. Ser. B Stat. Methodol.* **2005**, *67*, 301–320. [CrossRef]
25. Jin, B.; Lorenz, D.; Schiffler, S. Elastic-net regularization: error estimates and active set methods. *Inverse Probl.* **2009**, *25*, 1595–1610. [CrossRef]
26. Mol, C.D.; Vito, E.D.; Rosasco, L. Elastic-net regularization in learning theory. *J. Complex.* **2009**, *25*, 201–230. [CrossRef]
27. Wang, J.; Han, B.; Wang, W. Elastic-net regularization for nonlinear electrical impedance tomography with a splitting approach. *Appl. Anal.* **2018**, *2017*, 1–17. [CrossRef]
28. Zhdanov, M.S.; Alfouzan, F.A.; Cox, L.; Alotaibi, A.; Alyousif, M.; Sunwall, D.; Endo, M. Large-Scale 3D Modeling and Inversion of Multiphysics Airborne Geophysical Data: A Case Study from the Arabian Shield, Saudi Arabia. *Minerals* **2018**, *8*, 271. [CrossRef]
29. Tillmann, A.; Stöcker, T. A new approach for the joint inversion of seismic and geoelectric data. In Proceedings of the 63rd EAGE Conference and Technical Exhibition, Amsterdam, The Netherlands, 11–15 June 2000.
30. Jones, A.G.; Fishwick, S.; Evans, R.L.; Spratt, J.E. Correlation of lithospheric velocity and electrical conductivity for Southern Africa. 2009. Available online: https://www.researchgate.net/publication/252588025_Correlation_of_lithospheric_velocity_and_electrical_conductivity_for_Southern_Africa (accessed on 2 July 2019).
31. Hoversten, G.M.; Cassassuce, F.; Gasperikova, E.; Newman, G.A.; Chen, J.; Rubin, Y.; Hou, Z.; Vasco, D. Direct reservoir parameter estimation using joint inversion of marine seismic AVA and CSEM data. *Geophysics* **2006**, *71*, C1–C13. [CrossRef]
32. Harris, P.; MacGregor, L. Determination of reservoir properties from the integration of CSEM, seismic, and well-log data. *First Break* **2006**, *25*, 53–59.
33. Gallardo, L.A.; Meju, M.A. Characterization of heterogeneous near-surface materials by joint 2D inversion of dc resistivity and seismic data. *Geophys. Res. Lett.* **2003**, *30*, 1658. [CrossRef]
34. Gallardo, L.A.; Meju, M.A. Joint two-dimensional DC resistivity and seismic traveltime inversion with crossgradients constraints. *J. Geophys. Res. Solid Earth* **2004**, *109*, 3311–3315. [CrossRef]
35. Colombo, D.; De Stefano, M. Geophysical modeling via simultaneous joint inversion of seismic, gravity, and electromagnetic data: Application to prestack depth imaging. *Lead. Edge* **2007**, *26*, 326–331. [CrossRef]
36. Hamdan, H.A.; Vafidis, A. Joint inversion of 2D resistivity and seismic travel time data to image saltwater intrusion over karstic areas. *Environ. Earth Sci.* **2013**, *68*, 1877–1885. [CrossRef]

37. Bennington, N.L.; Zhang, H.; Thurber, C.H. Joint inversion of seismic and magnetotelluric data in the park field region of california using the normalized cross-gradient constraint pure and applied. *Geophysics* **2015**, *172*, 1033–1052.
38. Li, T.L.; Zhang, R.Z.; Pak, Y.C. Joint Inversion of magnetotelluric and first-arrival wave seismic traveltime with cross-gradient constraints. *J. Jilin Univ. Earth Sci. Ed.* **2015**, *45*, 952–961.
39. Li, T.L.; Zhang, R.Z.; Pak, Y.C. Multiple joint inversion of geophysical data with sub-region cross-gradient constraints. *Chin. J. Geophys.* **2016**, *59*, 2979–2988.
40. Zhang, R.Z.; Li, T.L.; Deng, H. 2D joint inversion of MT, gravity, magnetic and seismic first-arrival traveltime with cross-gradient constraints. *Chin. J. Geophys.* **2019**, *62*, 2139–2149.
41. Moorkamp, M.; Heincke, B.; Jegen, M.; Roberts, A.W.; Hobbs, R.W. A framework for 3D joint inversion of MT, gravity and seismic refraction data. *Geophys. J. Int.* **2011**, *184*, 477–493. [CrossRef]
42. Gallardo, L.A.; Fontes, S.L.; Meju, M.A.; Buonora, M.P.; de Lugao, P.P. Robust geophysical integration through structure-coupled joint inversion and multispectral fusion of seismic reflection, magnetotelluric, magnetic, and gravity images: Example from Santos Basin, offshore Brazil. *Geophysics* **2012**, *77*, B237–B251. [CrossRef]
43. Linde, N.; Binley, A.; Tryggvason, A.; Pedersen, L.B.; Revil, A. Improved hydrogeophysical characterization using joint inversion of cross-hole electrical resistance and ground-penetrating radar travel-time data. *Water Resour. Res.* **2006**, *42*, W12404. [CrossRef]
44. Linde, N.; Tryggvason, A.; Peterson, J.E.; Hubbard, S.S. Joint inversion of crosshole radar and seismic traveltimes acquired at the South Oyster Bacterial Transport Site. *Geophysics* **2008**, *73*, G29–G37. [CrossRef]
45. Zhdanov, M.S.; Gribenko, A.; Wilson, G. Generalized joint inversion of multimodal geophysical data using Gramian constraints. *Geophys. Res. Lett.* **2012**, *39*, L09301. [CrossRef]
46. Haber, E.; Holtzman Gazit, M. Model Fusion and Joint Inversion. *Surv. Geophys.* **2013**, *34*, 675–695. [CrossRef]
47. deGroot-Hedlin, C.D.; Constable, S.C. Occam's inversion to generate smooth, two-dimensional models from magnetotelluric data. *Geophysics* **1990**, *55*, 1613–1624. [CrossRef]
48. Wannamaker, P.E.; Stodt, J.A.; Rijo, L. A stable finite element solution for two-dimensional magnetotelluric modeling. *Geophys. J. Int.* **1987**, *88*, 277–296. [CrossRef]
49. Singh, B. Simultaneous computation of gravity and magnetic anomalies resulting from a 2D object. *Geophysics* **2002**, *67*, 801–806. [CrossRef]
50. Tarantola, A. Inversion problem theory: Methods for data fitting and model parameter estimation. *Phys. Earth Planet. Inter.* **1987**, *57*, 350–351.
51. Byrd, R.H.; Payne, D.A. *Convergence of the Iteratively Re-Weighted Least Squares Algorithm for Robust Regression*; Johns Hopkins University: Baltimore, MD, USA, 1979.
52. Wolke, R.; Schwetlick, H. Iteratively reweighted least squares: Algorithms, convergence analysis, and numerical comparisons. *SIAM J. Sci. Comput.* **1988**, *9*, 907–921. [CrossRef]
53. Chartrand, R.; Yin, W. Iteratively reweighted algorithms for compressive sensing. In Proceedings of the 33rd International Conference on Acoustics, Speech, and Signal Processing, Las Vegas, NV, USA, 28 March–4 April 2008; pp. 3869–3872.
54. Last, B.J.; Kubik, K. Compact gravity inversion. *Geophysics* **1983**, *48*, 713–721. [CrossRef]
55. Zhdanov, M.S. *Geophysical Inverse Theory and Regularization Problems*; Elsevier: Amsterdam, The Netherlands, 2002.
56. Rudin, L.; Osher, S.; Fatemi, E. Nonlinear total variation based noise removal algorithms. *Phys. D* **1992**, *60*, 259–268. [CrossRef]

Article

An Exploration Study of the Kagenfels and Natzwiller Granites, Northern Vosges Mountains, France: A Combined Approach of Stream Sediment Geochemistry and Automated Mineralogy

Benedikt M. Steiner *, Gavyn K. Rollinson and John M. Condron

Camborne School of Mines, University of Exeter, Penryn Campus, Penryn TR10 9FE, UK; G.K.Rollinson@exeter.ac.uk (G.K.R.); jmc258@exeter.ac.uk (J.M.C.)

* Correspondence: b.steiner@exeter.ac.uk

Received: 1 November 2019; Accepted: 30 November 2019; Published: 3 December 2019

Abstract: Following a regional reconnaissance stream sediment survey that was carried out in the northern Vosges Mountains in 1983, a total of 20 stream sediment samples were collected with the aim of assessing the regional prospectivity for the granite-hosted base and rare metal mineralisation of the northern Vosges magmatic suite near Schirmeck. A particular focus of the investigation was the suspected presence of W, Nb and Ta geochemical occurrences in S-type (Kagenfels) and I-S-type (Natzwiller) granites outlined in public domain data. Multi-element geochemical assays revealed the presence of fault-controlled Sn, W, Nb mineralisation assemblages along the margins of the Natzwiller and Kagenfels granites. Characteristic geochemical fractionation and principal component analysis (PCA) trends along with mineralogical evidence in the form of cassiterite, wolframite, ilmenorutile and columbite phases and muscovite–chlorite–tourmaline hydrothermal alteration association assemblages in stream sediments demonstrate that, in the northern Vosges, S-type and fractionated hybrid I-S-type granites are enriched in incompatible, late-stage magmatic elements. This is attributed to magmatic fractionation and hydrothermal alteration trends and the presence of fluxing elements in late-stage granitic melts. This study shows that the fractionated granite suites in the northern Vosges Mountains contain rare metal mineralisation indicators and therefore represent possible targets for follow-up mineral exploration. The application of automated mineralogy (QEMSCAN®) in regional stream sediment sampling added significant value by linking geochemistry and mineralogy.

Keywords: Vosges; Variscan orogeny; Natzwiller; Kagenfels; granite; lithium; tungsten; niobium; exploration targeting; stream sediments; QEMSCAN®

1. Introduction

Since 2010, the global drive for clean energy and industrial metals led to the delineation and shortlisting of a number of "critical" metals by the European Union [1]. Of the 26 metals listed in the 2017 report, high-tech metals, such as W, Nb and Ta, are of significant importance to the European manufacturing industry. Recent exploration for these metal deposits has been ongoing for more than two decades and has been predominantly focused around known prospective European Variscan belts, such as the Erzgebirge and Cornwall, and characteristic S-type granite provinces therein [2]. A recent regional geochemical reconnaissance sampling campaign, however, identified I-type granites of the lesser known central Vosges Mountains near Sainte-Marie-aux-Mines (France) to be prospective for W and Li-Cs-Ta [3]. Similarly, an investigation of Bureau de Recherches Géologiques et Minières (BRGM) public domain data [4,5] led to the delineation of a distinct W anomaly in the Natzwiller

Granite, which is located in the northern domain of the Vosges Mountains. The northern Vosges Mountains comprise a series of late Devonian–Permian intrusions which show a distinct development from primitive mantle to highly fractionated peraluminous melts and the emplacement of S and I–S-type granites. Detailed studies into the petrology and geochronology of the magmatic suites allowed the reconstruction of a complex magmatic system and evolution of tectonic processes during the Variscan Orogeny [6,7]. The present study aims to investigate the granite-hosted W, Li, Nb and Ta mineralisation potential of the northern Vosges Mountains and provides a link between chemical and mineralogical analyses of regional stream sediment samples and magmatic processes outlined in previous studies. The application of automated mineralogical techniques (QEMSCAN®) to stream sediment samples furthermore aims to demonstrate the usability of this technique in early stage mineral exploration and therefore adds to the limited literature available on the subject [8–10]. Consequently, this paper provides an economic perspective to previous research carried out on the northern Vosges magmatic suite.

2. Geology and Magmatic Pulses of the Northern Vosges Mountains

2.1. Regional Geological Setting

The Vosges Mountains (NE France) are part of the European Variscan orogenic belt and consist of three primary geological domains: the Southern Paleozoic basin (Southern Vosges), the central high grade gneiss and granulite domain (Central Vosges), and the northern low-grade Palaeozoic sedimentary succession (Northern Vosges) [11] (Figure 1a). In this context, the Late Carboniferous Lalaye–Lubine suture, a dextral shear zone, represents the boundary between the Saxothuringian domain of the northern Vosges and the Moldanubian domain of the central and southern Vosges [12]. Throughout the Vosges, a complex suite of Carboniferous magmatic rocks intruded into the early Palaeozoic–Carboniferous basement, predominantly comprising clastic metasedimentary rocks. Over the last three decades, these distinct magmatic suites have instigated significant interest and research into the study of Variscan tectonics and magmatic evolution [13].

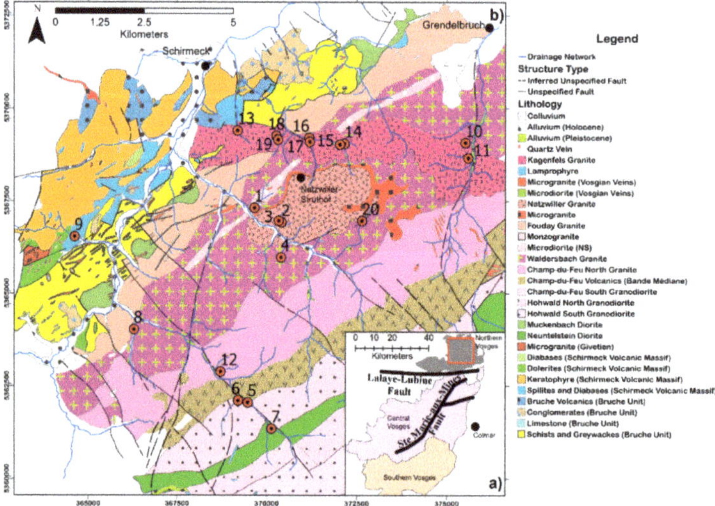

Figure 1. (**a**) Regional geological setting of the Vosges Mountains modified from Tabaud et al. (2015) [12] and (**b**) geological map of the northern Vosges Mountains with focus on the Champ du Feu area, NE France (modified from Bureau de Recherches Géologiques et Minières (BRGM) InfoTerre [4]). Projected coordinate system used: WGS 1984 UTM Zone 32N. Stream sediment sample numbers are indicated as 1–20.

2.2. Magmatic Suites: Geochronology, Geochemistry and Source Rocks

The northern Vosges Mountains (Figure 1b) consist of a Devonian–Carboniferous NE–SW striking succession of (volcano-) sedimentary belts (Bruche Unit, Schirmeck Volcanic "Massif") and weakly metamorphosed sediments of principally clastic origin (Steige Unit, Villé Unit), which were intruded by a Middle Devonian–Permian magmatic suite indicating an overall fractionation trend from tholeiitic to peraluminous S-type melts [6]. The evolution of the magmatic suite is a result of a complex interaction between mantle and crustal source rocks and therefore shows a variety of petrological and geochemical characteristics. The geochronological and geochemical development of these intrusions has been discussed in multiple publications [6,7,14]; however, an overall consensus about the origin and evolution of the granite suites appears to not have been reached yet as other authors question the peraluminous nature of Late Carboniferous granites [15]. In this paper, the geochronological and geochemical framework of the northern Vosges magmatic suites along with the implications for the overall tectonic setting and nature of source rocks will principally be based on the U-Pb zircon age and whole-rock geochemical data provided by Tabaud et al. (2014) [6].

The first magmatic episode took place during the Middle Devonian when continental back-arc tholeiitic to calc-alkaline (sub-aluminous to peraluminous) volcanic rocks of the Schirmeck–Rabodeau massifs were formed as a result of partial melting of a depleted mantle source. These volcanic rocks are generally depleted in Rb and Th. At ca. 330 Ma, calc-alkaline (metaluminous to weakly peraluminous) volcanic rocks of the Hohwald suite (Bande Médiane and Neuntelstein diorites and Hohwald granodiorites) were produced by the partial melting of an enriched mantle wedge followed by fractional crystallisation, crustal assimilation and metasomatism processes. The high-K calc-alkaline Belmont granite suite, comprising the Champ du Feu North, Waldersbach and Fouday granites, was intruded 10 Ma later at ca. 318 ± 3 Ma, and probably formed as a product of the mixing of enriched mantle-derived melt and felsic magma originating from dehydration fluids of subducted continental crust. The "Younger" granite suite, comprising the Andlau, Natzwiller and Senones granite stocks, was intruded at 312 ± 2 Ma and displays typical geochemical properties of felsic Mg-K magmatism, whereas the chemical composition is sub-aluminous to weakly peraluminous. Younger granites are generally enriched in Ba, Sr, Cr, Th, U and $\sum$REE (Rare Earth Elements). The Younger granites are considered to be derived from the partial melting of an enriched mantle source followed by interaction with subducted, young crustal material and therefore represent a hybrid I-S-type granite. At ca. 290 Ma, in-situ radiogenic heat production resulting from K, Th and U-enriched Younger granites led to crustal anatexis of metasedimentary country rocks and the emplacement of peraluminous to felsic peraluminous S-type Kagenfels granites. Kagenfels two-mica granites are the least ferro-magnesian and the most sodium–potassic rocks of the study area. The rocks are depleted and show negative trends in Ba, Sr, Ti, U and $\sum$REE, but are enriched in Al, Zr and Hf. The occurrence of metaluminous and peraluminous granites, in the form of the Younger suite and the Kagenfels granite, mirror the overall evolution of the Central Vosges Mg-K (CVMg-K) and Bilstein–Brézouard–Thannenkirch (BBTC) granites of the central Vosges domain. The granite suites in both domains share geochemical similarities and a prominent time lag of 10–15 Ma between the intrusion of I-type CVMg-K (337.2 ± 1.8 Ma) and predominantly S-type BBTC (330–325 Ma) granites [12].

2.3. Mineralisation in the Northern Vosges Mountains and Review of Historic BRGM Regional Geochemical Data

The overall metallogenic setting of the Vosges Mountains has been described in Dekoninck et al. (2017) [16], Fluck and Stein (1992) [17], Fluck (1977) [18] and Fluck and Weil (1976) [19]. Whilst the Central Vosges, in particular, are famous for previously mined post-Variscan polymetallic Pb-Zn-Cu-As-Co-Bi veins along the "Val d'Argent" trend near Sainte-Marie-aux-Mines [20,21], there are limited studies available that explicitly describe the mineralisation patterns of the northern Vosges magmatic suite. Billa et al. (2016) [22] evaluated the regional geochemical trends of historic BRGM data across French basement "massifs". Whilst this report outlined a number of polymetallic and Sn-W anomalies in the southern and central-northern Vosges Mountains, the Belmont Granite Suite of the northern

Vosges around Schirmeck has not been taken into consideration. On the other hand, Weil (1936) [23] noted the presence of adularia, beryl, fluorine and molybdenite in a pegmatite vein ("Grotte des Partisans") located in Kagenfels Granite between Schirmeck, Rothau and Natzwiller. The mineralogical observations confirm the peraluminous nature of S-type Kagenfels Granite and imply that late stage fractionation processes led, locally, to the enrichment of incompatible elements. Leduc (1984) [5] reported results of 905 regional stream sediment samples (data available from BRGM, 2019 [24]) collected across a 330 km^2 extensive area of the northern Vosges Mountains. This survey employed a 125 μm stream sediment fraction along with plasma emission spectroscopy analysis for 22 elements; however, the report does not specify details on the acid leach and analytical precision of the analyses, although these are the only data available for this area. The study resulted in the delineation of a number of polymetallic anomalies in the Fouday, Natzwiller and Kagenfels Granites. For example, distinctive anomalies—i.e., >98th percentile—of W (20–56 ppm) and Sn (27–35 ppm) can clearly be recognised in the eastern parts of the Natzwiller and Kagenfels Granites, respectively (Figure 2a,b, Table 1). Furthermore, elevated concentrations—i.e., >90th percentile—of Nb (41–57 ppm) are present in the Kagenfels Granite and streams draining the Kagenfels area to the north and east (Figure 2c). An analysis of the spatial distribution of Nb is not discussed in Leduc's report. Boron generally shows subdued concentrations of <11 ppm throughout the Natzwiller and Kagenfels Granites, whereas elevated concentrations of 22–63 ppm can be observed in the southeastern margin of the Kagenfels Granite, Ville and Steige schists (Figure 2d).

Table 1. Summary statistics of 579 out of 905 historic BRGM samples located in the study area (Figure 1). The 579 samples were selected from the larger dataset so that the northern Vosges magmatic suite around Hohwald, Champ du Feu, Natzwiller, and imminently adjacent portions of the Bruche Unit and Schirmeck Volcanic Massif, were covered. In total, 22 elements were analysed using Inductively Coupled Plasma Atomic Emission Spectroscopy (ICP-AES); however, only selected available elements of the typical granite-related mineralisation suite are presented here. Note the distinct anomalies of Sn, W (>99th percentile) and Nb (>98th percentile).

Summary Statistics	B (ppm)	Be (ppm)	Cu (ppm)	Nb (ppm)	Sn (ppm)	W (ppm)
Lower Limit of Detection (LoD)	10	1	10	10	20	10
Minimum (LoD/2)	5	0.5	5	5	10	5
Maximum	63	43	155	57	35	56
Mean	10	3.2	17.5	30.3	10.2	5.4
Median	5	3	15	30	10	5
5 percentile	5	0.5	5	16	10	5
10 percentile	5	0.5	5	20	10	5
25 percentile	5	2	5	25	10	5
30 percentile	5	2	11	26	10	5
60 percentile	5	3	16	32	10	5
75 percentile	12	4	21	36	10	5
80 percentile	14	4	23	37	10	5
90 percentile	22	5	31	41	10	5
95 percentile	30	6	42	44	10	5
98 percentile	44	9	63.4	48	10	5
99 percentile	47	22.5	86.4	51.4	27.6	24

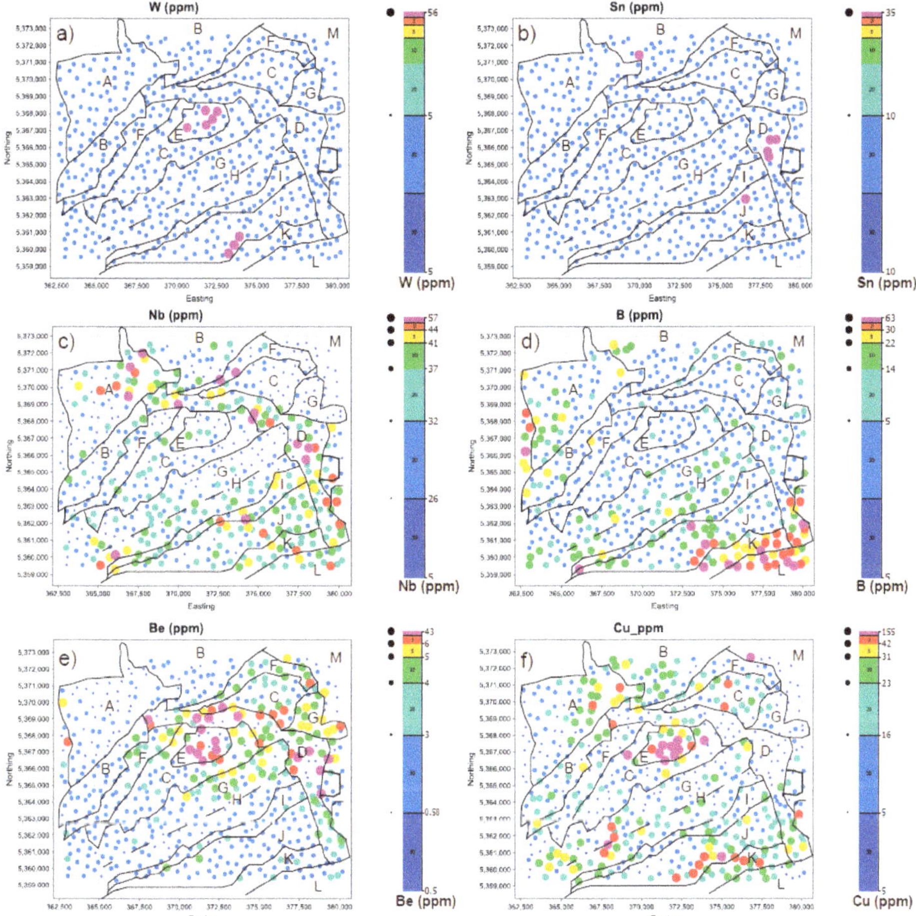

Figure 2. Graduated point symbol maps for (**a**) W, (**b**) Sn, (**c**) Nb, (**d**) B, (**e**) Be, (**f**) Cu using the 30th, 60th, 80th, 90th, 95th, 98th and 99th percentile distribution from Table 1. Principal geological domains are highlighted. A detailed geological map is shown in Figure 1. A, mafic and intermediate volcanic rocks of the Schirmeck Massif; B, metasedimentary rocks of the Bruche Unit; C, Waldersbach Granite; D, Kagenfels Granite; E, Natzwiller Granite; F, Fouday Granite; G, Champ du Feu granite suite; H, Bande médiane mafic volcanics; I, Neuntelstein Diorite; J, Hohwald granite suite; K, Steige schists; L, Ville schists; M, Lower Buntsandstein.

Elevated Be concentrations are confined to both the Natzwiller Granite (10–25 ppm) and the Kagenfels Granite (6–43 ppm), highlighting the enrichment of late-stage, incompatible elements in these granites when compared to the surrounding, less-evolved granites of the Belmont Granite Suite (Figure 2e). A prominent Cu anomaly is centred on the Natzwiller Granite (63–136 ppm, Figure 2f). The W, Cu, Nb and Be anomalies are located several hundred meters from the historic Natzwiller–Struthof prison camp quarry which produced "red granite" aggregate during World War II [25]. Whether the metallogenic significance of the Natzwiller and Kagenfels Granites was known to geologists at the time is unclear. Leduc (1984) [5] attributes these anomalies to molybdenite, scheelite, and beryl-bearing pegmatite stringers in Kagenfels Granite, alluvial cassiterite in streams, and chalcopyrite–molybdenite–scheelite (Cu-Mo-W-Ag) veins in Natzwiller Granite, essentially representing a porphyry mineralisation signature. However, whilst the 22 elements of the analytical suite seemed adequate for the exploration of base metal mineralisation at the time, resistate elements, such as Sn, and also W and Nb, are underestimated by the weak digest and plasma emission technique [5]. Furthermore, the 1980s analytical suite did not contain pathfinder elements (K, Rb, Zr, Hf, Li, Cs, Ta) that are utilised on a routine basis to assess the prospectivity of rare metal granites and pegmatites [26]. For example, K/Rb, Nb/Ta, and Zr/Hf ratios are commonly utilised to support the determination of magmatic fractionation trends, magmatic–hydrothermal interaction during fractional crystallisation, and the enrichment of incompatible elements of possible economic value [27–29]. For this reason, the present investigation aimed to collect a number of independent stream sediment samples in previously identified prospective lithologies, as well as to employ a comprehensive suite of pathfinder elements obtained by Inductively Coupled Plasma Mass Spectroscopy (ICP-MS), along with automated mineralogy techniques, to fingerprint the fractionation and mineralisation processes of the northern Vosges magmatic suite.

3. Methodology

In order to obtain an independent dataset and a comprehensive geochemical analysis suite characterised by low detection limits, 20 stream sediment samples (samples 1–20) were collected from first and second-order streams in the Hohwald, Natzwiller, Schirmeck and Grendelbruch areas (Figure 1b). The sample locations were principally designed to test distinctive structural and geochemical trends identified from previous BRGM investigations [5,24] as well as to characterise the different intrusive units across the northern Vosges. The sampling strategy mirrors the workflow and orientation study described in [3]. Samples were collected from stream traps, such as in the lee of large boulders or on point bars, and sieved to retain the <2 mm fraction in the field. The resulting material yielded average weights of 500 g per sample. Samples were stored in plastic bags, zip-tied and labelled with sample ID, coordinates and elevation information. The fine fraction was allowed to settle in the bag before excess water was poured back into the stream. After each sampling location, the equipment was thoroughly cleaned to prevent cross-contamination. A detailed list of stream sample attribute data (colour, grain and mesh size, contamination, trap type, etc.) was recorded on an iPad using ESRI's 'Collector for ArcGIS' app (Version 19.0, ESRI, Redlands, CA, USA). Daily data quality checks and synchronisation with a master database ensured that the data quality was consistent throughout the sampling campaign. Detailed observations and comparisons of drainage sediment composition, outcropping adjacent lithologies and heavy minerals present in pans were noted and supported the lithological classification of samples, along with the identification of fractionated lithologies; for example, quartz or pegmatite-rich rock. Linking observations of stream sediments and adjacent outcrops confirmed that stream sediments appeared unweathered and accurately represent the overall bedrock geology of respective catchment areas, and therefore can be used for further representative lithogeochemical investigations. While glaciation occurred in central–western Europe and affected parts of the central–southern Vosges Mountains, the only glacial debris related to the last glacial event (Weichselian) are recorded in the Bruche River valley. No evidence of glacial sediment was encountered during fieldwork.

Samples were returned to Camborne School of Mines, University of Exeter (UK) laboratories and dried in laboratory ovens at a constant temperature of 40 °C. The samples were then sieved using a sieve stack and a Pascal Sieve Shaker to isolate the <75 μm fraction which was previously identified to be host to W, Li and Cu anomalies in the Vosges [3]. Individual fraction weights were determined to assess if sample loss had occurred. All samples underwent ICP-MS analysis using a standard four acid digest (HCl-HF-HNO_3-$HClO_4$) and Agilent 7700 ICP-MS instrument (Agilent Technologies, Santa Clara, CA, USA), so that, compared to historic studies, a wider range of trace and pathfinder elements, including Nb, Ta, Li, Hf, could be obtained, aiding the determination of fractionation trends (Table 2). Quality control was ensured by inserting silica blanks, duplicate samples and OREAS 147 and 751-certified reference materials into the sample stream, with no accuracy issues noted outside +/− 1 standard deviation of the certified mean value. In a subsequent geochemical interpretation, following a previously published approach in the central Vosges [3] and southeast Ireland [30], the additional ICP-MS assays supported the usage of multi-element major and trace geochemistry in classifying lithological populations and petrogenetic and mineralisation processes. The classification of lithological units was achieved by delineating population clusters in bivariate geochemical plots. Each sample point was assigned a lithology, which was refined using geological observations in upstream catchment areas, outcrops and float in order to better represent subtle nuances in geochemical composition, such as "Natzwiller/Belmont Granite with visible cassiterite".

Of the 20 stream sediment samples (Samples 1–20) undergoing ICP-MS analysis, nine samples (2, 3, 14, 15, 16, 17, 18, 19, 20) of the <75 μm sample fraction were selected for mineralogical analysis. In addition, two rock samples (17A and 18A) were collected within a 10 m distance upstream of their corresponding stream sediment sample locations (17 and 18), and a mineralogical analysis (QEMSCAN®) was performed on uncovered polished thin sections of these samples. Mineralogical analysis was conducted on the fine <75 um stream sediment fraction, which is considered homogeneous and allowed a direct comparison between mineralogy and geochemistry. The samples were prepared into 30 mm diameter epoxy resin mounts and mixed with pure graphite powder to reduce settling bias and separate particles. The sample surface of the cured mounts was carefully ground to expose the particles and polished to a 1 micron finish using diamond media, then carbon-coated to 25 nm thickness. Samples were analysed using a QEMSCAN® 4300 [31–34] at Camborne School of Mines, University of Exeter, UK. Sample measurement used iMeasure version 4.2SR1 software for data collection and iDiscover 4.2SR1 and 4.3 software for data processing. The Particle Mineral Analysis (PMA) measurement mode was used to map particles at a resolution (pixel spacing) of 2 μm, field size of 600 μm (300 × 300 square, magnification of ×111), default of 1000 X-ray counts per analysis point and a target of 10,000 particles per sample. The final number of particles mapped per sample was higher than this (up to 14,556) due to the system completing the particles in the field it was on when it reached its 10,000-particle target. The number of analysis points per sample varied from 900,000 to 4 million.

The data collected during measurement were processed using a modified version of the standard LCU5 SIP (database), following and building upon details outlined in Section 7 of Rollinson et al. (2011) [35]. Both mineral area-% and mineral mass-% (density weighted) data were produced, and it was decided to use the mineral mass-% data as they better reflect the economic mineral content of the samples. As these were stream sediments, the focus was on both the major minerals and trace or unusual minerals, with the SIP customised to reflect the mineralogy of samples. This included checking the mineral identification not just from the measured chemical spectra, but also against in-house mineral reference standards. Moreover, checks were also completed for possible Li minerals following the method developed at Camborne School of Mines during the FAME EU Horizon 2020 project [36]. In the case of identified Ta-Nb minerals, independent SEM-EDS checks were also conducted.

Table 2. Summary statistics of elements and elemental ratios obtained for the present geochemical study. The stream sediment grain size fraction analysed was <75 μm. All 20 samples were analysed by ICP-MS using a four-acid digest (HCl-HF-HNO_3-$HClO_4$).

	As (ppm)	B (ppm)	Be (ppm)	Cs (ppm)	Cu (ppm)	Hf (ppm)	K (ppm)	Li (ppm)	Nb (ppm)	Pb (ppm)	Rb (ppm)
Lower Limit of Detection	0.0003	4.14	0.0003	0.0007	0.00005	0.0003	0.006	0.0002	0.0003	0.0003	0.0003
Minimum	7.28	9.9	1.25	3.59	7.88	8.65	12,513	12.13	9.93	8.72	47.3
Maximum	125.97	60.25	34.22	27.42	375.63	64.86	25,376	105.1	198.29	99.56	366.6
Mean	40.13	16.11	10.73	12.62	96	29.45	18,210	53.51	67.41	43.94	140.69
Median	27.1	14.08	10.21	11.93	77.71	30.51	18,226	50.57	58.41	39.39	136.4
5 percentile	7.37	9.9	1.27	3.79	8.74	8.7	12,533	12.83	10.11	9.29	67.1
10 percentile	9.07	9.97	1.79	7.78	26.84	9.66	13,087	26.96	13.59	20.73	77.9
25 percentile	15.62	12.04	2.8	9.79	50.18	14.4	15,382	41.4	16.03	27.3	112.75
30 percentile	16.75	12.28	3.08	10.56	54.09	17.73	15,536	41.65	19.06	29.79	116.3
60 percentile	38.63	15.06	11.82	12.08	87.94	35.13	19,110	53.9	82.25	40.44	144.9
75 percentile	64.57	15.43	13.61	14.13	119.21	39.51	20,371	66.09	97.96	53.73	162.55
80 percentile	72.9	15.45	18.08	16.08	122.23	44.12	21,099	75.96	100.08	54.28	175.2
90 percentile	91.78	21.34	23.83	19.37	175.49	50.68	23,542	81.37	185.66	92.04	192.5
95 percentile	124.31	58.34	33.71	27.03	365.67	64.15	25,289	103.92	197.99	99.27	215.7
98 percentile	125.97	60.25	34.22	27.42	375.63	64.86	25,376	105.1	198.29	99.56	316.76
99 percentile	125.97	60.25	34.22	27.42	375.63	64.86	25,376	105.1	198.29	99.56	366.6
	Sn (ppm)	**Sr (ppm)**	**Ta (ppm)**	**Th (ppm)**	**Ti (ppm)**	**W (ppm)**	**Zr (ppm)**	**K/Rb**	**Nb/Ta**	**Zr/Hf**	
Lower Limit of Detection	0.0003	0.0003	0.0003	0.0003	0.0007	0.0006	0.0003				
Minimum	5.48	21.41	0.76	8.87	2171	1.8	346.08	64.5	8.03	19.13	
Maximum	41.55	291.31	11.77	93.73	10,503	102.56	1923.08	273.2	17.75	45.01	
Mean	18.89	132.38	4.55	41.5	4676	21.84	962.67	134.74	13.41	34.75	
Median	18.39	133.64	3.93	39.7	4171	13.2	857.43	131	12.88	36.93	
5 percentile	5.52	22.15	0.78	8.93	2194	1.8	347.69	100.25	8.04	19.43	
10 percentile	6.33	36.91	1.183	10.1	2652	2.05	378.59	104.2	8.52	25.26	
25 percentile	10.87	59.5	1.83	16.82	3721	7.4	588.84	111	11.19	27.48	
30 percentile	13.39	84.617	1.87	20.21	3857	9.26	702.17	102.38	11.67	31.07	
60 percentile	19.82	144.036	5.12	47.2	4531	21.8	902.79	125	13.99	38.3	
75 percentile	25.2	182.54	7.06	64.56	5386	29.01	1368.48	141.2	16.641	40.01	
80 percentile	29.81	196.41	7.53	66.27	5531	33.67	1434.81	146.76	16.82	41.03	
90 percentile	33.74	257.58	10.56	85.67	8344	48.52	1819.55	228.35	16.87	42.57	
95 percentile	41.18	289.86	11.72	93.37	10,410	99.92	1918.6	271.4	17.71	44.89	
98 percentile	41.55	291.31	11.77	93.73	10,503	102.56	1923.08	273.2	17.75	45.01	
99 percentile	41.55	291.31	11.77	93.73	10,503	102.56	1923.08	273.2	17.75	45.01	

4. Results

4.1. Univariate Anomaly Analysis of Ore and Incompatible Elements

Graduated point symbol maps of incompatible elements (Figure 3) show an overall enrichment above the 95th percentile of W (99.92 ppm), Nb (197.99 ppm), Ta (11.72 ppm) and Be (33.71 ppm) in Kagenfels Granite. Compared to average concentrations of W (2 ppm), Nb (20 ppm), Ta (3.5 ppm) and Be (5 ppm) in granites [37], the Kagenfels Granite is locally characterised by an enrichment factor of more than 50 (W), 10 (Nb), 3.4 (Ta) and 6.7 (Be). The anomalous samples are located approximately 600–900 m north of the Natzwiller–Struthof camp at confluences with the Barembach stream (samples 13–19, Figure 1b) as well as the Magel stream to the east (samples 10–11). The anomalous zones principally correlate with regional NNE–SSW trending fault zones, where this study additionally located occurrences of 2–3 m wide, partially overburden-covered quartz–feldspar pegmatitic granite stringers trending parallel to these fault zones, particularly at sample location 18. In addition, the Kagenfels Granite is characterized by distinct magmatic fractionation ratios of $64.5 < K/Rb < 110$, $13.98 < Nb/Ta < 17.75$ and $19.12 < Zr/Hf < 37.91$. On the other hand, the southwestern part of the Natzwiller Granite displays an enrichment of Li (105.1 ppm), W (34.81 ppm) and B (21.99 ppm) corresponding to enrichment factors of 3.5 (Li), 17.4 (W) and 1.5 (B), respectively. Fractionation ratios of $101 < K/Rb < 128.8$, $12.35 < Nb/Ta < 12.7$, $33.85 < Zr/Hf < 36.45$ are slightly more elevated than the Kagenfels Granite, indicating generally less fractionation. Sample 20, located on a fault-controlled boundary between the Natzwiller Granite and Belmont Granite Suite, shows visible cassiterite in a stream pan, but only limited concentrations of Sn (17.98 ppm) and the highest K/Rb (273.2) ratio of the sampling campaign. The presence of coarse cassiterite visible in a pan and limited concentrations of Sn in the < 75 um fraction suggest a strong partitioning of Sn by size fraction. Cassiterite was not abraded and comminuted to finer size fractions due to the limited fluvial transport and therefore is thought to originate from a proximal source. The Belmont Granite Suite, comprising the Fouday, Waldersbach, Champ du Feu North Granites along with three samples collected in the older Neuntelstein Diorite and Hohwald Granodiorites, is typically characterised by an absence of notable concentrations of incompatible elements, whereby fractionation ratios of $113.5 < K/Rb < 237.1$, $37.42 < Zr/Hf < 45$ imply a less fractionated nature than the Natzwiller and Kagenfels Granites. The calculated Nb/Ta ratio of 8–13.1, however, is considerably lower than the Natzwiller and Kagenfels Granites.

4.2. Geochemical Classification of Principal Rock Units and Magmatic Fractionation Using Stream Sediment Data

The analysis of trace element concentrations, supported by geological observations in outcrops and stream catchment areas, led to the classification of four major regional lithology types (Figure 4). Stream sediment geochemistry was classified per lithology unit, and the link between the stream sediment geochemistry and lithology was investigated. The lithologies principally reflect the regional BRGM geological map of the northern Vosges Mountains. Major and trace elements and ratios used in this study are predominantly Ti, Th, B, K/Rb, Nb/Ta, and Zr/Hf. A combination of these elements supports the determination of significant petrological and fractionation processes characteristic for each lithology of the northern Vosges magmatic suite.

Kagenfels Granite was macroscopically distinguished in outcrop and stream sediments from other lithologies by the presence of muscovite, biotite, ilmenite, and minor beryl. The lithology forms distinct population clusters in Ti vs. Th, K/Rb vs. Th and Ti vs. B bivariate plots with concentrations of Th (38.33–86.53 ppm), Ti (2170–5683 ppm), B (9.9–13.85 ppm), $64.5 < K/Rb < 110$, and $19.12 < Zr/Hf < 37.91$. Whilst the K/Rb and Zr/Hf ratio of samples 10–11, 13–14, and 18 represent the lowest, and therefore most fractionated, magmatic fractionation indicator values in the studied samples, Nb/Ta ratios of 13.98–17.75 are unexpectedly high.

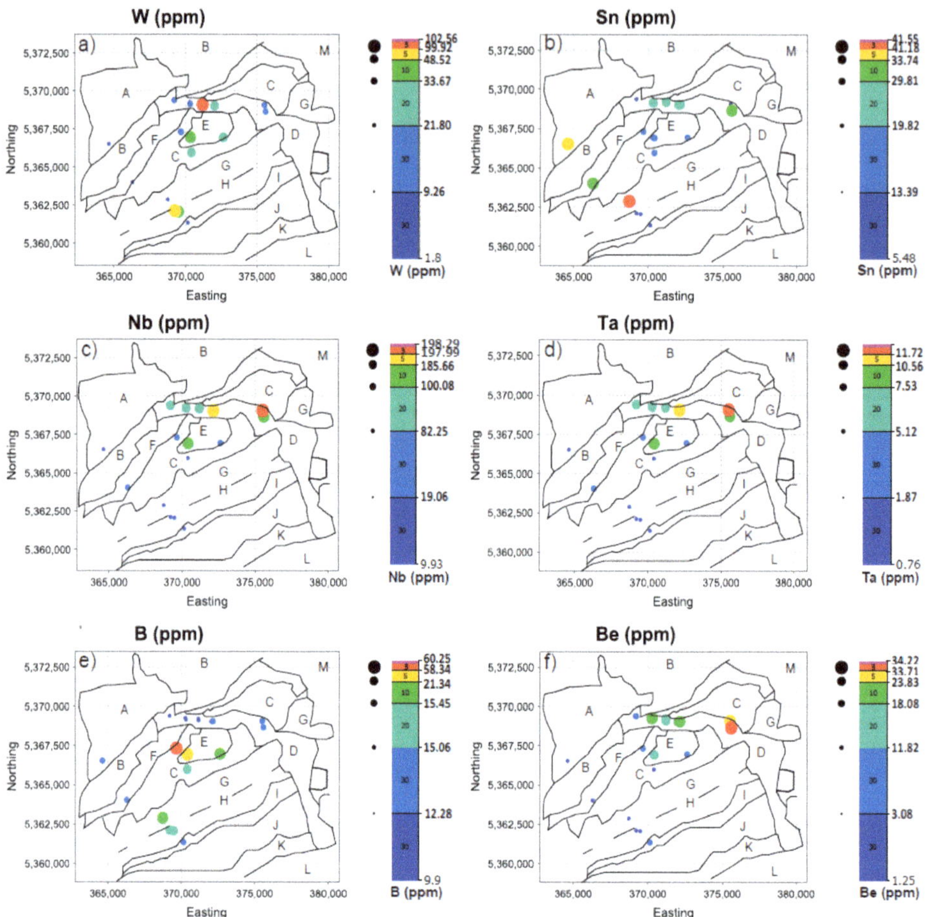

Figure 3. *Cont.*

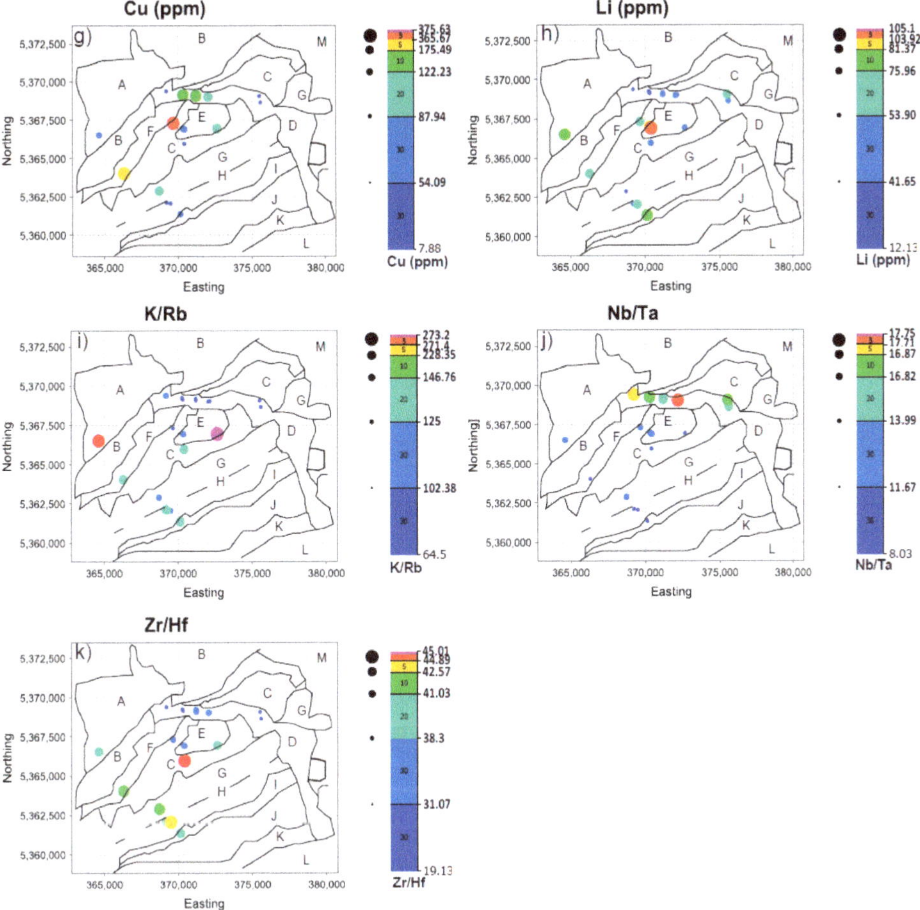

Figure 3. Graduated point symbol maps for (**a**) W, (**b**) Sn, (**c**) Nb, (**d**) Ta, (**e**) B, (**f**) Be, using the 30th, 60th, 80th, 90th, 95th, 98th and 99th percentile distribution from Table 1. Principal geological domains are highlighted in panel (**a**). A, mafic and intermediate volcanic rocks of the Schirmeck Massif; B, metasedimentary rocks of the Bruche Unit; C, Waldersbach Granite; D, Kagenfels Granite; E, Natzwiller Granite; F, Fouday Granite; G, Champ du Feu granite suite; H, Bande médiane mafic volcanics; I, Neuntelstein Diorite; J, Hohwald granite suite; K, Steige schists; L, Ville schists; M, Lower Buntsandstein. Graduated point symbol maps for (**g**) Cu, (**h**) Li, (**i**) K/Rb, (**j**) Nb/Ta and (**k**) Zr/Hf.

Natzwiller Granite contains macroscopic evidence of biotite, amphibole, hematite, ilmenite and rutile and is generally darker in appearance. This population can be geochemically distinguished from Kagenfels Granite and forms two distinct clusters in Ti vs. Th and Ti vs. B bivariate plots, with higher concentrations of Ti (8639–10,502 ppm), Th (77.95–93.73 ppm), and B (15.41–21.99 ppm). The K/Rb ratio varies from 101–128.8; i.e., it is reasonably similar to the Kagenfels Granite. The abundant visual presence of rutile, ilmenite, and magnetite in the stream sediment and outcrop supports the elevated contents of Ti, implying a more intermediate composition than the Kagenfels Granite. Natzwiller Granite plots on the southern and eastern margins of the corresponding BRGM lithology (Figure 1). Sample 20 contained abundant panned cassiterite which occurs along the faulted contact between the Natzwiller and Waldersbach granites. The sample was classified as Natzwiller/Belmont Granite with visible cassiterite. This K-feldspar and quartz-rich stream sediment sample has the highest K/Rb value of 273.2 and is generally depleted of trace elements.

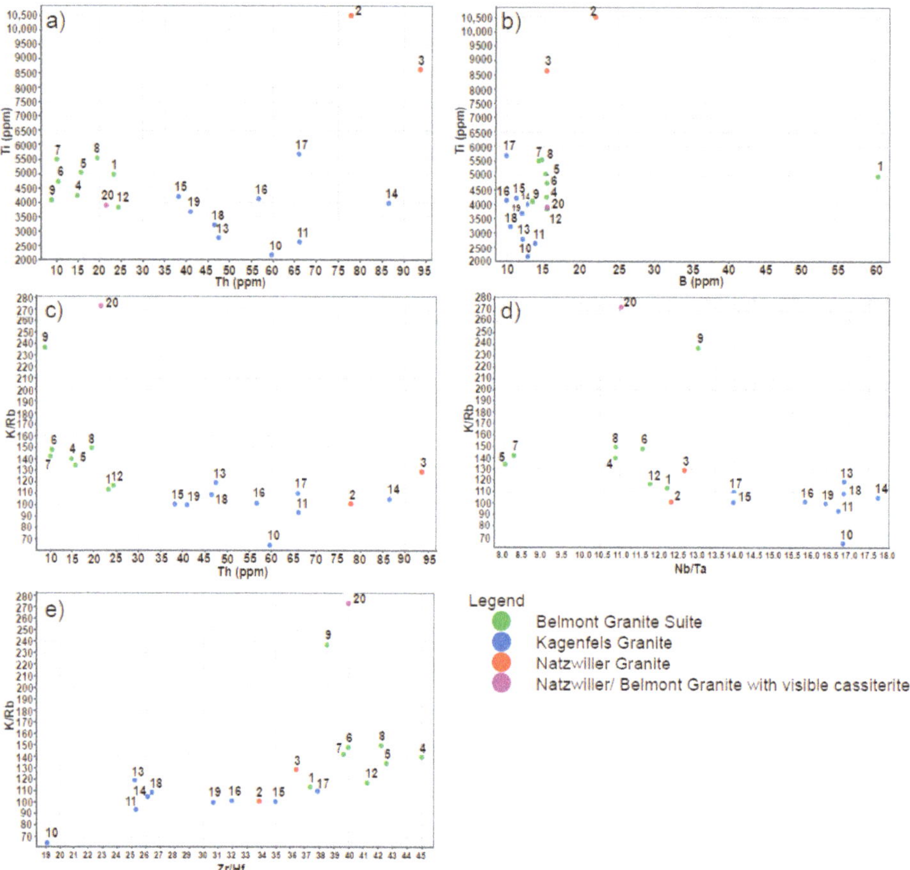

Figure 4. Binary plots of stream sediment geochemistry represented by broad lithological units in the drainage basin. (**a**) Ti vs. Th, (**b**) Ti vs. B, (**c**) K/Rb vs. Th, (**d**) Nb/Ta, and (**e**) Zr/Hf. Sample identification numbers are plotted in the graphs. Each population displays a characteristic clustering trace element assemblage. The stream sediment samples obtained downstream of the Belmont Granite suite generally has a less fractionated nature than the Natzwiller and Kagenfels Granites expressed in comparatively higher K/Rb and Zr/Hf ratios and lower Th concentrations. Decreasing fractionation indices (K/Rb, Nb/Ta and Zr/Hf) indicate an increasing influence of magmatic–hydrothermal interactions during fractional crystallization [27]. Stream sediment sample numbers are indicated as 1–20.

The Belmont Granite Suite comprising the Fouday, Champ du Feu and Waldersbach granites [6,7] forms a distinct population cluster in Ti vs. Th, Ti vs. B, and K/Rb vs. Th bivariate plots, and can be clearly distinguished from the Kagenfels and Natzwiller granites by lower concentrations of the aforementioned elements, particularly Th (8.87–24.45 ppm), and has K/Rb and Zr/Hf values ranging from 105–165 and 37.42–45, respectively. The Hohwald Granodiorite (samples 5–6) and the Neuntelstein Diorite (sample 7) plot within this lithogeochemical population have therefore been included in the present Belmont Granite suite classification.

A principal component analysis (PCA) of the geochemical dataset was carried out using a log10 transformation (to eliminate the closure effect) of B, Be, Cu, Li, Nb, Rb, Sn, Ta, Th, Ti, W and Zr input variables (Table 3, Figure 5). The PCA revealed that PC1 to PC4 represent 85.68% of the total variance of the dataset, whereas the remaining eight PCs can be attributed to random processes or noise. The eigenvector table and combined variable-sample analysis (RQ) plots, in which samples plot as points and variables as vectors with the length of the vectors proportional to the variability of the two displayed principal components, outline two major correlations and elemental trends. Firstly, previously classified Kagenfels and Natzwiller Granites are principally characterised by negative RQ1 loadings, whereas Natzwiller Granite can be distinguished from Kagenfels Granite by positive RQ2 and RQ4 loadings. On the other hand, there is a noteworthy variability of the Kagenfels Granite samples 10–11, 13–14 and 18, expressed by generally negative RQ2 and RQ4 loadings of Rb and Zr, respectively. In the K/Rb vs. Zr/Hf plot (Figure 4e), these samples represent the most fractionated population with K/Rb < 120 and Zr/Hf < 32. Therefore, the PCA trends are distinctive for the previously outlined fractionated lithogeochemical populations and characterised by a Ta-Nb-Be-Rb-Th-W and Zr geochemical association. In particular, whilst PC1 broadly characterises fractionated lithologies of the Kagenfels and Natzwiller Granites in the study area, PC2 and PC4 allow the determination of local variation of fractionation in the Kagenfels Granite. In contrast, the Belmont Granite Suite and the Natzwiller/Belmont Granite with visible cassiterite (sample 20) show fundamentally different trends with positive RQ1 loadings, and a Cu-Ti and Li geochemical association. Distinctive negative and positive variability in RQ2 and RQ4, respectively, allow the determination of potential sub-populations, such as samples 6 (Hohwald Granodiorite) and 7 (Neuntelstein Diorite), which are generally less evolved and have intruded approximately 10 Ma earlier than the Belmont Suite [6]. In summary, the analysis of petrogenetic indicator elements in bivariate plots along with principal component analysis provide a robust tool to determine different litho-geochemical populations and element associations in the study area.

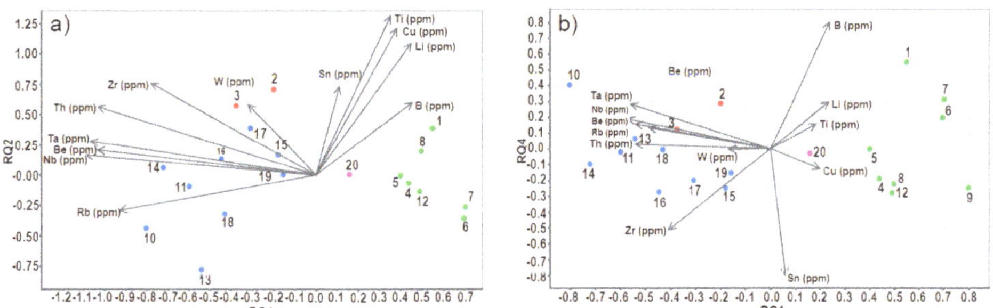

Figure 5. Principal component analysis of the stream sediment geochemical dataset showing (**a**) RQ1–RQ2 plot and (**b**) RQ1–RQ4 plots. In combined variable-sample analysis (RQ) plots, samples plot as points and variables as vectors, with the length of the vectors proportional to the variability of the two displayed principal components. The legend is as seen in Figure 4. The data clearly demonstrate that the Kagenfels and Natzwiller Granite are characterised by negative RQ1 loadings, whilst the Belmont Granite Suite has fundamentally positive RQ1 loadings. Natzwiller Granite can be distinguished by positive RQ1 and RQ4 loadings. The local fractionation of Kagenfels Granite is obvious from variable RQ2 and RQ4. Stream sediment sample numbers are indicated as 1–20.

Table 3. Principal component analysis of the stream sediment geochemical dataset showing (a) correlation of elements, (b) eigenvalues, and (c) scaled coordinates. Grey colours display an average correlation range (0.5–0.7), light brown a good correlation range (0.7–0.9) and purple a very good correlation range (0.9–1).

(a) Correlation	B (ppm)	Be (ppm)	Cu (ppm)	Li (ppm)	Nb (ppm)	Rb (ppm)	Sn (ppm)	Ta (ppm)	Th (ppm)	Ti (ppm)	W (ppm)	Zr (ppm)
B (ppm)	1	−0.22	0.43	0.38	−0.23	−0.26	0.07	−0.19	−0.2	0.32	−0.1	−0.45
Be (ppm)	−0.22	1	−0.11	−0.26	0.93	0.75	−0.04	0.93	0.87	−0.28	0.17	0.57
Cu (ppm)	0.43	−0.11	1	0.49	−0.23	−0.36	0.52	−0.21	−0.08	0.44	−0.02	−0.02
Li (ppm)	0.38	−0.26	0.49	1	−0.32	−0.23	0.19	−0.22	−0.2	0.64	0	−0.1
Nb (ppm)	−0.23	0.93	−0.23	−0.32	1	0.81	−0.1	0.98	0.94	−0.23	0.21	0.62
Rb (ppm)	−0.26	0.75	−0.36	−0.23	0.81	1	−0.01	0.77	0.67	−0.55	0.02	0.35
Sn (ppm)	0.07	−0.04	0.52	0.19	−0.1	−0.01	1	−0.15	0.07	0.08	−0.19	0.21
Ta (ppm)	−0.19	0.93	−0.21	−0.22	0.98	0.77	−0.15	1	0.94	−0.13	0.24	0.63
Th (ppm)	−0.2	0.87	−0.08	−0.2	0.94	0.67	0.07	0.94	1	0	0.31	0.77
Ti (ppm)	0.32	−0.28	0.44	0.64	−0.23	−0.55	0.08	−0.13	0	1	0.26	0.21
W (ppm)	−0.1	0.17	−0.02	0	0.21	0.02	−0.19	0.24	0.31	0.26	1	0.51
Zr (ppm)	−0.45	0.57	−0.02	−0.1	0.62	0.35	0.21	0.63	0.77	0.21	0.51	1

(b)	Eigenvalues	Percent	Cumulative %
PC1	5.321	44.35	44.35
PC2	2.386	19.88	64.23
PC3	1.432	11.93	76.16
PC4	1.143	9.524	85.68
PC5	0.6511	5.426	91.11
PC6	0.5012	4.177	95.28
PC7	0.3569	2.974	98.26
PC8	0.09674	0.8062	99.06
PC9	0.07267	0.6056	99.67
PC10	0.02147	0.1789	99.85
PC11	0.0125	0.1042	99.95
PC12	0.005843	0.04869	100

(c) Scaled Coordinates	PC1	PC2	PC3	PC4	PC5	PC6	PC7	PC8	PC9	PC10	PC11	PC12
B (ppm)	0.408	0.3572	0.3255	0.6315	0.3509	0.01225	0.2645	0.04051	0.07987	0.00297	0.00218	0.00289
Be (ppm)	−0.917	0.1213	0.1951	0.1241	0.04045	−0.04277	−0.1874	0.2209	−0.03386	−0.02367	0.04212	−0.00409
Cu (ppm)	0.3415	0.72	0.3555	−0.09954	0.2485	−0.03138	−0.3918	−0.1048	0.03856	−0.00284	0.00383	−0.00283
Li (ppm)	0.4033	0.6475	0.07263	0.2348	−0.4995	0.3126	−0.06785	0.06441	0.00516	0.0317	−0.02365	0.00966
Nb (ppm)	−0.9695	0.09748	0.1066	0.1512	0.01134	−0.0767	−0.00449	−0.05642	−0.04153	−0.03102	−0.0283	0.05989
Rb (ppm)	−0.8252	−0.1733	0.3217	0.1054	−0.1527	0.3506	0.06946	−0.1358	0.03502	−0.03331	0.05013	−0.01149
Sn (ppm)	0.1	0.4327	0.5633	−0.6311	0.0861	0.1112	0.2403	0.0451	−0.0835	−0.01961	−0.01414	−0.00206
Ta (ppm)	−0.9489	0.1672	0.05228	0.2242	−0.04822	−0.09056	−0.00587	−0.02147	−0.00787	−0.03192	−0.0687	−0.04341
Th (ppm)	−0.917	0.3374	0.04044	0.02026	0.02975	−0.1245	0.08018	−0.05258	−0.06128	0.1162	0.01846	−0.00596
Ti (ppm)	0.316	0.7821	−0.3607	0.1243	−0.2014	−0.2717	0.1259	−0.05419	−0.07271	−0.05353	0.03931	−0.00605
W (ppm)	−0.2828	0.3478	−0.7064	−0.00598	0.371	0.3977	−0.01112	0.007	−0.0623	−0.00659	−0.00518	−0.00106
Zr (ppm)	−0.6948	0.4521	−0.3004	−0.4066	−0.05321	−0.06105	0.09301	0.03585	0.2018	−0.00142	0.00119	0.00592

4.3. Bulk Mineralogy, Indicators for Magmatic Fractionation and Link to Stream Sediment Geochemistry

4.3.1. Correlation between Outcrop and Stream Sediment Sample Mineralogy

In order to ascertain the representativity of stream sediment samples and previously mapped lithological units, a comparison between the modal mineralogy of stream sediment and adjacent outcrop samples has been undertaken. Stream sediment samples 17 and 18 are located in the northern part of the Kagenfels Granite unit and are enriched in W (102.56 ppm and 6.94 ppm) and Nb (63.67 ppm and 92.76 ppm), respectively. Visual field mapping and description of drainage areas and sediments confirm the presence of a muscovite–biotite(–tourmaline) granite, along with sporadically occurring pods of a finer, hornblende-bearing variety, which was previously described as the less evolved "rhyolite facies" of the Kagenfels Granite [14]. Abundant outcrops of the granite are present several hundred metres upstream of the sampled locations, whilst pods of the "rhyolite facies" are encountered in the vicinity of the stream sources along the margins of the granite. Sample 18 yielded minor columbite grains during routine pan inspection. The QEMSCAN modal mineralogy of stream sediment samples 17 and 18 (Table 4) confirms the peraluminous monzogranitic nature of the samples containing tourmaline (1.43%, 0.82%), muscovite (1.34% and 1.37%), and spessartine (0.14% and 0.65%) as abundant accessory Al-rich mineral phases. Furthermore, wolframite (0.03% in 2060) and ilmenorutile (0.01% and 0.05%) represent the hosts for W and Nb-rich accessory mineral phases, which correlate with the geochemical anomalies outlined in this study.

The mineralogical analysis of the collected samples confirms that major silicate phases (e.g., quartz, feldspars, micas, and tourmaline) of the −75 μm fraction are not or very weakly weathered and generally do not show weathering rims or the complete replacement of the mineral grains with kaolinite (Figure 6). This implies a rapid erosion of outcrops and (re)deposition in stream traps. Variable amounts of Fe-Mn oxides are present in the samples and indicate that the weathering of primary Fe-rich minerals can to some extent affect stream sediments of the study area.

The outcropping muscovite–biotite(–tourmaline) host granite bedrock samples 17B and 18B display a comparable monzogranitic composition and variably contain more abundant K-Feldspar (10–14%) and quartz (5–10%) than the stream sediments. Importantly, both rock samples are characterised by similar, yet slightly subdued, amounts of key accessory minerals, such as tourmaline (0.29% and 0.22%), muscovite (0.46% and 0.42%), ilmenite (0.09% and 0.18%) and rutile (0.04%). Due to local variations in bedrock geology and the nature of sample origin and preparation techniques (i.e., a thin section of a singular outcrop chip vs. well-mixed stream sediment samples), trace and heavy minerals will naturally be more abundant in samples obtained from stream traps where the accumulation of denser mineral grains is more pronounced. Conversely, minerals with a lower specific gravity, such as quartz, mica and feldspar, will likely be less abundant in stream sediments than in outcrop. The mineralogical analysis of stream sediments and outcropping bedrock, however, demonstrates that the mineralogical composition of stream sediments is generally indicative of its corresponding source lithology. As a result, it can be assumed that the largely unweathered stream sediment samples reflect the overall lithological composition and are representative of the stream catchments.

Table 4. Modal mineralogy (mineral mass-%) of selected stream sediment and outcrop samples of the Kagenfels (17A and 18B) and Natzwiller Granites (2 and 3). The four stream sediment samples show a broadly similar monzogranitic composition; however, the Natzwiller Granite has generally more abundant B and Ti-rich mineral phases—e.g., tourmaline and rutile. Outcropping rocks and stream sediments of the Kagenfels granite have a comparable mineralogical composition; however, with a subdued heavy mineral signature. Please refer to Table A1 for a complete list of sample mineralogy tables.

Stream Sediment		Stream Sediment		Stream Sediment		Stream Sediment		Rock		Rock	
2		3		17		18		17A		18A	
Plagioclase	27.08	Plagioclase	29.01	Plagioclase	29.24	Plagioclase	36.97	K-Feldspar	35.78	Quartz	36.55
Quartz	19.6	K-Feldspar	20.96	Quartz	24.10	K-Feldspar	25.9	Plagioclase	32.23	K-Feldspar	31.35
K-Feldspar	18.53	Quartz	19.65	K-Feldspar	22.14	Quartz	24.34	Quartz	29.44	Plagioclase	29.65
Chlorite	13.61	Chlorite	6.88	Fe-Ox (Mn)/CO3	5.89	Chlorite	2.56	Fe-Ox (Mn)/CO3	0.77	Fe-Ox (Mn)/CO3	0.8
Tourmaline	8.48	Fe-Ox (Mn)/CO3	6.73	Chlorite	3.52	Fe-Ox (Mn)/CO3	1.91	Biotite	0.72	Biotite	0.53
Biotite	3.26	Hornblende	3.86	Zircon	3.10	Biotite	1.79	Muscovite	0.46	Muscovite	0.42
Fe-Ox (Mn)/CO3	2.07	Tourmaline	3.39	Hornblende	3.03	Muscovite	1.37	Tourmaline	0.29	Tourmaline	0.22
Hornblende	1.92	Zircon	2.3	Biotite	2.01	Ilmenite	0.91	Chlorite	0.12	Ilmenite	0.18
Muscovite	1.12	Biotite	1.98	Tourmaline	1.43	Tourmaline	0.82	Ilmenite	0.09	Chlorite	0.12
Rutile	1.09	Apatite	1.26	Muscovite	1.34	Spessartine	0.65	Rutile	0.04	Kaolinite	0.05
Kaolinite	0.76	Rutile	1.11	Apatite	0.97	Zircon	0.56	Zircon	0.02	Rutile	0.04
Zircon	0.57	Muscovite	0.71	Rutile	0.70	Hornblende	0.48	Ti-Magnetite	0.01	Zircon	0.03
Ca-Fe-Al silicates	0.38	Ca-Fe-Al silicates	0.42	Ilmenite	0.60	Rutile	0.41	Monazite	0.01	Ti-Magnetite	0.02
Titanite	0.17	Titanite	0.37	Ca-Fe-Al silicates	0.50	Mn Oxides	0.33	Others	0.02	Monazite	0.02
Ilmenite	0.14	Ilmenite	0.33	Kaolinite	0.44	Kaolinite	0.31			Thorite	0.02
Spessartine	0.12	Kaolinite	0.26	Titanite	0.20	Calcite	0.23			Others	0
Chrome spinel	0.11	Spessartine	0.11	Ti-Magnetite	0.18	Ca-Fe-Al silicates	0.19				
Ti-Magnetite	0.05	Ti-Magnetite	0.08	Wolframite	0.03	Ilmenorutile	0.05				
Monazite	0.02	Monazite	0.04	Ilmenorutile	0.01	Others	0.22				
Ilmenorutile	0.01	Ilmenorutile	0.01	Spessartine	0.14						
Thorite	0.01	Columbite	0.01	Others	0.43						
Cu sulphides	0.01	Cu sulphides	0.01								
Others	0.89	Others	0.52								

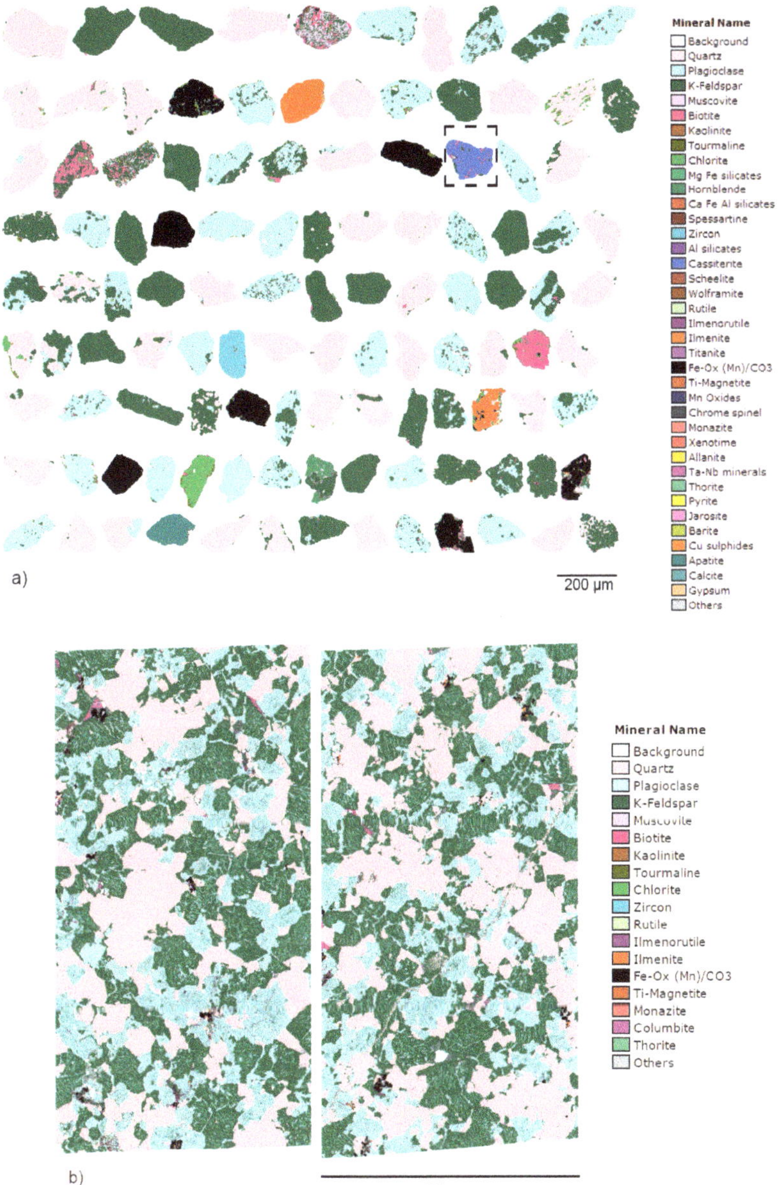

Figure 6. (**a**) Extract of studied composite QEMSCAN® mineral maps sorted by area and large to small particle size. The false colour image of the particles show their general mature appearance with a distinct lack of weathering rims and complete replacement. A cassiterite–columbite particle is indicated with a dashed rectangle and shown in Figure 7. (**b**) QEMSCAN® mineral maps of outcrop samples 17A (left) and 18A (right). Note the similarities in composition to composite mineral maps in a).

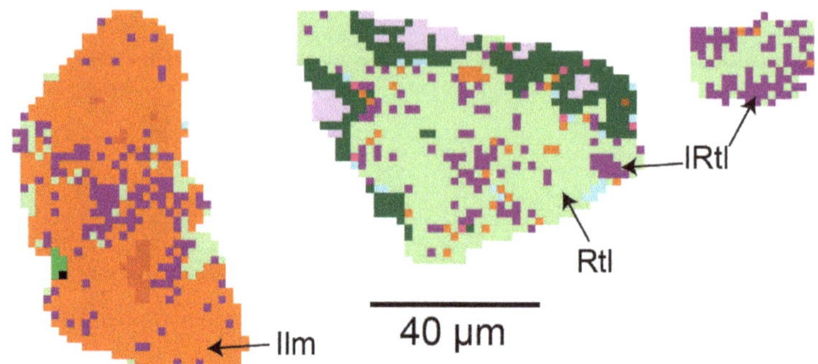

Figure 7. Stream sediment particles illustrating the intergrowth of rutile (rtl), ilmenorutile (irtl) and ilmenite (ilm). The legend is shown in Figure 6.

4.3.2. Petrogenetic Indicators

Other samples obtained and analysed from the Kagenfels Granite (Appendix A) show similar modal mineralogical compositions compared to stream sediment samples 17 and 18 and therefore provide evidence for peraluminous two-mica monzogranite. In particular, stream sediment samples 14–16, 19 and 20 are characterised by the presence of muscovite (0.80–1.97%), biotite (1.26–2.01%), tourmaline (0.51–1.43%), chlorite (1.47–3.52%), ilmenite (0.28–0.98%), ilmenorutile (0.01–0.05%), zircon (0.54–3.1%), monazite (0.01–0.4%), cassiterite (0.01% in sample 16) and columbite (0.01%). Ilmenorutile is commonly intergrown with rutile and ilmenite (Figure 7) and columbite can be found as inclusions in cassiterite (Figure 8). Further follow-up by SEM confirmed the presence of various Nb-rich phases (Mn-columbite, U-rich euxenite-(Y), and Nb-rich titanium oxide/ ilmenorutile) in sample 17 (Figure 9), which, together with a high geochemical Nb/Ta ratio of 14, implies that Nb-rich mineral phases are distinctively more abundant than Ta phases in the western part of the Kagenfels Granite. Sample 18 (Figure 10) shows that chlorite is associated with tourmaline and muscovite and therefore represents evidence for granite-related magmatic–hydrothermal alteration [38]. However, due to the size of the fragments and the absence of quartz, it is not entirely conclusive whether this mineral association is related to veins or wall-rock alteration.

Stream sediment samples 2 and 3, obtained from the two principal drainages on the southwestern flank of the Natzwiller Granite, generally show a similar monzogranitic composition as the Kagenfels Granite. Accessory Th- and Ti-bearing mineral phases in Natzwiller and Kagenfels Granites (Table 4) indicate a comparable content of Th-rich mineral phases, with the highest content (1.91 mass-%) of Ti-phases in sample 3. This generally corresponds to the Th vs. Ti bivariate plot (Figure 5a), where the Natzwiller Granite samples have Ti concentrations of >8640 ppm. However, given the notable enrichment of Ti in the geochemical analyses, a higher abundance of Ti mineral phases would be expected in the mineralogical analysis. However, this may be explained by the occurrence of trace Ti occurring in other minerals, such as micas [39]. A distinctive feature of samples 2 and 3, on the other hand, is the abundance of tourmaline (8.48% and 3.39%) and chlorite (13.61% and 6.88%), which in Figure 5b correspond to a distinctive B signature of 21.99 ppm and 15.41 ppm, respectively, and associated Nb (100.33 ppm and 94.64 ppm), Ta (8.12 ppm and 7.45 ppm), Nb/Ta (12.35 and 12.7) and W (21.55 ppm and 34.81 ppm).

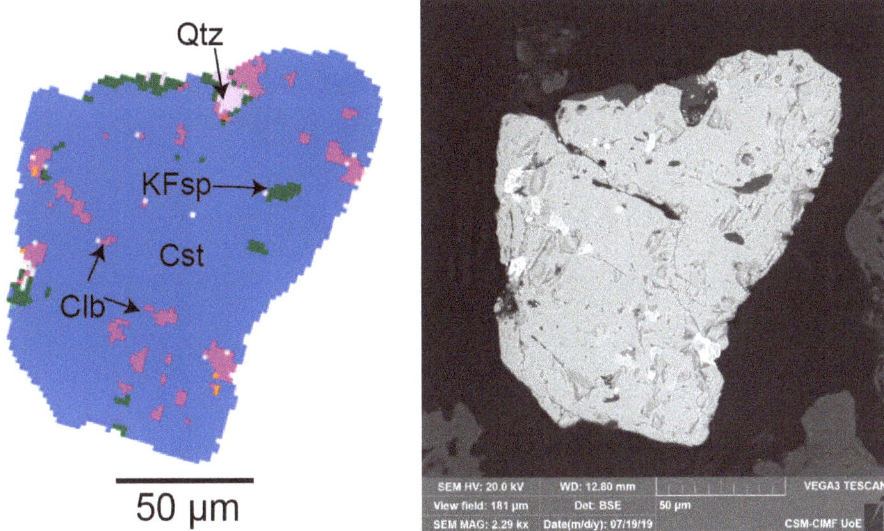

Figure 8. QEMSCAN® and SEM mineral map of a stream sediment particle (sample 16) showing the intergrowth of cassiterite (cst)–columbite (clb)–K–Feldspar (KFsp) and quartz (qtz).

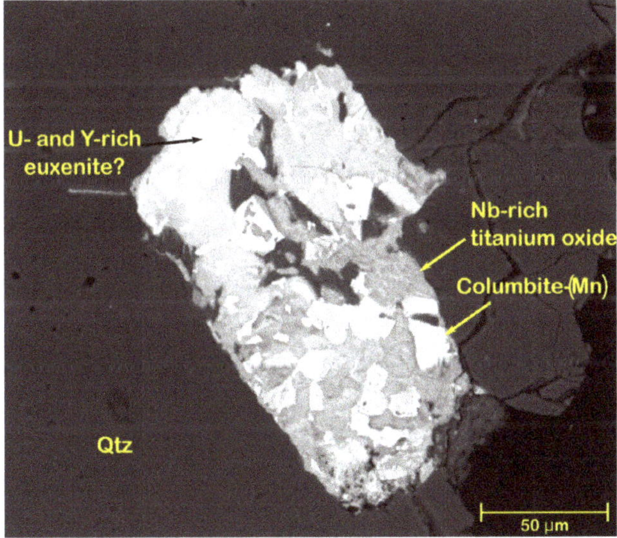

Figure 9. SEM image of stream sediment particle (sample 17) showing the intergrowth of Nb-rich titanium oxide (ilmenorutile), columbite (Mn) and euxenite (Y).

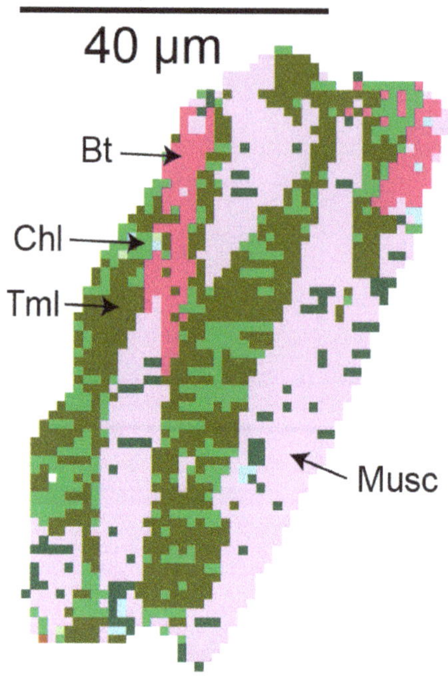

Figure 10. QEMSCAN® mineral map of stream sediment particle (sample 18) showing the intergrowth of chlorite (chl)–tourmaline (tml)–muscovite (musc)–biotite. This mineral association is interpreted to represent evidence for a granite-related magmatic–hydrothermal alteration.

5. Discussion

The graduated point symbol and bivariate fractionation plots demonstrate a distinctive endowment of base metals and high field strength elements (HFSE) in the northern Vosges magmatic suite, with a particular emphasis on the Natzwiller and Kagenfels Granite suites. The geochemical trends observed in these plots are comparable with previous regional reconnaissance sampling [5] and whole-rock geochemical data presented in Tabaud et al. (2014) [6], particularly for Th, Sr, Rb and Ti, which provide a tool to distinguish intermediate and felsic magmatic rocks. The application of univariate anomaly mapping and the lithogeochemical classification of catchment sediments, therefore, supports the delineation of areas of increased enrichment of economically sought-after metals and corresponding magmatic lithologies. In particular, the present data confirm the known presence of Nb and Be and outline additional, previously unrecognised W anomalies in Kagenfels Granite.

Earlier empirical mineral and whole-rock geochemical studies have successfully demonstrated the use and application of magmatic fractionation ratios in defining late-stage magmatic melts prospective for Sn-W and Li-Cs-Ta mineralization [27,28,40–43]. Decreasing K/Rb, Nb/Ta, and Zr/Hf ratios indicate the increasing fractionation of the granitic melt and a transition to hydrothermal alteration [27]. Numerical changes in these ratios during late-stage magmatic fractionation are a result of the substitution of K with Rb in micas and feldspars [44], fractionation of Nb over Ta due to secondary muscovitisation and hydrothermal sub-solidus reactions enriching Ta in F-rich residual melts [29], and increasing Kd values of Hf in zircon, which are only weakly influenced by secondary fluid-related processes [28,45,46]. Recent geochemical studies of the Leinster Granite (Ireland) and Central Vosges Mg-K granites [3,30], along with mineralogical results of this study, have shown that these petrogenetic ratios are equally applicable to determine highly fractionated lithologies using geological materials affected by secondary dispersion processes, such as stream sediments. The lack of significant weathering of K, Rb, Sn, Nb and Zr-bearing silicate mineral phases in the analysed stream sediment fraction that were collected as well as the representativity of these stream sediments in relation to mapped and sampled outcrops of the catchment area confirm that these petrogenetic ratios can be employed to fingerprint fractionation patterns in the study area. In this context, the Natzwiller Granite shows fractionation ratios of 101 < K/Rb < 128.8, 12.3 < Nb/Ta < 12.7, and 33 < Zr/Hf < 36 (Figure 3i–k), along with the generally highest Ti concentrations of 8640–10,500 ppm (Figure 5) evidenced by abundant rutile, ilmenite and titanite in the samples. Of particular interest are the elevated concentrations of incompatible elements, such as Li (105.1 ppm), W (34.81 ppm), Ta (8.1 ppm), B (21.99 ppm) and Be (13.47 ppm) in sample 2 along with abundant tourmaline (8.48%). In addition, despite a relatively low concentration of Sn (17.98 ppm) being measured in sample 20, automated mineralogical techniques identified multiple cassiterite grains in the stream sediment sample. This evidence suggests that the Natzwiller Granite has experienced, at least locally around NE–SW trending fault zones, the introduction of a highly fractionated melt enriched in fluxing elements, which allows incompatible and HFS elements to remain in late-stage, low-temperature melts [29]. Tabaud et al. (2014) [6] previously described the Natzwiller Granite as sub-aluminous to weakly peraluminous in nature, resulting from partial melting of an enriched mantle source and subsequent interaction with subducted metasedimentary and metaigneous crustal source material. Therefore, despite the comparably lower fractionation grade, the Natzwiller Granite was able to retain incompatible elements of possible economic interest.

In contrast, the peraluminous S-type Kagenfels Granite typically displays low values of 64.5 < K/Rb < 119, 14 < Nb/Ta < 17.7, and 19 < Zr/Hf < 38 (Figure 3i–k), and therefore demonstrates a high degree of magmatic fractionation and secondary muscovitisation, characteristic of peraluminous S-type granites [27,42]. Geochemical fractionation trends (Figure 5), PC analysis (Figure 6) and mineralogical evidence in the form of tourmaline, muscovite, chlorite, wolframite, cassiterite, columbite and ilmenorutile imply a peraluminous evolution and confirm a highly fractionated and locally hydrothermally altered nature of the Kagenfels Granite. Strong fractionation of the melt, along with a predominant NE–SW structural control in the Kagenfels Granite, led to the emplacement of pegmatitic quartz-feldspar (-beryl) veins at "Grotte des Partisans" [23] and the Barembach stream confluences observed in the present study. The same process also resulted in characteristic elemental concentrations of As (25.52–92 ppm), Cu (7.88–166.61 ppm) and incompatible elements, such as Be (10–21 ppm), Sn (7.19–26.29 ppm), Li (12.13–52.61 ppm), and W (6.94–102.56 ppm). However, distinctive high values of Nb/Ta > 17 and the general absence of Li concentrations of >45 ppm across the western part of the Kagenfels Granite (Figure 3h,j) suggest that, locally, magmatic fractionation and hydrothermal alteration processes were not as pronounced as in the eastern part of the Kagenfels Granite, which yields higher concentrations of Be (24–34.22 ppm), and Li (49.35–54.77 ppm), at 16.4 < Nb/Ta < 16.8, 64.5 < K/Rb < 93, and 19 < Zr/Hf < 25 (Figure 3h–k). The comparatively higher Nb/Ta ratios in the western Kagenfels Granite are a result of the predominant occurrence of Mn-columbite, euxenite-(Y), and Nb-rich titanium oxide/ ilmenorutile as evidenced in stream sediment sample 17 and 18. Consequently, the western part of the Kagenfels Granite predominantly produced a mineral assemblage with Nb > Ta-rich minerals, indicating the preferential fractionation of Nb over Ta-rich minerals and, therefore, the locally lower fractionation of the granitic melt. In a regional context, these observations imply that the Kagenfels Granite is the most fractionated granite suite in the northern Vosges Mountains, and consequently represents a prime target to explore for granite-hosted mineralisation.

The application of automated mineralogical techniques, such as QEMSCAN® and manual SEM EDS, supported the routine collection of stream sediment samples in a mineral exploration context. These techniques were not only able to identify the bulk mineralogical composition of the stream sediment samples and therefore link geochemical signatures to source mineralogy, but also provide potential information about element deportment characteristics (ilmenorutile and columbite as principal Nb hosts, wolframite as principal W host, cassiterite as principal Sn host) for mineral processing-related studies. Therefore, this study, along with previous investigations into heavy mineral ilmenite deposits in India [8,9], demonstrates the usefulness of automated mineralogical techniques in early-stage exploration campaigns, which often predominantly involve the routine collection and analysis of stream sediment, soil and till samples and do not necessarily involve the link between sample geochemistry and mineralogical response and element deportment.

6. Conclusions and Implications for Mineral Exploration

This regional follow-up stream sediment exploration study has outlined geochemical and mineralogical evidence to support the previously established presence of Sn-W mineralization [5] and has provided new insights into this mineralisation suite, particularly regarding potential Li-Cs-Ta-Nb mineralisation in the S-type Kagenfels and I–S-type Natzwiller Granite suites of the northern Vosges Mountains. The occurrence of late-stage, incompatible elements in strongly fractionated lithologies and minerals implies the presence of a local mineralisation system. The Natzwiller and Kagenfels granites of the northern Vosges therefore warrant further exploratory work, particularly in areas of known structural control and occurrence of pegmatitic quartz–feldspar (-beryl) vein systems. The study has shown that automated mineralogical techniques can routinely be used in conjunction with univariate, multivariate and litho-geochemistry to determine the nature of stream sediments, source rock and generated exploration targets. A combined geochemical and mineralogical approach in early-stage grassroots exploration is therefore beneficial and useful when screening large areas during routine surveys.

Author Contributions: B.M.S. is the principal investigator on this project and compiled the data from fieldwork, geochemical and software analysis. B.M.S. also compiled the report. G.K.R. performed the QEMSCAN laboratory work and contributed to the methodology section of this paper. J.M.C. assisted in the fieldwork. Conceptualization, B.M.S.; methodology, B.M.S. and G.K.R.; software, B.M.S.; validation, B.M.S. and G.K.R.; formal analysis, B.M.S.; investigation, B.M.S.; resources, B.M.S.; data curation, B.M.S. and G.K.R.; writing—original draft preparation, B.M.S.; writing—review and editing, B.M.S.; visualization, B.M.S.; supervision, B.M.S.; project administration, B.M.S.; funding acquisition, B.M.S. and J.M.C.

Funding: The fieldwork was partly funded by a research grant from the McKinstry Fund of the Society of Economic Geologists Foundation Inc.

Acknowledgments: Sharon Uren and Stephen Pendray at Camborne School of Mines are thanked for providing assistance with the analytical work. Matthew Burford contributed to the field sampling in the Vosges.

Conflicts of Interest: The authors declare no conflict of interest.

Appendix A

Table A1. Modal mineralogy (mineral mass-%) of stream sediment and outcrop samples of the Kagenfels (14–19, yellow header) and Natzwiller Granites (2–3, 20, grey header).

Stream Sediment		Stream Sediment		Stream Sediment		Stream Sediment		Stream Sediment		Stream Sediment	
2		3		14		15		16		17	
Plagioclase	27.08	Plagioclase	29.01	Plagioclase	35.30	Plagioclase	32.18	Plagioclase	32.38	Plagioclase	29.24
Quartz	19.6	K-Feldspar	20.96	K-Feldspar	28.96	K-Feldspar	25.12	K-Feldspar	26.44	Quartz	24.10
K-Feldspar	18.53	Quartz	19.65	Quartz	22.12	Quartz	24.30	Quartz	22.27	K-Feldspar	22.14
Chlorite	13.61	Chlorite	6.88	Fe-Ox (Mn)/CO3	4.58	Fe-Ox (Mn)/CO3	3.35	Fe-Ox (Mn)/CO3	6.98	Fe-Ox (Mn)/CO3	5.89
Tourmaline	8.48	Fe-Ox (Mn)/CO3	6.73	Chlorite	1.47	Chlorite	3.14	Zircon	2.27	Chlorite	3.52
Biotite	3.26	Hornblende	3.86	Biotite	1.26	Hornblende	2.17	Chlorite	2.10	Zircon	3.10
Fe-Ox (Mn)/CO3	2.07	Tourmaline	3.39	Zircon	1.20	Biotite	1.52	Biotite	1.48	Hornblende	3.03
Hornblende	1.92	Zircon	2.3	Ilmenite	0.98	Tourmaline	1.33	Hornblende	1.48	Biotite	2.01
Muscovite	1.12	Biotite	1.98	Muscovite	0.83	Zircon	1.20	Ilmenite	0.87	Tourmaline	1.43
Rutile	1.09	Apatite	1.26	Rutile	0.77	Rutile	1.10	Muscovite	0.80	Muscovite	1.34
Kaolinite	0.76	Rutile	1.11	Tourmaline	0.51	Muscovite	0.84	Tourmaline	0.74	Apatite	0.97
Zircon	0.57	Muscovite	0.71	Hornblende	0.49	Ca-Fe-Al silicates	0.43	Rutile	0.73	Rutile	0.70
Ca-Fe-Al silicates	0.38	Ca-Fe-Al silicates	0.42	Monazite	0.40	Kaolinite	0.40	Apatite	0.40	Ilmenite	0.60
Titanite	0.17	Titanite	0.37	Kaolinite	0.36	Ilmenite	0.28	Kaolinite	0.32	Ca-Fe-Al silicates	0.50
Ilmenite	0.14	Ilmenite	0.33	Spessartine	0.21	Thorite	0.21	Ca-Fe-Al silicates	0.17	Kaolinite	0.44
Spessartine	0.12	Kaolinite	0.26	Ca-Fe-Al silicates	0.17	Apatite	0.18	Ti-Magnetite	0.13	Titanite	0.20
Chrome spinel	0.11	Spessartine	0.11	Ilmenorutile	0.04	Spessartine	0.17	Spessartine	0.11	Ti-Magnetite	0.18
Ti-Magnetite	0.05	Ti-Magnetite	0.08	Others	0.35	Mn Oxides	0.10	Monazite	0.11	Wolframite	0.03
Monazite	0.02	Monazite	0.04			Columbite	0.01	Ilmenorutile	0.02	Ilmenorutile	0.01
Ilmenorutile	0.01	Ilmenorutile	0.01			Cu sulphides	0.01	Cassiterite	0.01	Spessartine	0.14
Thorite	0.01	Columbite	0.01			Ilmenorutile	0.01	Others	0.19	Others	0.43
Cu sulphides	0.01	Cu sulphides	0.01			Others	1.95				
Others	0.89	Others	0.52								

Table A1. *Cont.*

Stream Sediment		**Rock**		**Rock**		**Stream Sediment**		**Stream Sediment**	
18		**17A**		**18A**		**19**		**20**	
Plagioclase	36.97	K-Feldspar	35.78	Quartz	36.55	Plagioclase	33.93	Plagioclase	31.05
K-Feldspar	25.9	Plagioclase	32.23	K-Feldspar	31.35	K-Feldspar	23.32	Quartz	24.65
Quartz	24.34	Quartz	29.44	Plagioclase	29.65	Quartz	29.01	K-Feldspar	22.23
Chlorite	2.56	Fe-Ox (Mn)/CO3	0.77	Fe-Ox (Mn)/CO3	0.8	Chlorite	2.88	Fe-Ox (Mn)/CO3	9.81
Fe-Ox (Mn)/CO3	1.91	Biotite	0.72	Biotite	0.53	Fe-Ox (Mn)/CO3	2.75	Chlorite	2.49
Biotite	1.79	Muscovite	0.46	Muscovite	0.42	Biotite	1.28	Biotite	1.82
Muscovite	1.37	Tourmaline	0.29	Tourmaline	0.22	Muscovite	1.46	Tourmaline	1.23
Ilmenite	0.91	Chlorite	0.12	Ilmenite	0.18	Tourmaline	0.92	Muscovite	1.16
Tourmaline	0.82	Ilmenite	0.09	Chlorite	0.12	Hornblende	0.75	Cassiterite	1.04
Spessartine	0.65	Rutile	0.04	Kaolinite	0.05	Ca-Fe-Al silicates	0.4	Zircon	0.93
Zircon	0.56	Zircon	0.02	Rutile	0.04	Spessartine	0.38	Calcite	0.93
Hornblende	0.48	Ti-Magnetite	0.01	Zircon	0.03	Zircon	0.54	Rutile	0.84
Rutile	0.41	Monazite	0.01	Ti-Magnetite	0.02	Rutile	0.62	Ilmenite	0.76
Mn Oxides	0.33	Others	0.02	Monazite	0.02	Ilmenorutile	0.01	Kaolinite	0.33
Kaolinite	0.31			Thorite	0.02	Ilmenite	0.59	Apatite	0.19
Calcite	0.23			Others	0	Titanite	0.13	Hornblende	0.14
Ca-Fe-Al silicates	0.19					Mn Oxides	0.2	Ti-Magnetite	0.12
Ilmenorutile	0.05					Columbite	0.01	Columbite	0.01
Others	0.22					Cu sulphides	0.01	Others	0.27
						Kaolinite	0.54		
						Others	0.27		

Table A2. Raw geochemical data for the present geochemical study. The stream sediment grain size fraction analysed was <75 µm. All 20 samples were analysed by ICP-MS using a four-acid digest (HCl-HF-HNC_3-$HClO_4$).

Sample ID	As (ppm)	B (ppm)	Be (ppm)	Cs (ppm)	Cu (ppm)	Hf (ppm)	K (ppm)	Li (ppm)	Nb (ppm)	Pb (ppm)	Rb (ppm)	Sn (ppm)	Sr (ppm)	Ta (ppm)	Th (ppm)	Ti (ppm)	W (ppm)	Zr (ppm)
1	15.35	60.25	6.35	14.41	375.63	10.11	12,513	64.24	31.12	54.29	110.2	18.4	214.44	2.54	23.37	4980	11.39	378.28
2	30.45	21.99	13.47	19.53	85.34	36.88	14,558	105.1	100.33	52.22	144.1	18.38	262.37	8.12	77.95	10,502.82	21.55	1248.59
3	18.67	15.41	10.2	12.65	86.53	50.3	18,186	81.55	94.64	33.29	141.2	13.66	291.31	7.45	93.73	8639.58	34.81	1833.39
4	9.38	15.38	2.33	10.63	51.75	22.96	19,889	51.79	15.16	36.07	142.3	19.48	138.48	1.39	14.99	4241.68	21.96	1033.49
5	13.69	15.26	2.82	16.5	49.65	19.83	15,368	66.7	14.62	77.18	114.5	13.28	177.75	1.82	15.89	5048.43	36.05	844.8
6	7.28	15.43	1.25	11.99	42.99	8.65	15,425	42.14	13.47	39.22	104.3	6.31	199.66	1.16	10.41	4723.44	49.9	346.08
7	16.41	14.31	3.68	17.92	62.55	9.61	16,553	78.27	15.28	93.69	116.4	5.48	132.92	1.85	10.07	5498.79	1.89	381.42
8	19.26	14.75	2.29	11.94	176.48	18.39	20,405	61.55	20.85	40.06	136.4	30.69	134.35	1.91	19.59	5539.26	6.92	776.97
9	50.27	13.48	1.73	27.42	70.08	12.41	18,473	79.65	9.93	28.29	77.9	34.06	98.89	0.76	8.87	4093.86	3.53	478.42
10	125.97	12.8	24.06	9.25	25.05	37.22	23,638	54.77	198.29	54.23	366.6	6.5	53.17	11.77	59.76	2170.97	11.83	711.86
11	73.95	13.85	34.22	10.88	48.77	32.66	18,266	49.35	126.27	46.11	195.5	30.82	49.41	7.55	66.3	2637.35	10.42	827.32
12	17.55	15.45	2.8	11.93	102.23	13.38	17,816	41.45	18.29	26.97	152.2	41.55	102.73	1.55	24.45	3839.98	1.8	552.45
13	44.09	12.1	10.32	3.59	7.88	35.85	22,682	12.13	99.06	8.72	190.2	7.19	21.41	5.87	47.58	2784.06	14.56	905.28
14	92.71	12.7	19.18	9.54	60.8	64.86	25,376	34.12	192.26	26.08	242	21.92	43.13	10.83	86.53	3994	8.76	1695.02
15	25.52	11.25	13.66	13.29	120.18	40.27	15,796	44.4	53.14	35.79	157	21.61	147.74	3.8	38.33	4208.34	29.11	1408.45
16	52.17	9.91	13.41	10.53	88.88	45.08	21,273	41.38	92.71	20.13	210.1	26.29	101.8	5.85	56.71	4133.23	28.7	1441.4
17	28.68	9.9	12.82	12.14	122.74	50.72	19,535	52.61	63.67	39.57	177.7	14.14	183.39	4.55	66.16	5683.34	102.56	1923.08
18	83.4	10.46	21.8	7.62	59.56	34.05	20,270	26.16	92.76	26.64	186.4	10.07	36.22	5.5	46.62	3223.2	6.94	899.06
19	68.71	12.02	10.22	11.36	166.61	28.35	15,251	35.5	66.55	40.69	152.7	20.05	78.5	4.06	41.08	3681.99	10.84	870.06
20	9.04	15.46	8.03	9.35	116.28	17.45	12,924	47.29	29.72	99.56	47.3	17.98	179.99	2.69	21.65	3896.3	23.26	698.02

References

1. European Commission. *Directorate-General for Internal Market, Industry, Entrepreneurship and SMEs (European Commission) Study on the Review of the List of Critical Raw Materials 2017*; European Commission: Brussels, Belgium, 2017.
2. Romer, R.L.; Kroner, U. Phanerozoic tin and tungsten mineralization—Tectonic controls on the distribution of enriched protoliths and heat sources for crustal melting. *Gondwana Res.* **2016**, *31*, 60–95. [CrossRef]
3. Steiner, B.M. W and Li-Cs-Ta signatures in I-type granites—A case study from the Vosges Mountains, NE France. *J. Geochem. Expl.* **2019**, *197*, 238–250. [CrossRef]
4. Bureau de Recherches Géologiques et Minières BRGM InfoTerre. Available online: http://www.infoterre.brgm.fr (accessed on 20 October 2019).
5. Leduc, C. *Paris-Vosges—Zones D et E. Prospection Géochimique Stratégique Cirey-sur-Vezouze et Molsheim. Interprétation des Résultats Analytiques*; Bureau de Recherches Géologiques et Minières: Orléans, France, 1984; p. 93.
6. Tabaud, A.-S.; Whitechurch, H.; Rossi, P.; Schulmann, K.; Guerrot, C.; Cocherie, A. Devonian–Permian magmatic pulses in the northern Vosges Mountains (NE France): Result of continuous subduction of the Rhenohercynian Ocean and Avalonian passive margin. *Geol. Soc. Lond.* **2014**, *405*, 197–223. [CrossRef]
7. Altherr, R.; Holl, A.; Hegner, E.; Langer, C.; Kreuzner, H. High-potassium, calc-alkaline I-type plutonism in the European Variscides: Northern Vosges (France) and northern Schwarzwald (Germany). *Lithos* **2000**, *50*, 51–73. [CrossRef]
8. Bernstein, S.; Frei, D.; McLimans, R.K.; Knudsen, C.; Vasudev, V.N. Application of CCSEM to heavy mineral deposits: Source of high-Ti ilmenite sand deposits of South Kerala beaches, SW India. *J. Geochem. Expl.* **2008**, *96*, 25–42. [CrossRef]
9. Keulen, N.; Frei, D.; Riisager, P.; Knudsen, C. Analysis of heavy minerals in sediments by computer-controlled scanning electron microscopy (CCSEM): Principles and applications. *Mineral. Assoc. Can. Short Course* **2012**, *42*, 67–184.
10. Mackay, D.A.R.; Simandl, G.J.; Ma, W.; Redfearn, M.; Gravel, J. Indicator mineral-based exploration for carbonatites and related specialty metal deposits—A QEMSCAN® orientation survey, British Columbia, Canada. *J. Geochem. Expl.* **2016**, *165*, 159–173. [CrossRef]
11. Piqué, A.; Fluck, P.; Schneider, J.L.; Whitechurch, H. The Vosges Massif. In *Pre-Mesozoic Geology in France and Related Areas*; Chantraine, J., Rolet, J., Santallier, D.S., Piqué, A., Keppie, J.D., Eds.; IGCP-Project 233; Springer: Berlin, Heidelberg, 1994; pp. 416–425.
12. Tabaud, A.-S.; Janoušek, V.; Skrzypek, E.; Schulmann, K.; Rossi, P.; Whitechurch, H.; Guerrot, C.; Paquette, J.-L. Chronology, petrogenesis and heat sources for successive Carboniferous magmatic events in the Southern–Central Variscan Vosges Mts (NE France). *J. Geol. Soc.* **2015**, *172*, 87–102. [CrossRef]
13. Schulmann, K.; Martínez Catalán, J.R.; Lardeaux, J.M.; Janoušek, V.; Oggiano, G. The Variscan orogeny: Extent, timescale and the formation of the European crust. *Geol. Soc. Lond.* **2014**, *405*, 1–6. [CrossRef]
14. Hess, J.C.; Lippolt, H.J.; Kober, B. The age of the Kagenfels granite (northern Vosges) and its bearing on the intrusion scheme of late Variscan granitoids. *Geol. Rundsch.* **1995**, *84*, 568–577. [CrossRef]
15. Elsass, P.; Eller, J.P.; Stussi, J.M. *Géologie du Massif du Champ du Feu et de ses Abords: Éléments de Notice Pour la Feuille Géologique 307 Sélestat.*; Bureau de Recherches Géologiques et Minières: Orléans, France, 2008; p. 187.
16. Dekoninck, A.; Rochez, G.; Yans, J.; Fluck, P. Mineralizing events in the Vosges massif: Insights from the Mn-W Haut-Poirot deposit (NE France). *Proc. Miner. Resour. Discov.* **2017**, *4*, 1519–1522.
17. Fluck, P.; Stein, S. Espèces minérales des principaux districts miniers du massif vosgien. *Pierres et Terre* **1992**, *35*, 107–115.
18. Fluck, P. *Metallogeny of Vosges*; Freiberger Forschungshefte: Salamanca, Spain, 1977; pp. 83–93.
19. Fluck, P.; Weil, R. *Géologie des Gîtes Minéraux des Vosges et des Régions Limitrophes*; Mémoires du, B.R.G.M., Ed.; Bureau de Recherches Géologiques et Minières: Orléans, France, 1976.
20. Mariet, A.L.; Bégeot, C.; Gimbert, F.; Gauthier, J.; Fluck, P.; Walter-Simonnet, A.V. Past mining activities in the Vosges Mountains (eastern France): Impact on vegetation and metal contamination over the past millennium. *Holocene* **2016**, *26*, 1225–1236. [CrossRef]

21. Forel, B.; Monna, F.; Petit, C.; Bruguier, O.; Losno, R.; Fluck, P.; Begeot, C.; Richard, H.; Bichet, V.; Chateau, C. Historical mining and smelting in the Vosges Mountains (France) recorded in two ombrotrophic peat bogs. *J. Geochem. Expl.* **2010**, *107*, 9–20. [CrossRef]
22. Billa, M.; Gloaguen, E.; Melleton, J.; Tourlière, B. *Consolidation des Anomalies Géochimiques et Géophysiques du Territoire Métropolitain*; Bureau de Recherches Géologiques et Minières: Orléans, France, 2016; p. 42.
23. Weil, R. Sur la présence de l'adulaire dans la Grotte des Partisans. *Sciences Géologiques Bulletins et Mémoires* **1936**, *3*, 27–28. [CrossRef]
24. Bureau de Recherches Géologiques et Minières SIG Mines France. Available online: http://sigminesfrance.brgm.fr/sig.asp (accessed on 20 October 2019).
25. Schneider, M. *Vogesengranit—Letter: Natzweiler-Struthof Memorial Museum Collection 1940*; Natzweiler-Struthof Memorial Museum Collection: Natzwiller, France, 1940.
26. Steiner, B.M. Tools and Workflows for Grassroots Li-Cs-Ta (LCT) pegmatite exploration. *Minerals* **2019**, *9*, 499. [CrossRef]
27. Ballouard, C.; Poujol, M.; Boulvais, P.; Branquet, Y.; Tartèse, R.; Vigneresse, J.L. Nb-Ta fractionation in peraluminous granites: A marker of the magmatic–hydrothermal transition. *Geology* **2016**, *44*, 231–234. [CrossRef]
28. Breiter, K.; Škoda, R. Zircon and whole-rock Zr/Hf ratios as markers of the evolution of granitic magmas: Examples from the Teplice caldera (Czech Republic/Germany). *Mineral. Petrol.* **2017**, *111*, 435–457. [CrossRef]
29. Linnen, R.L.; Keppler, H. Columbite solubility in granitic melts: Consequences for the enrichment and fractionation of Nb and Ta in the Earth's crust. *Contrib. Mineral. Petrol.* **1997**, *128*, 213–227. [CrossRef]
30. Steiner, B.M. Using Tellus stream sediment geochemistry to fingerprint regional geology and mineralisation systems in southeast Ireland. *Irish J. Earth Sci.* **2018**, *36*, 45–61. [CrossRef]
31. Gottlieb, P.; Wilkie, G.; Sutherland, D.; Ho-Tun, E.; Suthers, S.; Perera, K.; Jenkins, B.; Spencer, S.; Butcher, A.; Rayner, J. Using Quantitative Electron Microscopy for Process Mineralogy Applications. *JOM* **2000**, *52*, 24–25. [CrossRef]
32. Pirrie, D.; Butcher, A.; Power, M.R.; Gottlieb, P.; Miller, G.L. Rapid quantitative mineral and phase analysis using automated scanning electron microscopy (QEMSCAN®); potential applications in forensic geoscience. *Geol. Soc. Lond.* **2004**, *232*, 123–136. [CrossRef]
33. Pirrie, D.; Rollinson, G.K. Unlocking the applications of automated mineral analysis. *Geol. Today* **2011**, *27*, 235–244. [CrossRef]
34. Rollinson, G.K. Automated Mineralogy by SEM-EDS. In *Earth Systems and Environmental Sciences*; Science Direct; John Wiley and Sons: Hoboken, NJ, USA, 1996.
35. Rollinson, G.K.; Stickland, R.J.; Andersen, J.C.Ø.; Fairhurst, R.; Boni, M. Characterisation of Supergene Non-Sulphide Zinc Deposits using QEMSCAN®. *Miner. Eng.* **2011**, *24*, 778–787. [CrossRef]
36. Simons, B.; Rollinson, G.K.; Andersen, J.C.Ø. Characterisation of lithium minerals in granite-related pegmatites and greisens by SEM-based automated mineralogy 2018. In Proceedings of the Mineral Deposits Study Group Winter Meeting, Brighton, UK, 3–5 January 2018.
37. Taylor, S.R. Abundance of chemical elements in the continental crust: A new table. *Geochim. Cosmochim. Acta* **1965**, *28*, 1273–1285. [CrossRef]
38. Andersen, J.C.Ø.; Stickland, R.J.; Rollinson, G.K.; Shail, R.K. Indium mineralisation in SW England: Host parageneses and mineralogical relations. *Ore Geol. Rev.* **2016**, *78*, 213–238. [CrossRef]
39. Fleet, M.E. *Sheet Silicates: Micas*; Deer, Howie and Zussman Rock Forming Minerals Series; Geological Society: London, UK, 2003.
40. Breiter, K.; Ďurišová, J.; Hrstka, T.; Korbelová, Z.; Vašinová Galiová, M.; Müller, A.; Simons, B.; Shail, R.K.; Williamson, B.J.; Davies, J.A. The transition from granite to banded aplite-pegmatite sheet complexes: An example from Megiliggar Rocks, Tregonning topaz granite, Cornwall. *Lithos* **2018**, *302*, 370–388. [CrossRef]
41. Simons, B.; Andersen, J.C.Ø.; Shail, R.K.; Jenner, F.E. Fractionation of Li, Be, Ga, Nb, Ta, In, Sn, Sb, W and Bi in the peraluminous Early Permian Variscan granites of the Cornubian Batholith: Precursor processes to magmatic-hydrothermal mineralisation. *Lithos* **2017**, *278*, 491–512. [CrossRef]
42. Selway, J.B.; Breaks, F.W.; Tindle, A.G. Pegmatite Exploration Techniques for the Superior Province, Canada, and Large Worldwide Tantalum Deposits. *Explor. Min. Geol.* **2005**, *14*, 1–30. [CrossRef]
43. Černý, P. Exploration strategy and methods for pegmatite deposits of tantalum. In *Lanthanides, Tantalum, and Niobium*; Moller, P., Černý, P., Saupe, F., Eds.; Springer: New York, NY, USA, 1989; pp. 274–302.

44. Shaw, D. A review of K-Rb fractionation trends by covariance analysis. *Geochim. Cosmochim. Acta* **1968**, *32*, 573–601. [CrossRef]
45. Bau, M. Controls on the fractionation of isovalent trace elements in magmatic and aqueous systems: Evidence from Y/Ho, Zr/Hf, and lanthanide tetrad effect. *Contrib. Mineral. Petrol.* **1996**, *123*, 323–333. [CrossRef]
46. Fujimaki, H. Partition-coefficients of Hf, Zr, and REE between zircon, apatite and liquid. *Contrib. Mineral. Petrol.* **1986**, *94*, 42–45. [CrossRef]

Article

Laser-Induced Breakdown Spectroscopy—An Emerging Analytical Tool for Mineral Exploration

Russell S. Harmon [1,*], Christopher J.M. Lawley [2], Jordan Watts [3], Cassady L. Harraden [4], Andrew M. Somers [5] and Richard R. Hark [6]

1 Department of Marine, Earth, and Atmospheric Sciences, North Carolina State University, Raleigh, NC 27695, USA
2 Natural Resources Canada, Geological Survey of Canada, 601 Booth Street, Ottawa, ON K1A 0E8, Canada; christopher.lawley@canada.ca
3 School of the Environment, University of Windsor, 401 Sunset Avenue, Windsor, ON N9B 3P4, Canada; watts114@uwindsor.ca
4 Corescan Pty Ltd., Suite 1900, 1055 West Hastings Street, Vancouver, BC V6E 2E9, Canada; cassady.l.harraden@gmail.com
5 SciAps Inc., 7 Constitution Way, Woburn, MA 01801, USA; asomers@sciaps.com
6 Institute for the Preservation of Cultural Heritage, Yale University, West Haven, CT 06516, USA; richard.hark@yale.edu
* Correspondence: rsharmon@ncsu.edu; Tel.: +1-919-588-0613

Received: 22 October 2019; Accepted: 13 November 2019; Published: 20 November 2019

Abstract: The mineral exploration industry requires new methods and tools to address the challenges of declining mineral reserves and increasing discovery costs. Laser-induced breakdown spectroscopy (LIBS) represents an emerging geochemical tool for mineral exploration that can provide rapid, in situ, compositional analysis and high-resolution imaging in both laboratory and field and settings. We demonstrate through a review of previously published research and our new results how LIBS can be applied to qualitative element detection for geochemical fingerprinting, sample classification, and discrimination, as well as quantitative geochemical analysis, rock characterization by grain size analysis, and in situ geochemical imaging. LIBS can detect elements with low atomic number (i.e., light elements), some of which are important pathfinder elements for mineral exploration and/or are classified as critical commodities for emerging green technologies. LIBS data can be acquired in situ, facilitating the interpretation of geochemical data in a mineralogical context, which is important for unraveling the complex geological history of most ore systems. LIBS technology is available as a handheld analyzer, thus providing a field capability to acquire low-cost geochemical analyses in real time. As a consequence, LIBS has wide potential to be utilized in mineral exploration, prospect evaluation, and deposit exploitation quality control. LIBS is ideally suited for field exploration programs that would benefit from rapid chemical analysis under ambient environmental conditions.

Keywords: laser-induced breakdown spectroscopy; LIBS; geochemical exploration; geochemical fingerprinting; micro-imaging; grain size analysis; mineral texture

---

1. Introduction

Mineral exploration plays an important role in society, as the continued discovery of new deposits is required to supply mineral and other natural resources for the more equitable and low-carbon economy of the future [1–4]. Rising global population and increased consumption by a growing middle class will apply new pressure to find additional resources, particularly for the elements that are required to make the specialized materials and components contained in advanced technologies [5,6]. The increased demand for these critical commodities led to significant investment by industry and government to search for additional mineral resources in green-field (i.e., remote) and brown-field (i.e., near mine) exploration environments [7–9]. However, the global trends of declining mineral reserves for many commodities and increasing discovery costs [3,4,10] suggest that exploration investment is insufficient and/or is not being deployed in the most effective manner possible. Both trends are also occurring at a time when new mineral deposit discoveries tend to be deeper, covered, and/or more remote, which are unlike near-surface mines that were often found, at least initially, by prospectors [4,7–9,11].

To address increasing demand and declining mineral reserves from deeper and more challenging deposits, the mineral exploration industry had to evolve and innovate by adopting new, cost-effective methodologies and technologies. For example, new conceptual models provide a predictive framework to identify the kinds of large-scale geological environments that should be considered the most prospective for finding additional mineral resources in greenfield areas of sparse geological data [12–15]. The ore system concept, which includes all of the geological processes required to transport and concentrate ore components from source to ore (i.e., drivers, sources, pathways, and traps), is one such predictive framework [12–14,16]. Because each of the required ore-forming components is manifested in the rock record as changes in mineralogy, texture, and/or composition, conceptual models can be transformed to mappable criteria to support mineral exploration [16]. Taken together, these mappable criteria are known as the mineral deposit's *footprint*.

Some mineral deposit footprints are spectacular, including massive sulfide mineralization or coarse visible gold. Such obvious, mineralogical, and/or textural indicators are a direct consequence of ore-forming fluids and/or magmas transporting and concentrating base, precious, and critical metals. Geochemical and mineralogical investigations form the backbone of most mineral exploration programs, as geologists analyze minerals in sediments and hydrothermally altered bedrock, catalog the lithology and mineralogy of drill core, and model their spatial distributions at different spatial scales using the latest three-dimensional (3D) visualization tools. The recent introduction of field-portable technologies (e.g., X-ray diffraction (XRD), short-wavelength infrared (SWIR) spectrometers, field-portable X-ray fluorescence (fp-XRF)) is making the process of mapping geochemical and mineralogical footprints more robust and quantitative. Laser-induced breakdown spectroscopy (LIBS) is the latest addition to this group of technologies. Advanced technologies in the form of field-portable analyzers, mobile core scanners, and other on-site sensor technologies are also being utilized at drilling locations or in the core shack. As discussed and demonstrated here, chemical, mineralogical, and textural studies also have implications for processing and recovery of mineral resources later on in the supply chain, and LIBS analysis has an important potential application in such studies, as well as during initial exploration.

Advances in geophysical and geochemical methods, coupled with the reduced cost of these technologies, are also allowing more cryptic footprints to be mapped at greater resolution and/or larger spatial scales. Such multi-parameter, mineral deposit footprints are now well documented at most ore system types and are widely used to vector toward ore zones within individual deposits and, more rarely, at continent to mineral district scales [17–20]. Geophysical imaging is proven particularly effective at mapping these deposit footprints because ore-forming processes tend to produce mineralogical changes that impact rock properties, e.g., density, magnetic susceptibility, conductivity [21]. However, some components of a mineral deposit footprint can be invisible to geophysical methods, either because the rock property changes between the host rock and the ore-forming mineral assemblage are relatively subtle and/or because the ore-forming process only resulted in trace element substitution into the main rock-forming minerals. Cryptic geochemical footprints such as these may not be not associated with any visual indicators, but tend to be larger than any single mineralogical or geophysical pathfinder at most ore systems [19,22].

The identification of geochemical footprints has the potential to greatly increase the size of the exploration target. A large variety of approaches were deployed for geochemical vectoring, such as lithogeochemistry, stable and radiogenic isotopes, indicator minerals, hyperspectral scanning, biogeochemistry, hydrochemistry, and sediment geochemistry [23,24]. Each of these approaches depends on a growing list of analytical techniques with improved sensitivity for more elements, including mass spectrometry (e.g., inductively coupled plasma mass spectrometry (ICP-MS), secondary ion mass spectrometry (SIMS)), X-ray fluorescence (XRF), optical emission spectroscopy (ICP-OES), and a suite of electron micro-beam techniques (e.g., energy-dispersive spectroscopy (EDS), wavelength-dispersive spectroscopy (WDS), synchrotron micro-XRF). Despite this proliferation of geochemical methods and tools, many challenges remain. Firstly, mapping cryptic geochemical footprints requires low analytical detection limits because the absolute abundance of some pathfinder elements may range from μg/g (ppm) to ng/g (ppb) concentrations. Quantifying the low abundance of pathfinder elements is typically addressed by using large, expensive, and sensitive instruments hosted within research laboratories. Laboratory analysis is expensive, takes time, and most often needs to be completed off-site from the mineral exploration camp, which, in the case of greenfield exploration, often leads to a disconnect between sample collection and drill target decision-making. Secondly, most ore systems are complex and often comprise multiple overprinting events, some of which may be unrelated to mineralization. Unraveling the complex geologic history from the host rocks of ore systems must often be done at the microscale with mineral chemistry. Unfortunately, micro-analytical geochemistry was until very recently restricted to research applications due to the sophisticated instrumentation required and the lengthy time and high cost of analysis. Thirdly, the geochemical expression of ore-forming fluids and magmas often comprises more than just the element of economic interest. Very few analytical methods are capable of detecting the complete range of elements comprising a mineral deposit's geochemical footprint. Therefore, mapping the complete multivariate geochemical footprint of mineral deposits often requires the use of multiple analytical techniques. Elements with low atomic mass (i.e., atomic number ≤12, or Mg), the so-called *light elements*, represent a specific challenge for conventional geochemical techniques despite their significant potential to be used as geochemical vectors. Some of these light elements (e.g., Li) also happen to be among the critical commodities that are required for green technologies in the low-carbon economy, which makes understanding their ore-forming processes particularly important.

Here, we focus on LIBS, an emerging technology for real-time chemical analysis in the field that has the potential to address each of the challenges of geochemical vectoring. We briefly introduce the theoretical and mechanistic basis of LIBS and follow that discussion with example applications relevant to mineral and ore deposit exploration, including element detection (i.e., geochemical fingerprinting), sample classification/discrimination, quantitative geochemical analysis, rock characterization (e.g., grain size analysis), and in situ geochemical imaging. The specific examples presented here illustrate how LIBS can (i) provide a geochemical signature for a large number of elements in minerals and rocks, including the light elements that cannot be analyzed by other in-field analytical techniques, (ii) support exploration decision-making with real-time and low-cost geochemical analysis in the field, and (iii) deliver micro-analytical geochemistry and textural analysis in support of mineral identification to unravel complex, multiply overprinting hydrothermal histories from the rock record.

2. Laser-Induced Breakdown Spectroscopy

Atomic emission spectroscopy is a technique for chemical analysis to determine either the presence or the mass fraction of an element present in a sample based on measurement of the intensity of light emitted from a flame, spark, arc, or plasma [25,26]. Laser-induced breakdown spectroscopy (LIBS) is an established, straightforward, reliable, and versatile form of atomic emission spectroscopy that has broad capability for rapid, in situ elemental detection in any material (solid, liquid, or gas), and quantitative analysis by LIBS is possible using either conventional calibration methods or calibration-free approaches [25–31]. Thus, LIBS has the potential for widespread use for rapid chemical detection and analysis outside the research laboratory. At its core, an LIBS analytical system (Figure 1) consists of just a few components: (i) a solid-state, short-pulsed, Q-switched laser operating at 1064 nm (or one of its frequency-multiplied harmonics) used to create a microplasma on the target, (ii) a set of optics to focus the laser light onto the target and to collect the light emitted as the plasma cools, (iii) a coupled fiber-optic and spectrometer/detector system for acquisition of the plasma light emission and spectral resolution of the light spectrum, and (iv) a computer for system control and data processing.

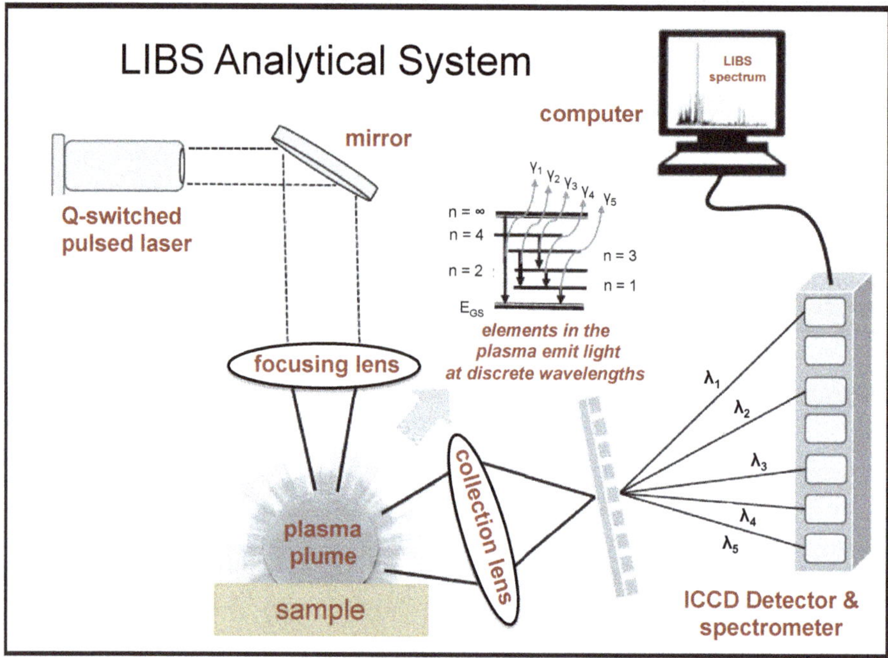

Figure 1. Schematic diagram of a typical laser-induced breakdown spectroscopy (LIBS) system consisting of a solid-state, short-pulsed, Q-switched laser used to create a micro-plasma on the sample, lenses to focus the laser light onto the sample and then collect and transmit the light produced during the micro-plasma event, a detector/spectrometer combination to receive and resolve the light spectrum, and a computer for system control and data processing.

2.1. The LIBS Analysis

In LIBS, a high-intensity pulsed laser beam is focused onto the surface of a sample, which creates a high-temperature plasma through the multi-stage breakdown process that occurs when laser energy couples to a material. Typically, a sub-mg amount of material is ablated, vaporized, and dissociated in the high-temperature plasma into a mixture of free electrons and weakly ionized molecular, atomic, and ionic species. The plasma cools down rapidly, causing species recombination, de-excitation, and the release of energy as photons when electrons return to lower energy levels. Relatively sharp spectral lines are produced at discrete wavelengths after a few hundred nanoseconds of plasma expansion and cooling that results in the decay of the continuum emission caused by radiative recombination and bremsstrahlung radiation. The detector can be time-gated to optimize collection of the LIBS emission for specific elements. Because all elements emit in the 200–900 nm spectral range and many electron orbital transitions occur for most elements, an LIBS emission intensity spectrum, such as that shown in Figure 2, consists of multiple peaks for the majority of elements and contains hundreds to thousands of spectral lines for most geological materials. Monitoring the wavelength and intensity of emission lines in the LIBS plasma provides information on both the chemical species present and their abundance. The spectral line wavelength documents the identity of an element, whereas its intensity is proportional to the number of atoms of the element present. Generally, elements on the left side of the periodic table, such as Li, Na, and Ca that have relatively low ionization energy, display strong emission and, therefore, can be detected at very small abundances. By contrast, non-metallic elements on the right side of the periodic table with high ionization potentials, such as the halogen elements, are more challenging to determine by LIBS and, consequently, have much higher limits of detection.

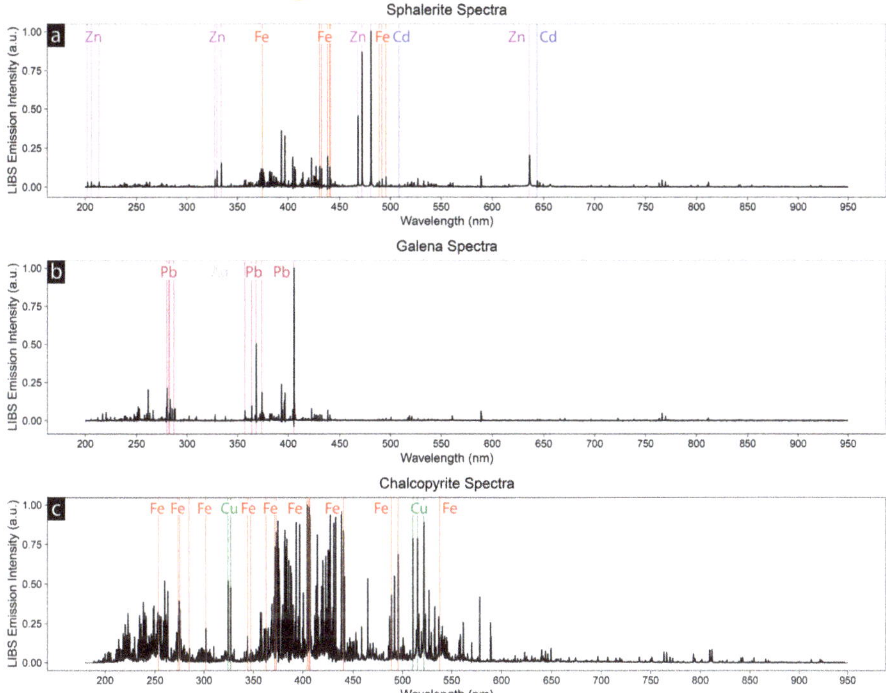

Figure 2. Example broadband LIBS spectra for (**a**) sphalerite showing emission lines for Zn, Fe, and Cd, (**b**) galena showing emission lines for Pb and Ag, and (**c**) chalcopyrite showing emission lines for Fe and Cu. Data were acquired in Ar with a SciAps Z-300 handheld LIBS analyzer and processed (i.e., baseline correction and normalization) using the internal software utility described in Section 3. The large number of Fe emission lines often results in complex LIBS spectra for Fe-rich minerals (e.g., chalcopyrite), particularly in the 350–450-nm spectral range.

Gases can be analyzed by forming the plasma within the gaseous environment of interest. Liquid analysis is accomplished by focusing the laser on the surface of the liquid or within the body of a liquid below its surface. The character of the micro-plasma created during solid sample analysis is complex and determined by several interrelated factors—the nature of the material being ablated (e.g., its composition, crystallinity, surface texture, and optical reflectivity), the operational characteristics of the laser (i.e., wavelength, energy, beam shape, and pulse duration), the degree of laser energy coupling to the sample surface, and the ambient environment in which the LIBS plasma is formed [32].

2.2. Specific Attributes of LIBS

LIBS has many attributes that make it an attractive analytical tool for analysis of a wide variety of geological materials both in a conventional laboratory setting and in the field. LIBS can analyze material in any state (solid, liquid, or gas) in real time with little to no sample preparation. All elements have optical emission lines within the 200–900-nm spectral range; thus, LIBS can detect all elements with typical limits of detection in the μg/g range. LIBS is especially sensitive to the light elements (H, He, Li, Be, B, C, N, O, Na, and Mg) that can be present in high abundance within natural materials but are difficult to determine by many other analytical techniques (e.g., fp-XRF). The LIBS signal generation of simultaneous multi-element emission can also be used for simple elemental detection or for quantitative analysis in real time. Despite the complications associated with the analysis of solid samples noted above, quantitative analysis is readily achieved by LIBS. Excellent precision and accuracy can be obtained utilizing an external calibration approach based on reference materials that are very closely matrix-matched to the type of samples being analyzed [33,34], and semi-empirical analysis is possible through calibration-free methodologies or through internal calibration-free techniques [35,36].

LIBS spectral analysis is capable of identifying and discriminating unknown materials when used in conjunction with chemometric techniques and pre-established databases. A particularly attractive feature of LIBS is its capability for chemical analysis outside the laboratory with little or no sample preparation for applications, such as geological prospecting, where rapid in-field analysis under ambient conditions is beneficial. Winterburn et al. [23] noted that the light-element detection capability of LIBS is particularly valuable in an exploration context. A recent increase in global demand for Li promoted the application of LIBS as a field analytical tool. As demonstrated herein, LIBS has the potential in the near future to become nearly as common a field tool in mineral exploration as fp-XRF.

There is the additional feature that LIBS sampling is highly spatially resolved, as the plasma forms over a limited spatial area of only tens to hundreds of microns on the sample surface, so that only a small amount of material (typically picograms to nanograms) is sampled by each laser pulse. This allows for in situ analysis of individual particles, mineral grains, or inclusions [33,37–41], as well as the fine-scale compositional mapping of a complex sample such as a chemically zoned mineral [42–47], the analysis of thin crusts, coatings, or surface alteration zones without substrate interference [48], or chemical analysis at highly spatially resolved spatial scales to below ~10 µm [47,49–54]. Stratigraphic analysis of a sample by depth profiling is also possible, as sequential ablation forms a vertical crater that progressively bores down into a sample with successive laser pulses [55–57].

2.3. LIBS for Ore Prospecting

LIBS followed a typical arc of technology development. Since its beginnings in the 1960s, and throughout most of its history, LIBS was a laboratory-based research technique. Typically, researchers would assemble LIBS systems in many different configurations to conduct experiments aimed at understanding some aspect of laser–material interaction, breakdown phenomena in different media, or plasma physics. This resulted in a multiplicity of different bespoke and often ad hoc system designs that would frequently be modified from one experiment to another. However, this situation began changing in the early 2000s, driven by advances in compact laser sources, optical-fiber light delivery and collection and compact beam optics designs, and the miniaturization of high-resolution spectrometers, microelectronics, and computers that in parallel led to commercial laboratory LIBS systems, mobile LIBS devices, and handheld LIBS analyzers for real-time chemical analysis outside the laboratory [58–67]. Because of recent technological advances, the current generation of LIBS handheld analyzers has many features present in laboratory systems, including variable gating, gas purging, sample rastering, video targeting, and on-board chemometric analysis, which substantially enhance the application of LIBS outside the laboratory. Researchers utilized each stage of technology development to understand and demonstrate how LIBS might be utilized in prospecting for economic minerals and ores. Some early studies that set the stage for the applications discussed here are reviewed in the paragraphs that follow.

It is well established that LIBS is an efficacious means for rapidly discriminating different types of minerals [68,69], and LIBS is widely applied to the analysis of ore samples in this context. Early application of LIBS to ore prospecting and processing examined the effects of particle size and ore mineralogy on the quantitative analysis of major, minor, and trace elements in iron ore and phosphates [70–76].

In a pioneering LIBS study, Kaski et al. [50] demonstrated in a laboratory study using a bespoke LIBS system that sulfide minerals in drill cores could be rapidly identified in situ. The spectral region between 170 and 210 nm was monitored using a Kr–F excimer laser operating at 248 nm to create the emission plasma. This spectral region was selected for interrogation because the most intense sulfur lines are located here, and all minerals analyzed have a characteristic emission spectrum in this region. Reference spectra of pyrite, pyrrhotite, chalcopyrite, and sphalerite were obtained by accumulating the emission signals generated by 10 laser pulses acquired in air, and then three mineralized drill core specimens were analyzed along 5 cm lines at 0.2 mm sampling intervals using five laser pulses at 15 Hz repetition rate, with unknown spectra matched with reference spectra for identification. In one drill core, some 33% of the sampling points were observed to be sphalerite, 27% of them were observed to be pyrrhotite, and 23% of them were observed to be barite; pyrite comprised >70% of the sulfides in a second core, and the third was composed of 43% dolomite and 25% pyrite.

Haavisto et al. [51] used a similar analytical approach with a bespoke LIBS system to demonstrate the potential of LIBS for on-line analysis of an Au-mineralized drill core from the Agnico-Eagle Kittilä Mine in northern Finland, where the main minerals in the ore are carbonates, quartz, albite, chlorite, sericite, and graphite, with Au in the ore contained primarily in arsenopyrite and pyrite. Elemental concentrations determined by LIBS corresponded to laboratory assay by X-ray fluorescence. Gold was not detected by LIBS because of the refractory nature of the ore, but measured As concentrations directly reflected the amount of Au in the ore. Chemometric data processing based on based on principal component analysis was able to extract mineralogical information from the LIBS spectra based on elemental ratios.

Kuhn et al. [46] used LIBS as a method to obtain chemical information for drill cores from tailing material in a former Pb–Zn deposit at Beythal near Aachen, Germany. This deposit was mined for 43 years, with 3.7 Mm^3 of flotation residues deposited over an area of about 45 ha. The tailing material was characterized by quartz and phyllosilicate sand plus minor amounts of carbonate, with drill cores 1 m long and 5 cm in diameter taken from a depth of about 7–9 m below surface. LIBS analyses were obtained utilizing a prototype LIBS core scanner system manufactured by LTB Lasertechnik GmbH (Berlin, Germany), capable of mapping an area of 1 m by 2.5 cm at a user-defined step size. Cores were mapped within an area of about 1 m by 1 cm at a step size of 400 µm, collecting the emission from five laser shots in air into each of 25 cumulate spectra along the y-direction and between 2478 and 2490 spectra along the x-direction for a total of approximately 62,000 spectra per meter of core. Quantitative analysis was undertaken on the basis of internal reference samples by partial least squares regression. Maximum concentrations of Pb, Zn, and Cu were observed between 785 and 789 cm sediment depth in the tailings pile, with average bulk concentrations of 0.95% Pb, 1.20% Zn, and 262 ppm Cu in that depth range. Distribution patterns of element content and variation predicted for whole drill cores exhibited a good correlation with determinations by WDS analysis.

Several recent studies examined gold ore using LIBS analysis as a means of improving ore processing efficiency and productivity through innovation in on-line analysis and control, automation, and real-time decision support. Monitoring and control of ore quality during mining operations utilizes laboratory analysis and, thus, requires sample collection and preparation for off-line analysis via traditional laboratory methodologies, an approach which is time-consuming and expensive, and which can also cause production delays to mining or processing that increase a mine's operating and production costs.

Harhira et al. [77] investigated the use of LIBS to assess gold ores and quantify gold content in an Au-bearing core from the Lapa Mine in Quebec, Canada. The bespoke LIBS system employed a pulsed neodymium-doped yttrium aluminum garnet (Nd:YAG) laser operating at the fundamental wavelength of 1064 nm, which delivered a maximum pulse energy of 300 mJ at a maximum repetition rate of 10 Hz. The plasma emission in air was recorded by a Czerny–Turner/intensified charge coupled device (ICCD) spectrometer/detector system that was optimized for a spectral window centered at 265 nm and was tuned in terms of spectral response, sensitivity, and resolution to meet the requirements of detecting the Au at the ppm level. The first part of this study documented that principal component analysis of LIBS spectra was able to differentiate between Fe-rich and Si-rich gold-bearing minerals by using the data as geochemical fingerprints of the mineralized-rock mineralogy. The second demonstrated quantitative Au analysis in drill core with limits of detection of 0.75 ppm for Si-rich samples and 1.5 ppm for Fe-rich samples. The core was mounted on a motorized stage for scanning across the sample surface, with LIBS spectra acquired based on 10,000 laser shots of ~600-µm spot size at a step interval of 800 µm over a scanned area of 5 × 20 cm^2 on the flat drill core surface. Calibration curves prepared from synthetic samples having Au concentrations between 0 and 120 ppm were used to quantify the Au content based on the measured intensity of the Au emission line at 267.59 nm. Averaging over the 10,000 laser shots gave a mean Au content in the core sample of 25.47 ppm, but the LIBS analysis also documented the heterogeneous distribution of Au in the core at the millimeter spatial scale, with high Au contents of up to a few thousand ppm recorded at some locations but very low concentrations at other sites. A quartz chlorite was determined to contain 31 ppm Au and 4.99% Fe, whereas 25.4 ppm Au and 39.55% Fe were measured for a pyrite sample.

Rifai et al. [47] determined the Au contents of 43 rock samples and 44 synthetic pressed powder reference materials of different composition, which had quasi-homogeneous Au concentrations between 0 and 1000 ppm based on a calibration curve developed using the Au spectral line at 267.59 nm. The bespoke LIBS system used for this study consisted of a 1064-nm Nd:YAG laser capable of delivering pulse energies of up to 600 mJ at a repetition rate of 10 Hz. The LIBS spectral emission in air was recorded by a Czerny–Turner spectrometer equipped with a grating of 2400 lines/mm blazed at 300 nm that covered the wavelength range of 180–650 nm. An ICCD camera with a resolution of 12 pm/pixel at 267 nm was coupled to the spectrometer to complete the LIBS system. Chemometric data processing using principal component analysis indicated that ~83% of the LIBS spectral variation was attributable to the presence of Fe in the samples. The calibration curve was characterized by two distinct branches, one for Si-rich samples and the other and for Fe-rich samples, with limits of detection of 0.8 ppm and 1.5 ppm, respectively. A detailed mapping of the Au content in a solid core sample revealed heterogeneity of the Au distribution at multiple spatial scales.

In a set of companion papers, Diaz et al. [53,78] first developed an analytical protocol for the quantification of Au and Ag in pressed pellets of surrogate samples prepared by the doping of an Au-free particulate matrix with Au and Ag standard solutions and naturally occurring Au- and Ag-bearing ore samples from a Colombian mine and then investigated the effects of laser wavelength and irradiance on gold analysis. The local geology in the vicinity of the mine consisted of mesothermal to epithermal gold-bearing quartz veins hosted in zones of faulting/shearing in the local granitoid intrusive and metamorphic rocks. Gold in ore samples was present as free native gold, electrum, and in association with the sulfide minerals pyrite, chalcopyrite, and galena. A bespoke laboratory LIBS system consisting of an Nd:YAG laser operating at 450 mJ, with a 6-ns pulse duration and a 5-Hz pulse repetition rate, was used to produce the LIBS plasma in air, with the plasma emission recorded by a 300-mm-focal-length, 10-µm-slit Czerny–Turner spectrometer. The effect on Au quantification was investigated at laser wavelengths and irradiances for 355 nm from 0.36–19.9 × 10^9 W/cm^2 and from 0.97–4.3 × 10^9 W/cm^2 at 1064 nm, respectively. Calibration curves behaved linearly for all wavelengths and irradiances, samples, and analytes for concentrations from 1–9 ppm Au in the surrogate samples and 0.7–47.0 ppm Ag in ore samples, but it was not possible to develop calibration curves for the Au-doped surrogates below 1 ppm and at any concentration for ore samples. Detection and intensity measurement of Ag was straightforward using the spectral line at 328.06 nm for concentrations from 0.4–43 ppm in ore samples. The spectral emission line for Au at 267.59 nm was not observed after accumulation of 100 single-shot LIBS spectra of ore samples with Au contents of up to 9.5 ppm, but quantification was successfully achieved at concentrations as low as 1 ppm after 5000 laser shots. It was noted that Ag quantification was accomplished at a limit of detection of 0.4 ppm without regard to matrix composition, but that the matrix effect is an important consideration for Au quantification, as different calibration slopes were observed for the surrogate samples that were prepared from quarry sands of similar chemical composition to the ore matrix at limits of detection between 0.8 and 2.6 ppm.

3. Analytical Methods

As noted for specific instances in the discussion that follows, with one exception, the results presented were obtained on either an RT100-HP laboratory LIBS system manufactured by Applied Spectra Inc. (West Sacramento, CA, USA), or one of the Z-series handheld analyzers manufactured by SciAps Inc. (Woburn, MA, USA). One of the premier attributes of LIBS analysis is that minimal to no sample preparation is required. Generally, it is common practice to use the first few laser shots to clean the surface of a mineral or section of core as it is scanned, but this is not necessary if a freshly exposed surface is analyzed and may not be desirable for soil analysis.

The RT100-HP is a versatile commercial laboratory LIBS system that consists of a 50-mJ Nd:YAG laser operating at 1064 nm with a 5-ns pulse width and 1–20-Hz variable repetition rate, a Czerny–Turner spectrograph, and a high-performance ICCD detector. Operational parameters that can be controlled include laser power, gate width, and signal acquisition delay time. The spectrograph/ICCD detector has a dual grating turret, involving 600 grooves/mm for low-resolution analysis and 2400 grooves/mm for high-resolution analysis, respectively providing 0.2–0.3-nm and 0.05–0.1-nm spectral resolution across a 190–1040-nm spectral window to produce composite LIBS spectra with over 12,000 data points. The wavelength coverage is adjustable, with the low-resolution option allowing for a 230–260-nm wavelength range and the high-resolution setting providing a 35–50-nm range. In addition, a broadband option is available in which the six gated Czerny–Turner spectrographs are coupled with high-performance CCD detectors to provide coverage from 187–1044 nm at a resolution of 0.055–0.068 nm. The RT100-HP has an automated 3D translational stage that permits data to be collected over a user-defined grid pattern at 0.5-mm spacing, with a guide laser for ablation spot location, complementary metal–oxide–semiconductor (CMOS) camera imaging, and active focusing system that automatically refocuses the laser onto the surface each time the stage is moved to a new surface location.

Handheld LIBS analyzers are well suited to geochemical applications [65,79,80], particularly for real-time analysis in the field. The work reported here utilized the Z-300 LIBS analyzer or its Z-500 predecessor. Both instruments have a built-in camera for beam targeting, a translational stage for 3D beam rastering across the sample surface, the capability to flow an inert gas across the sample surface, and broad spectral ranges of 190–950 nm for the Z-300 series instruments or the more limited 180–675 nm range for the Z-500 series. Typical detection limits for the Z-300 are in the tens to hundreds of ppm range for most elements when averaged across the field of sample collection points, which is possible with its rastering capability. The binary presence or absence of specific elements, such as Au, which are typically heterogeneously distributed within samples, can be observed at far lower concentrations especially when using high-density rastering [65].

The Z-300 uses a proprietary PULSAR™ 1064-nm Nd:YAG pulsed laser with a 50-μm focused beam size. This diode-pumped solid-state laser delivers 6 mJ to the sample at a 1-ns pulse duration and 10-Hz firing rate. The instrument is capable of delivery of Ar or He purge gas directly to the sample surface. Small, replaceable gas cartridges that fit in the handle of the Z-300 can be used for field-based analyses. Alternatively, the Z-300 can be connected to a larger gas cylinder for stationary use. It is also possible to operate the analyzer in air without purge gas. If the purge gas is selected, flow is started shortly before the laser pulse train to displace any air present and ceases after the last shot to minimize gas consumption. The Z-300 analyzer is equipped with a 3D translation stage that is computer-controlled for automatically adjusting the laser focus on each sample location. Automated stage movements allow analysis over a raster pattern of up to a 2 × 2 mm area. The raster pattern, its internal spacing, and number of laser shots can be customized by the user. Multiple laser shots can also be collected at each location to improve data quality. Non-analytical cleaning shots, i.e., firing of the laser during which no data are collected, can be performed prior to the collection of data to remove surface coatings or tarnish. The resultant light emission signal from the micro-plasma is collected and the light is passed by fiber-optic cable into three internal spectrometers covering the wavelength intervals of 190–365 nm, 365–620 nm, and 620–950 nm. The Z-300 is equipped with time-gated CCD detectors with a resolution of 0.1 nm full width at half maximum (FWHM) below 365 nm and 0.3 nm above 365 nm. Data are typically collected with 1-µs delay times (i.e., the time between the laser pulse and beginning of emitted light collection and integration) over a 1-ms integration time, although data can be collected in a non-gated state or with variable gate delays from 250 ns up to 100 ns in 20.8 ns increments. This produces composite LIBS spectra comprising more than 23,400 data points.

The SciAps LIBS analyzers used in our work employ an Android operating system with a graphic user interface (GUI), and they are operated in the field by rechargeable Li-ion batteries that provide up to 10 h of operation or in the laboratory by connecting to alternating current (AC) power. Acquired spectral data can be stored on the analyzer and accessed via GUI, downloaded to a local computer via Universal Serial Bus (USB) or Wi-Fi, or emailed using a mail account set-up on the instrument when Wi-Fi is connected. There are two different methods for geochemical analysis: (i) a GeoChem mode that provides a quantitative analysis based upon a previously developed empirical calibration curve, and (ii) a GeoChem Pro mode for qualitative analysis that can both identify the spectral peaks of specific elements and generate estimated elemental concentration maps based upon the relative intensities of selected elemental peaks across a raster pattern.

4. Results and Discussion

4.1. Elemental Detection and Qualitative Analysis

Geological materials have an extremely wide range of composition, grain size, texture, and surface roughness. These features can be highly variable in the spatial scale and, thus, affect the LIBS analysis. For example, whether a sample surface is polished and smooth versus naturally rough determines the degree of laser energy coupling to the surface and consequent strength of the LIBS signal produced. Most rocks consist of multiple minerals. These minerals are of different hardness with variable grain size, and the minerals themselves may have interstitial phases, micro-fracture fillings, grain-boundary precipitates, or solid and liquid inclusions of different composition. Many minerals, particularly those in silicate rocks, are compositionally zoned. Soils and poorly indurated rocks may be aggregates of particles whose response to the laser pulse is unlikely to yield a reproducible plasma from one laser pulse to another. Variations in the intrinsic characteristics of these materials affect the development of the plasma and the resultant spectral emission intensity. For example, variations in the grain size or moisture of a soil sample can influence the detection limit of an element [69]. Thus, in many applications, where elemental abundance information is sought, it is common practice to aggregate the light emission signal from hundreds or thousands of laser pulses as a means of mitigating shot-to-shot variations in emission signal intensity arising for the reasons previously discussed. However, there is an increasing interest within the geosciences community to use single-pulse, qualitative LIBS elemental detection and material identification and discrimination.

Elemental detection is readily accomplished with LIBS by monitoring the spectral position of emission lines in the LIBS spectrum, so that the chemical species present in a sample can be readily ascertained by identifying the different peaks in an LIBS emission spectrum (Table 1, Figure 2). A broadband LIBS spectrum can be considered a *geochemical fingerprint* for a sample because all elements emit over the 200–900-nm range of typical LIBS signal monitoring. The concept of geochemical fingerprinting holds that minerals form in certain structures according to sets of well-understood rules (i.e., mineral stoichiometry) and that the chemical composition of a mineral or rock reflects the geological environment and processes associated with its formation [81]. In an LIBS context, geochemical fingerprinting uses the totality of chemical information contained in a broadband emission spectrum to provide a qualitative compositional comparison and discrimination amongst a group of related samples through chemometric analysis of their LIBS spectra [69,82–86]. As noted by Harmon et al. [79], although elemental identification and quantification is a primary LIBS capability, geochemical fingerprinting by LIBS can also be readily used (i) to discriminate between rocks and minerals of similar appearance but different composition, (ii) for stratigraphic correlation of volcanic, sedimentary, or metamorphic rocks, and (iii) to determine geomaterial provenance when used in conjunction with statistical chemometric data processing techniques.

Table 1. Examples of useful spectral emission lines (wavelengths in nm) for elemental identification in common rock-forming minerals (blue columns) and ore deposits (green columns).

Element	Wavelength	Wavelength	Wavelength	Element	Wavelength	Wavelength	Wavelength
O	777.42	794.76	844.64	Ag	328.07	520.91	338.29
Si	288.16	251.61	390.55	Au	267.59	242.80	312.28
Al	309.30	394.40	396.15	Co	238.89	389.41	258.04
Mg	2795.5	383.33	279.80	Cr	425.44	427.48	284.33
Fe	259.94	259.84	438.35	Cu	521.82	324.75	578.21
Ca	393.37	396.85	422.67	Mn	478.34	482.35	602.48
Na	589.99	589.59	330.24	Mo	267.28	379.89	268.41
K	766.49	769.90	404.72	Ni	239.45	241.63	300.25
Li	670.78	610.35	812.62	Pb	405.78	438.65	363.96
B	249.77	249.68	208.96	Pt	265.95	214.42	224.55
Rb	780.03	794.76	-	Sn	380.10	283.99	317.50
Sr	430.54	407.78	460.73	Ti	334.94	375.93	376.13
Ba	452.49	614.17	389.18	Zn	472.22	481.05	328.23

Sample discrimination using qualitative LIBS data is a fundamentally different task than using LIBS to quantify the absolute abundance of an element of interest in a sample. The discrimination/classification task relies on chemometric approaches using multivariate statistics to develop a set of mathematical *features* to characterize each sample in a population. For LIBS, these features are the plasma emission intensities at each wavelength, as illustrated in the broadband spectra of Figure 2. For classification/discrimination, it is required that that each feature must be the same for every input to the classifier. Therefore, for example, if the 100th feature in an LIBS spectrum for one sample is the Pb emission line intensity at 405.78 nm in the spectrum, then the 100th feature for all samples must be that same emission line intensity. Classification approaches develop statistical models for the behavior of each feature in a dataset and, then, based on these models, determine which features are most important for the discrimination between the classes of samples to be labeled, rather than using a priori knowledge of the elements that are likely to distinguish samples collected from different sources. Once a classifier is trained on a dataset that links samples to their respective classes, the features the classifier used to make its decisions can be identified, and knowledge can be obtained about the compositional differences by which samples were discriminated. Then, by creating representative libraries of LIBS spectra from samples of known origin, one can readily develop classifier models that allow for identification and provenance attribution of samples of unknown origin. This is the approach utilized in the examples for the carbonate minerals, garnet, cassiterite, columbite–tantalite, and gold discussed below.

4.1.1. Chemometrics

Principal component analysis (PCA) is an unsupervised statistical analysis technique that reduces the complexity (i.e., dimensionality) of multidimensional compositional data by finding linear combinations of variables (i.e., principal components) that explain the differences between samples. This type of exploratory data analysis provides a graphical representation of the natural grouping of samples and highlights which variables (i.e., emission wavelengths) most strongly influence sample class differentiation. PCA scores are linear combinations of the original variables and describe how the samples relate to each other, whereas PCA loadings contain information about how the variables relate to each other. The first few PCA loadings usually explain most of the covariance observed between samples.

Partial least squares discriminant analysis (PLSDA) is a supervised inverse least squares technique in which an algorithm with predictive latent variables is created that maximizes the variance between input variables. In the case of LIBS analysis, the object is to differentiate samples on the basis of their emission spectra. Once the algorithm is trained on a sample of known origin, the model can then be used to predict the probability of an LIBS spectrum from a new sample belonging to a previously identified class. PLSDA results can be displayed in the form of a "classification matrix", in which the model classes represent the different categories (e.g., a geographic identifier such as the sample, mine, or area location), to which each LIBS spectrum was assigned prior to building the PLSDA model. The algorithm then places each of the spectra being evaluated into one of the model classes. Correct classification corresponds to the values found along the diagonal, whereas values in off-axis cells correspond to misclassifications. PLSDA was proven to be a particularly effective technique for geochemical fingerprinting because it maximizes the inter-class variance (i.e., provenance) whilst minimizing the intra-class variance (i.e., shot-to-shot variability).

The necessary foundation for turning a successful geochemical fingerprinting study for lithologic correlation or mineral provenance determination into a practical application utilizing supervised chemometric analysis (e.g., PLSDA) for use in the field during an exploration campaign is the prior development of a robust spectral library of the geomaterial of interest, be it a specific mineral, soil class, or fresh/altered rock type. The spectral library would be pre-loaded onto a handheld LIBS analyzer, and then unknown samples could be interrogated in the field in real time to validate or refute a correlation or ascertain provenance through spectral matching.

4.1.2. Metal Carbonates

LIBS analysis of carbonate minerals, a class of minerals in which a divalent metal ion is coordinated by the CO_3^{2-} carbonate molecule, provides an illustrative example of the concept of geochemical fingerprinting. Carbonate minerals have a flexible crystal structure that is hexagonal rhombohedral, orthorhombic, or monoclinic. Hexagonal rhombohedral forms result when anionic CO_3^{2-} groups are combined with small divalent cations with ionic radii <1 Å, e.g., $MgCO_3$, $ZnCO_3$, and $MnCO_3$, whereas large divalent cations with ionic radii >1 Å produce orthorhombic forms (e.g., $BaCO_3$, $SrCO_3$, and $PbCO_3$). Ca^{2+} has an ionic radius of ~1 Å and, as a consequence, occurs in both hexagonal rhombohedral and orthorhombic polymorphs as calcite and aragonite. The crystal form of dolomite, $CaMg(CO_3)_2$, is similar to that of calcite, $CaCO_3$, except that there is regular alternation of Ca and Mg ions in a rhombohedral structure. The hydrous copper carbonate minerals azurite and malachite, $Cu_3(CO_3)_2(OH)_2$ and $Cu_2CO_3(OH)_2$, have a monoclinic structure. Carbonate minerals are commonly formed during sedimentary, hydrothermal, metamorphic, and weathering processes because of the nearly ubiquitous presence of the carbonate molecule in the Earth's surface and near-surface environments.

Broadband LIBS spectra for a suite of 11 carbonate minerals were obtained in air using a bespoke laboratory LIBS system similar to that shown in Figure 1. As illustrated in Figure 3, the compositionally different carbonate minerals are not only clearly differentiated in the 10-shot average spectra acquired across the spectral range of 180–680 nm on the basis of their primary cationic constituent (Mg, Zn, Mn, Ca, Sr, Pb, Ba, and Cu), but also exhibit a high degree of compositional variability due to the broad range of ionic substitutions possible in both the rhombohedral and orthorhombic isostructural groups. As expected, the geochemical fingerprints for the copper carbonate minerals azurite and malachite are largely indistinguishable. Although Cu is the dominant cation in these two minerals, Cu was also detected in smithsonite in low abundance, which is not surprising because smithsonite occurs in weathered hydrothermal ore deposits, which are common sources of both Cu and Zn. Although the LIBS spectra for the $CaCO_3$ polymorphs calcite and aragonite are quite different in terms of signal intensity, the same peaks are present in both spectra and, therefore, the two samples are characterized by the same overall geochemical fingerprint. As expected, the C lines at 193.09 and 247.86 nm are ubiquitous, but characterized by very reduced intensities because C has a relatively high ionization energy.

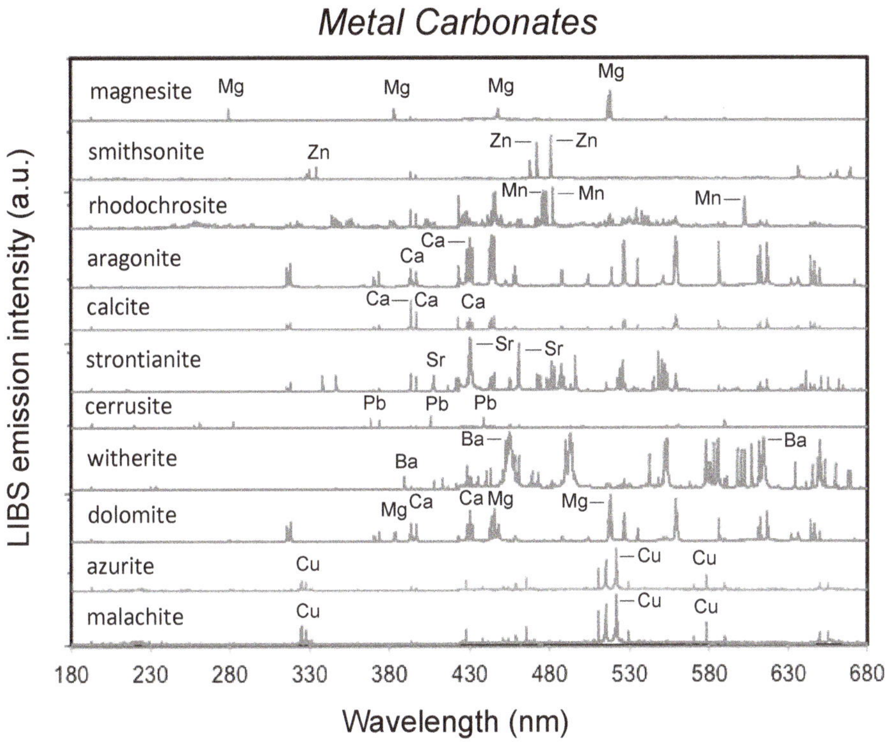

Figure 3. Single-pulse LIBS spectra acquired with a SciAps Z-500 handheld analyzer for a suite of divalent metal carbonate minerals (data from Reference [79]). From top to bottom: magnesite—$MgCO_3$, smithsonite—$ZnCO_3$, rhodochrosite—$MnCO_3$, aragonite—orthorhombic $CaCO_3$, calcite—hexagonal $CaCO_3$, strontianite—$SrCO_3$, cerrusite—$PbCO_3$, witherite—$BaCO_3$, dolomite—$CaMg(CO_3)_2$, azurite—$Cu_3(CO_3)_2(OH)_2$, and malachite—$Cu_2CO_3(OH)_2$.

4.1.3. Garnet

The study of the gem mineral garnet by Alvey et al. [83] provides a straightforward demonstration of multivariate statistical analysis for LIBS spectral data for classification and discrimination. Garnets, which occur in a broad spectrum of mantle and crustal lithologies and can be a pathfinder mineral for diamond kimberlites, are orthosilicate mineral silicates of widely varying major element composition based upon the general chemical formula $X_3Y_2Si_3O_{12}$, where "X" is a divalent metal (typically Mg, Fe, Mn, or Ca) and "Y" is a trivalent metal (typically Al, Fe, or Cr). The principal mineral end-member compositions are pyrope ($Mg_3Al_2Si_3O_{12}$), almandine ($Fe_3Al_2Si_3O_{12}$), spessartine ($Mn_3Al_2Si_3O_{12}$), grossular ($Ca_3Al_2Si_3O_{12}$), andradite ($Ca_3Fe_2Si_3O_{12}$), and uvarovite ($Ca_3Cr_2Si_3O_{12}$). Members of the garnet mineral group have the same cubic crystal structure, but vary widely in chemical composition and, therefore, also exhibit ranges in many of their physical properties including color. Garnet is rarely found in nature in a colorless state, but commonly occurs in any and all colors. Garnet is allochromatic, which means that color variations in different species of garnet are due to their highly variable trace element impurities rather than to the major, mineral-forming elements that define their bulk composition. Garnets of the same type can have different colors, and different types of garnet can have the same color; thus, attempting to discriminate on the basis of color alone is not feasible, although it is the most common basis for classifying garnets in the field or when chemical analysis is not available.

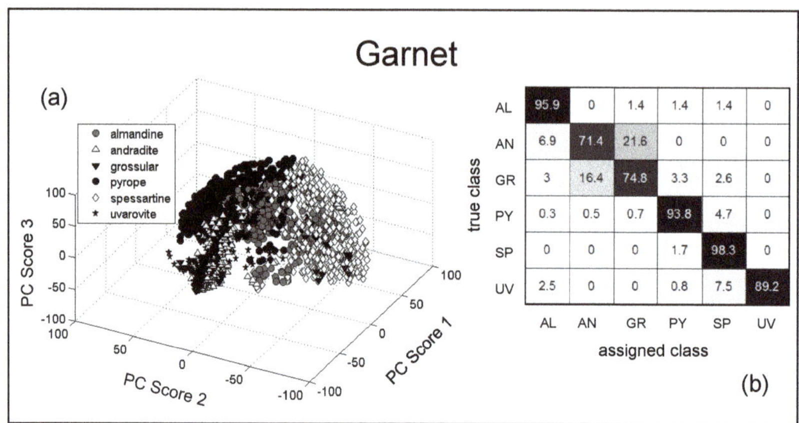

Figure 4. Principal component analysis (PCA) score plot (**a**) and partial least squares discriminant analysis (PLSDA) classification matrix (**b**) for the analysis of the common garnet groups using a SciApS Z-500 handheld LIBS analyzer (data from Reference [79]). AL = almandine ($Fe_3Al_2Si_3O_{12}$), AN = andradite ($Ca_3Fe_2Si_3O_{12}$), GR = grossular ($Ca_3Al_2Si_3O_{12}$), PY = pyrope ($Mg_3Al_2Si_3O_{12}$), SP = spessartine ($Mn_3Al_2Si_3O_{12}$), and UV = uvarovite ($Ca_3Cr_2Si_3O_{12}$).

Alvey et al. [83] acquired single-shot broadband LIBS spectra in air over the spectral range of 200–960 nm for a suite of 157 garnets of different composition collected from 92 locations worldwide using a bespoke LIBS system employing a Big Sky CFR200 laser (Big Sky Laser Technologies Inc., Bozeman, MT, USA) and Ocean Optics LIBS2500 (Ocean Optics, Inc., Delray, FL, USA) spectrometer system. Discrimination of garnet compositional groups was accomplished by PLSDA analysis at an overall success rate of >95% based on sets of 25 broadband LIBS spectra for each sample. The LIBS spectral wavelengths of the main cations that differentiated the different garnet species (Mg, Fe, Mn, Al, Ca, and Cr) determined the garnet compositional classification. A cross-validation procedure was also used to ensure the robustness and performance of the classification algorithm, so that it could be used in the future as a spectral library to determine the type and geographic location of an unknown garnet specimen. In a subsequent study, Harmon et al. [79] using a SciAps Z-500 handheld analyzer and a much larger sample set of 288 garnets consisting of 66 almandines, 49 andradites, 60 grossulars, 35 pyropes, 56 spessartines, and 17 uvarovites, observed that the compositional differences characterizing the six garnet compositional groups were manifest on a three-component PCA plot (Figure 4a) and that the overall success for garnet type discrimination by PLSDA was >90% (Figure 4b). Fe–Al almandine, Mg–Al pyrope, and Mn–Al spessartine were distinguished from the other garnet groups at a success level better than 95% and Ca–Cr uvarovite at 89%. Significant misclassification only occurred for Ca–Fe andradite and Ca–Al grossular garnets, which commonly form a solid solution series in crystalline rock of intermediate compositions. Some 22% of the andradites were classified as grossular and, conversely, about 12% of grossular were classified as andradite. Metamorphic garnets could be distinguished from igneous garnets at a success of better than 80% through PLSDA (Figure 5a) and, within the igneous garnet type, garnet from granitic pegmatites and rhyolites, chromitite pods, and kimberlites were readily differentiated at a classification success of 92.7% (Figure 5b). Additionally, LIBS analysis of a small suite of garnets from South Africa supported the idea that pyrope from kimberlites might be distinguished by locality. Pyrope from South Africa was correctly distinguished from all other pyropes at about 90% success based on differences in Na, Mg, Ca, Al, Mn, Fe, and Cr emission line intensities. Likewise, as shown in Figure 5c, diamond-bearing kimberlites from within South Africa were confidently differentiated at an overall success of >96%, with the same seven elements responsible for the discrimination. This result has important implications for diamond kimberlite exploration, as it offers the potential for in-field analysis of garnet and, by analogy, clinopyroxene, which would permit the rapid identification kimberlite pathfinder minerals on a daily basis by a field exploration team.

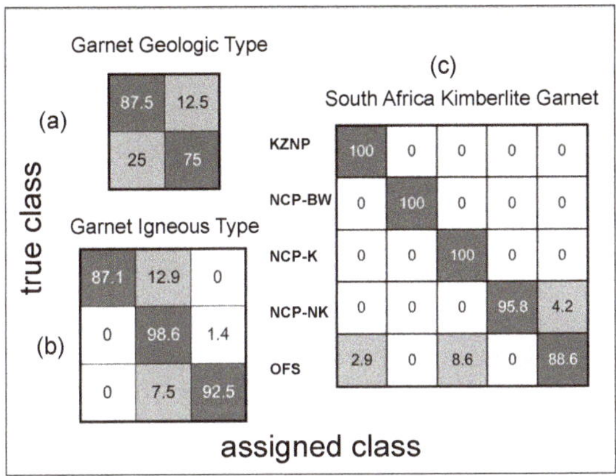

Figure 5. PLSDA classification matrices from handheld LIBS analysis of garnets by (**a**) geological type and (**b**) igneous host type, and (**c**) for garnets from five South African kimberlite locations (after Harmon et al. [79]): IGN = igneous, MET = metamorphic; CP = chromitite pods, GPR = granitic pegmatites and rhyolites, and K = kimberlites; KZNP = KwaZulu-Natal Province, NCP-BW = Barkley West, in Northern Cape Province, NCP-K = Kimberley in Northern Cape Province, NK = North of Kimberley in Northern Cape Province, and OFS = Orange Free State (data from Reference [79]). See text for discussion.

4.1.4. Oxide Minerals

The identification of oxide minerals can pose a challenge to the field exploration geologist. The two examples below demonstrate how LIBS can be used for their rapid identification and source discrimination.

Cassiterite (SnO_2) is the primary ore for tin, an element with high industrial demand for use as a corrosion-resistant coating on other metals, in electrically conductive coatings, as a fire-retardant in plastics, in anti-fouling paint for marine vessels, and in the alloys used for soft solder, pewter, bronze, and superconducting magnets. Cassiterite occurs naturally in hydrothermal veins associated with S-type granites, but is most commonly found as alluvial and placer accumulations because of its resistance to weathering. The increase in demand and relatively large supply of cassiterite in the underdeveloped parts of central Africa contributed to its illicit mining and trading as a "conflict mineral". Determining the chemical composition of an ore is one means of illuminating its provenance, and ascertaining whether or not it comes from an area of civil conflict. Geochemical fingerprinting is a means of rapidly ascertaining the provenance of an ore based on the unique crustal signature associated with the location of its formation.

Hark and Harmon [84] analyzed 38 cassiterite ore samples from South America and southeastern Asia using a Applied Spectra RT100-HP laboratory LIBS system to determine if the geographic origin of cassiterite could be identified using LIBS geochemical fingerprinting. LIBS spectra were obtained in air and at low laser power (~9 mJ) at a spectral resolution 0.2–0.3 nm over three wavelength regions (220–440, 460–700, and 680–910 nm) using a laboratory RT100-HP LIBS system. Four laser pulses at each location on a 5 × 5 grid were acquired to generate 100 spectra for each sample, which were chemometrically processed by a PLSDA classifier that employed 120 components and 10-fold cross-validation to build a robust model that provided an overall correct sample-level classification rate of 97%. When the sample suite was grouped according to the 11 locations from which the ore concentrate was mined, the PLSDA model gave a correct overall classification rate of 87% (Figure 6).

CASSITERITE

True Class / PLSDA Model Class

	Brazil	Bolivia	Peru	Indonesia	Malaysia	Thailand
Brazil	99	0	0	0	0	1
Bolivia	3	88	0	2	6	1
Peru	0	3	94	0	2	1
Indonesia	0	0	0	99	0	1
Malaysia	0	1	0	0	99	0
Thailand	0	1	0	0	0	99

Figure 6. PLSDA classification matrix for LIBS spectral analysis of 38 cassiterite samples from six countries generated using 80 components and 10-fold cross-validation (after Hark and Harmon [84]).

The rare metals Nb and Ta are currently elements of high economic importance, as the high demand for these elements as components in modern electronics and medical devices drives exploration programs to locate new ore deposits. As the most important carriers of Nb and Ta in granites and granitic pegmatites, minerals of the columbite group, $(Fe,Mn)(Nb,Ta)_2O_6$, were long considered economic commodities. Columbite, the niobium-dominant member of the group ($FeNb_2O_6$ to $MnNb_2O_6$), and tantalite, its tantalum-rich analogue ($Fe(Ta,Nb)_2O_6$ to $MnTa_2O_6$), are commonly grouped together under the appellation *coltan*. Coltan ore occurs in many areas worldwide, such as Brazil, Australia, and Central Africa, but its exploitation became politically problematic as one of the most prominent of the African "conflict minerals", because some 60% of the world's coltan reserves are located in the eastern portion of the Democratic Republic of the Congo and adjacent areas, from which its illicit export and sale to the North American, European, and Asian markets is thought to be an important means by which civil conflicts in Central Africa are financed [87].

Harmon et al. [88] used a laboratory RT100-HP LIBS system to acquire LIBS spectra in air over the wavelength range of 250–490 nm and high-resolution spectra over the more limited 235–285-nm and 313–358-nm wavelength ranges for 14 columbite–tantalite samples from three pegmatite fields in North America formed in widely distinct space and time (Figure 7a). These spectral ranges were chosen to encompass many of the intense emission lines for the major elements known to commonly substitute in the columbite-group minerals (Table 1). Before statistical processing, each spectrum was normalized by the sum of its emission intensity values so that each spectrum exhibited unity when summed across all wavelengths. Partial least squares discriminate analysis was used to successfully discriminate the provenance of the coltan samples at a success rate of >90%. In a follow-up validation study, Hark et al. [84] also used the same RT100-HP laboratory LIBS system to analyze a larger geographically diverse set of 57 samples from 37 granite pegmatite fields in Africa, Asia, Australia, North America, and South America, which is representative of the natural range of compositions for columbite-group minerals. Each sample group was unique in its geologic environment, mineralogical make-up, and geochemical character. This heterogeneity is reflected in the variety of major elements (Ta, Nb, Fe, Mn) and significant trace elements (e.g., W, Ti, Zr, Sn, U, Sb, Ca, Zn, Pb, Y, Mg, and Sc) known to commonly substitute in the crystal structure of the columbite-group minerals ferrocolumbite ($FeNb_2O_6$), manganocolumbite ($MnNb_2O_6$), ferrotantalite ($FeTa_2O_6$), and manganotantalite ($MnTa_2O_6$) [89]. Chemometric analysis using a PLSDA classification model with k-fold cross-validation achieved a correct place-level geographic classification at success rates between

90% and 100%, highlighting the potential of LIBS as a real-time field tool to discriminate the different provenance of coltan ore.

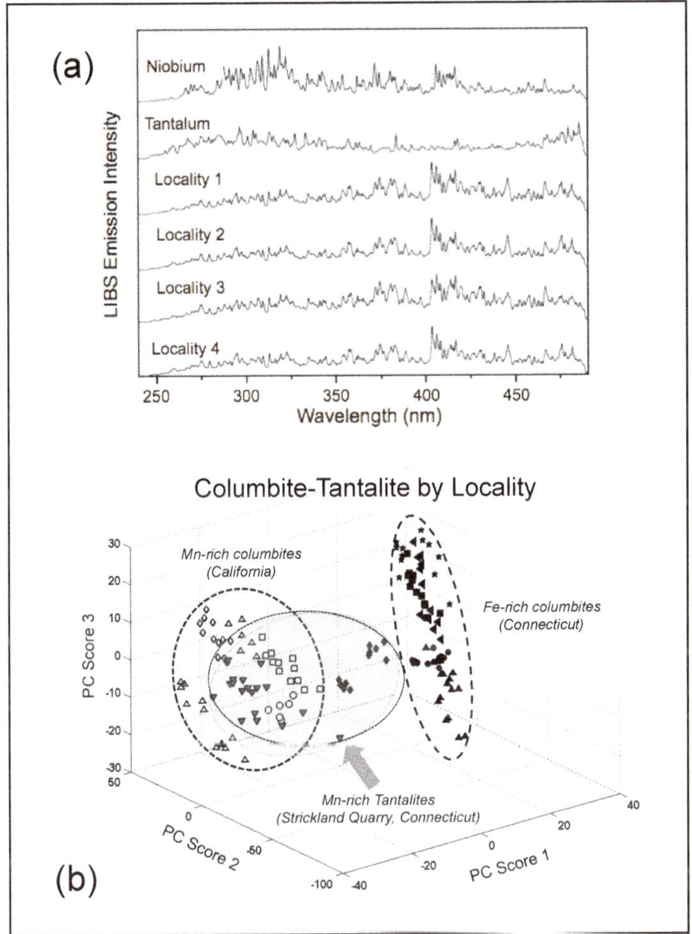

Figure 7. (**a**) LIBS spectra from 250–490 nm of pure niobium and tantalum, together with representative spectra of columbite–tantalite samples from (1) the Starrett pegmatite in southern Maine, (2) the Pack Rat pegmatite in San Diego County, California, (3) the Moose pegmatite in the east–central Northwest Territories, Canada, and (4) the BeeBee pegmatite in San Diego County, California. (**b**) PCA score plot for columbite–tantalite samples from LCT (Li, Cs, and Ta; see text) pegmatite locations at Branchville, Glastonbury, Haddam, Middletown, Portland, and the Strickland Quarry in Connecticut and mines at El Molino, Ingram, Katerina, and Olla in California (modified from Harmon et al. [79]). See text for discussion.

Columbite-group minerals occur most frequently in granitic pegmatites of the three geochemical families defined by Černý and Ercit [89]—(i) the NYF pegmatites characterized by progressive accumulation during magmatic crystallization and fractionation of subaluminous to metaluminous A- and I-granites of Nb, Y, and F in addition to Be, rare-earth element (REE), Sc, Ti, Zr, Th, and U, (ii) a peraluminous LCT pegmatite family typified by prominent accumulation of Li, Cs, and Ta in addition to Rb, Be, Sn, B, P, and F predominantly associated with S-type granites, and (iii) and a mixed family of diverse origins that can involve contamination by undepleted supracrustal lithologies. Pegmatites of the LCT family tend to host the greatest abundance of columbite-group minerals.

Based upon the two successful studies using the laboratory LIBS instrumentation, Harmon et al. [79] examined two suites of columbite–tantalite samples from LCT-type pegmatites in Connecticut and California (United States of America (USA)) by handheld LIBS using a SciAps Z-500 analyzer. These samples originated from three pegmatite districts, with those from Connecticut (CT) originating from the Middletown (Glastonbury, Haddam, Middletown, Portland mines, and Strickland quarry) and Redding (Branchville mine) pegmatite districts, whereas all of the California specimens (El Molino, Katerina, Ingram, Olla mines) were from the Pala district. The California sample suite consisted of Mn-rich columbites, whereas the Connecticut sample suite comprised Fe-rich columbites, with one exception. The samples from the Strickland location in Connecticut were distinct, comprising a range of Mn-rich tantalite compositions. PLSDA classification success for these samples was >99% with a single Middletown, CT spectrum incorrectly assigned to the Glastonbury, CT class. Loading weights indicated that the main elements responsible for discrimination of the samples, in relative order of importance, were Na, Ca, Li, Mg, Mn, Be, and Ta. The ability of the handheld LIBS instrument to identify the presence of elements with very low atomic number such as Li and Be is especially noteworthy as this capability does not exist in any other field-portable technique. Interestingly, all of the spectra for one Connecticut sample (Strickland #1) clustered with the four California classes in the PCA score plot (Figure 7b). Examination of the chemical data for these samples revealed that the classes were defined by their differences in Fe and Mn content. Qualitative analysis scanning electron microscopy/energy-dispersive X-ray spectroscopy established that all of the California samples were Mn-columbite whereas most of the Connecticut samples were Fe-columbite. The composition of the Strickland #1 sample was Mn-tantalite and, therefore, it was not surprising that it plotted in the domain of the Mn-rich California samples (Figure 7b). Conversely, chemical heterogeneity with compositions ranging from Mn-columbite to (Fe, Mn)-tantalite was recognized in the Strickland #2 sample, which explained why its LIBS data plotted between the Fe-rich and Mn-rich classes. These results reinforce the idea that LIBS has the potential to be utilized in the field as a real-time screening tool to discriminate columbite–tantalite ores from different granites and associated pegmatites based on chemical composition.

4.1.5. Gold

Native gold occurs as lode deposits in a wide range of geological settings, having well-defined characteristics and environments of formation [90]. Gold is common in placer accumulation deposits that develop during weathering and fluvial transport processes because it is dense and resistant to chemical weathering. In many areas, such as Yukon region of Canada and the eastern Alaska region of the United States, placer gold is common and widely distributed in river sediments, although the lode sources for most deposits remain unknown [91,92]. Because natural gold typically contains a variety of elements as trace constituents, for example Ag, As, Bi, Ca, Cu, Fe, Hg, Mg, Mn, Pb, Pt, Pd, Sb, Si, Te, Ti, and V [93,94], handheld LIBS could be an expeditious and efficacious means for elucidating the provenance of placer gold or differentiating between gold in a placer deposit derived from one or more lode sources.

Samples from native gold from 18 placer locations in New Zealand, Australia, and the USA (Alaska, Idaho, California, Colorado, Virginia, and North Carolina) were analyzed by Harmon et al. [79] using a SciAps Z-500 handheld LIBS analyzer to ascertain the potential for geochemical fingerprinting by LIBS. Initially, samples were consolidated into 11 groups based on geological affinity, and these domains differentiated by handheld LIBS with an overall success of 98.4%. From the PLSDA analysis, the most important spectral lines for the discrimination were the Au lines at 312.28, 523.03, 583.73, and 479.26 nm plus the Ag lines at 546.55, 520.91, 328.07, and 338.29 nm. The stacked spectral plot of Figure 8 displays clear differences in Au and Ag line intensities, with the portion of the spectra from 305–345 nm for pure Au and Ag and three representative samples illustrating how LIBS could be used to identify the relative amounts of Ag in a placer gold deposit and potentially determine provenance on this basis.

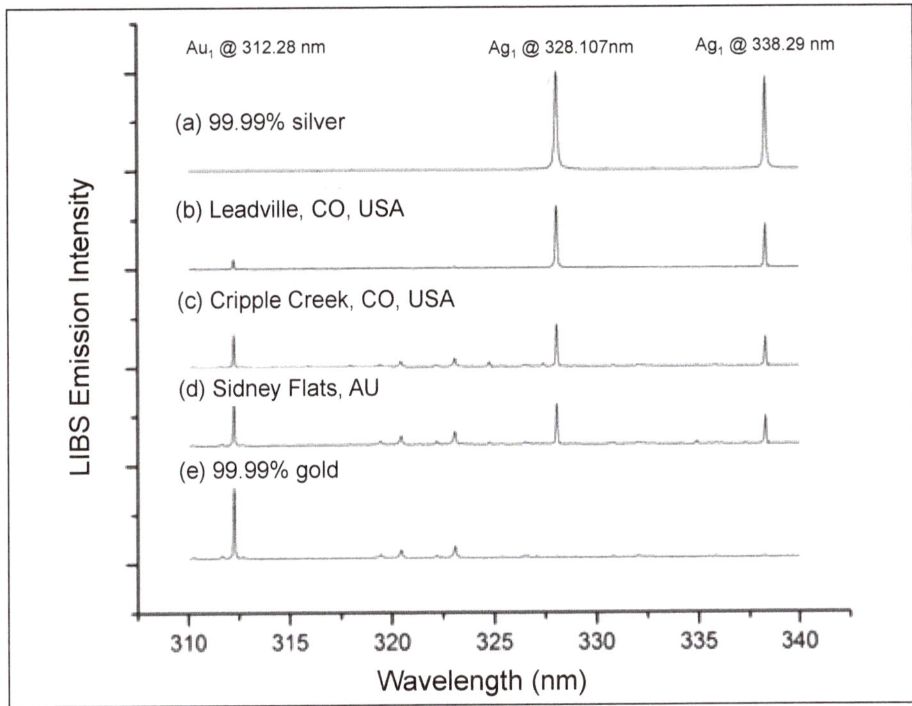

Figure 8. LIBS emission spectra between 305 and 345 nm for pure silver (**a**), placer gold samples from Leadville, Colorado, USA (**b**), Cripple Creek, Colorado, USA (**c**), and Sidney Flats, Australia (**d**), and pure gold (modified from Harmon et al. [79]). See text for discussion.

Chemometric analysis of the full LIBS broadband spectra from 180–675 nm resulted in >98% classification success. By contrast, consideration of just Au and Ag compositions using the 16 most prominent Au and Ag spectral lines resulted in a discrimination success for the 11 gold samples of better than 92.5%. These results indicate that discrimination was based on the relative amount of silver present. Not only does the classifier consider the presence or absence of loadings for Au and Ag, it also takes into account the magnitudes of the emission lines relative to eac h other. Geochemical fingerprinting is frequently based on variations in trace element composition; however, in this case, the major elements Au and Ag were sufficient for discrimination. These results suggest that handheld LIBS could be expeditiously utilized in the field to determine the fineness of alluvial gold, to differentiate gold from different lode deposits, and to match samples to their geological source after creation of a provenance library.

4.2. Quantitative Analysis

Quantitative analysis is possible using LIBS, as the intensity of the LIBS plasma emission is proportional to the concentration of an element in a sample. For solid samples, the character of the microplasma created is determined by the physical nature of the material being ablated, i.e., its composition, crystallinity, optical reflectivity, optical transmissivity, and surface morphology [68], as well as the operational characteristics of the laser (i.e., wavelength, energy, and pulse duration), the degree of laser energy coupling to the sample surface, and the ambient environment in which the

LIBS plasma is formed [95]. The elemental abundance in a sample can be quantified by measuring the intensity of the light captured at specific spectral wavelengths.

The use of LIBS for quantitative elemental analysis relies on some fundamental assumptions that must be verified, particularly that the plasma is optically thin and that a condition of local thermodynamic equilibrium is established within the plasma [26]. Quantitative measurement by LIBS is also complicated as a consequence of the transient nature of the LIBS plasma and matrix effects. Nonetheless, it is possible to quantify the elemental abundance in a sample by LIBS through a measurement of intensity of the light recorded at specific spectral wavelengths and then undertaking calibration against a reference material of the same type. As spectral emission intensity is not only influenced by the elemental concentration in the sample, but also by both the laser properties and the physical character of the sample, it is necessary to know how much mass is sampled by each laser pulse for quantitative analysis. Thus, standards must be homogeneous, and both the standard and the sample must be equally affected by the ablation, as it cannot be assumed that the standard and the desired element in the sample exhibit the same mass ablation rate behavior.

In the ideal case, when matrix-matched reference materials exist, LIBS can provide excellent quantitative results using single-element or multivariate calibration procedures. Typically, this involves the development of calibration curves for an element in the material of interest, and then LIBS measurements for the unknown samples are acquired under identical experimental conditions. Like all analytical methodologies, the ability to perform such quantitative analysis and its optimization are based on the quality of the standards. However, this becomes particularly problematic when reference standards do not exist, as is often the case with geological materials. There is also the possibility to use calibration-free (CF) approaches for semi-empirical analysis based on theoretical plasma models when matrix-matched standards do not exist [35,36]. However, there is still much research to be done on CF-LIBS before it can be routinely applied to the analysis of compositionally diverse and heterogeneous samples and, at present, calibration-free approaches are only possible for laboratory or mobile LIBS systems and not field-portable handheld LIBS analyzers.

The simplest form of quantitative LIBS analysis is the use of element ratios, which are readily calculated from pre-processed LIBS emission spectra because elemental concentrations in a sample are manifest in the observed spectral peak intensities. Gold deposits exhibit a wide range of Ag/Au ratios [96]. Harmon et al. [79] described an example of how this characteristic of placer gold deposits might be used in an exploration prospecting application.

As noted above, quantitative analysis using handheld LIBS is possible for many elements, provided that calibration curves for the type of geological material to be analyzed can be developed from fit-for-purpose calibration materials. Calibration standards must encompass the same element suite and target element concentration ranges that the analytical investigation is expected to encounter in the field. Ideally, the calibration standards would be matrix-matched to address the complicating phenomena discussed in the previous paragraph. Where matrix-similar certified reference materials (CRMs) cannot be used for the entirety of the calibration set, then reliably assayed samples from the specific investigation site or from another site with similar material can be used as secondary reference materials to augment the CRM set or for calibration in the absence of suitable CRMs. For example, a calibration of the kind described above was reported by Afgan et al. [97], who used a B&W Tek (Newark, DE, USA) NanoLIBS-Q handheld LIBS analyzer to develop calibration curves for C, Cr, Mn, Mo, Mo, Si, V, and Cu in steel (Figure 9). Using dominant factor-based partial least squares regression with spectral standardization resulted in respective average absolute measurement errors of 0.019%, 0.039%, 0.013%, and 0.001% for Si, Cr, Mn, and Ni, and the overall average relative standard deviation for quantitative analysis of the eight elements was less than 5%.

Unless the CRMs are pure metals (e.g., as in the example above), calibration materials should be very fine-grained, preferably <200 mesh (≤75 μm), and be pressure-pressed into pellets that remain fully consolidated during the ablation process (Figure 10). Binders should not be used, as incomplete mixing of the sample produces heterogeneous results. Typically, the LIBS analyzer would be moved to multiple locations across the pellet surface for analysis, and the data from each laser shot would be averaged in order to obtain a representative result. For example, to enhance the representativeness when using the SciAps Z-300 handheld analyzer, the default raster setting is for analysis of 12 individual locations in a 3 × 4 grid pattern at 500 m location spacing (Figure 10). Multiple shots can be collected at a sample site to improve the quality of analytical results. A set of similar matrix-matched quality control samples should be reserved for post-calibration validation to document satisfactory performance of the calibration.

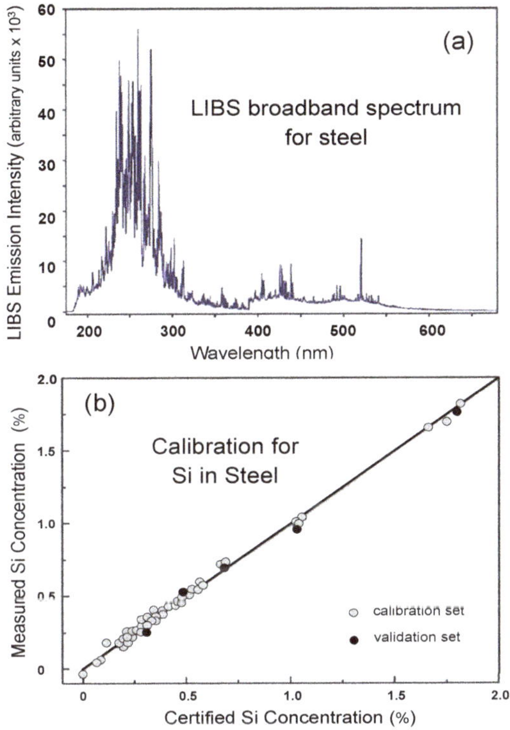

Figure 9. LIBS emission spectrum for a steel sample between 200–550 nm (**a**) and calibration/validation data for Si in different steels (**b**) using a dominant factor based PLS with spectrum standardization (modified from Reference [97]).

The basic approach to empirical calibrations using LIBS utilizes a concentration versus intensity ratio approach. Firstly, emission line intensities measured in a CRM are used to develop a calibration curve for each element of interest (Figure 11). The intensity values for each element are generated from selecting specific wavelength ranges in the LIBS spectrum to define regions of interest (ROIs), as shown in Figure 11a for sulfur in a sulfide mineral matrix. Secondly, a ratio is calculated for each ROI of the selected element. Wavelength intensity ratios can be calculated on an element-by-element basis for the calibration standards and unknowns or against a relatively flat background portion of the sample spectra. Following this ratio-based approach, the target element in the unknowns represents the numerator, and the calibration element or local spectrometer background region represents the denominator. Whilst concentration values for each of the target elements are required, ROIs in the denominator are not; thus, other elements of approximately constant composition (e.g., O, S, Si, Al, Ca, or Fe) may be used for spectral intensity normalization. Once a set of calibration curves are constructed (Figure 11b), each LIBS analysis displays elemental concentrations in a test sample in real time (Figure 12).

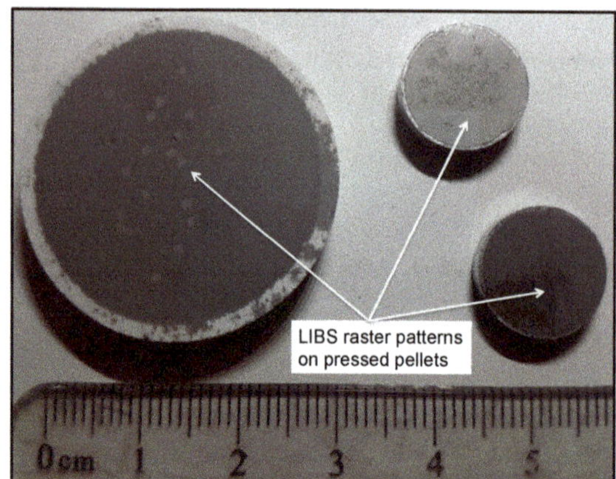

Figure 10. Example of pressed powder pellets showing a typical 12-position 3 × 4 LIBS analysis raster pattern.

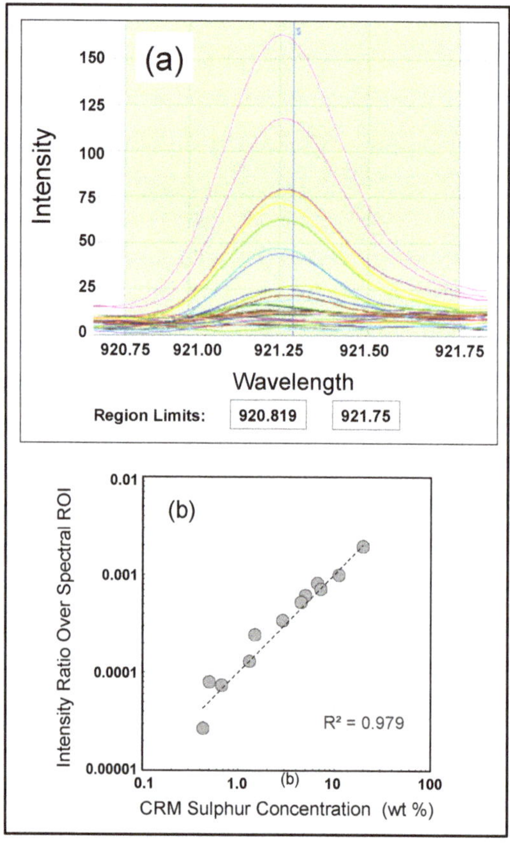

Figure 11. (**a**) SciAps Z-300 Geochem Pro screen display showing the LIBS spectra for S acquired from analysis of 15 Ore Research and Exploration Pty Ltd Assay Standards (OREAS) certified reference materials (CRMs) of known sulfur content over the spectral range of 920.82–921.75 nm, and (**b**) the resulting calibration curve. See text for discussion.

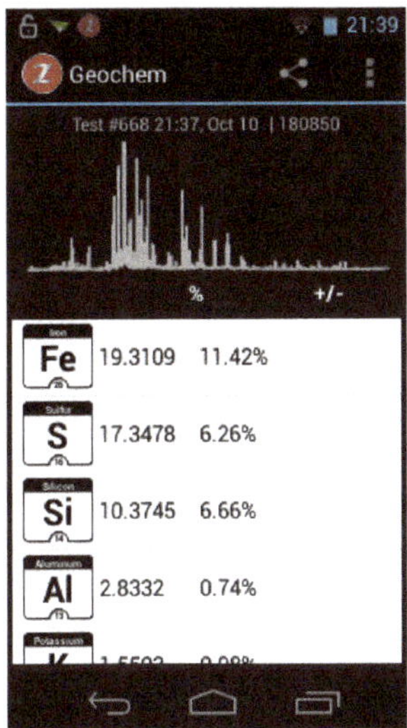

Figure 12. SciAps Z-300 screen display showing the quantitative analytical results for Fe, S, Si, and Al obtained after analyzer calibration.

Quantitative calibration curves can be established using the SciAps Z-300 handheld LIBS analyzer and its SciAps proprietary Profile Builder personal computer (PC)-based software package. This software utilizes a graphic user interface to facilitate the building of calibration curves by an operator through a pre-established workflow (e.g., Figure 11) Calibrations can be based upon existing factory templates, where much of the spectral pre-processing and ROIs are pre-defined for common applications or established by the user for a bespoke application. Selection of pre-loaded ROIs for many elements uses the SciAps proprietary LIBS spectral line library; alternatively, users can select from the National Institute of Standards and Technology (NIST) list of optical emission spectrometry lines, create custom element line lists, or create custom spectrometer regions. Users are guided through the following LIBS workflow: (i) selection of the element suite of interest; (ii) standard definition; (iii) standard acquisition; (iv) modeling of intensity ratio versus concentration plots for calibration. These models can then be saved to the analyzer using a USB or Wi-Fi connection to allow quantitative testing to be performed on the handheld analyzer. Once loaded onto the analyzer, basic slope and intercept corrections can be made to the calibration to make minor adjustments for field conditions. Finally, concentrations can be calculated for the target elements using the user-selected calibration method.

It should be noted that the use of univariate empirical calibrations comes with an inherent matrix specificity that needs to be considered when testing materials in the field using calibrations established under controlled conditions. For example, calibrations where S is not defined in the denominator would generally be unsuitable for the quantitative analysis of elements within sulfide minerals such as pyrite, arsenopyrite, chalcocite, or galena. LIBS analyzers often allow the choice of several different ROIs for elements so that different ROIs can be used for different matrix types to avoid interferences caused by constituents within the specific matrix other than those for which the calibration was developed.

The example below describes a field application of the process described above, including sample preparation, testing procedure, and results obtained using a SciAps Z-300 handheld LIBS analyzer during an exploration campaign at the Li-bearing hectorite clay-hosted Agua Fria prospect in the Sonora region of Mexico. The project campaign, as described by Griffin [98], was undertaken in 2017 by Lithium Australia NL (ASX: LIT) and Alix Resources Corp (TMX: AIX) and included examination of 16 reverse circulation drill holes, each completed to a depth of somewhat over 100 m.

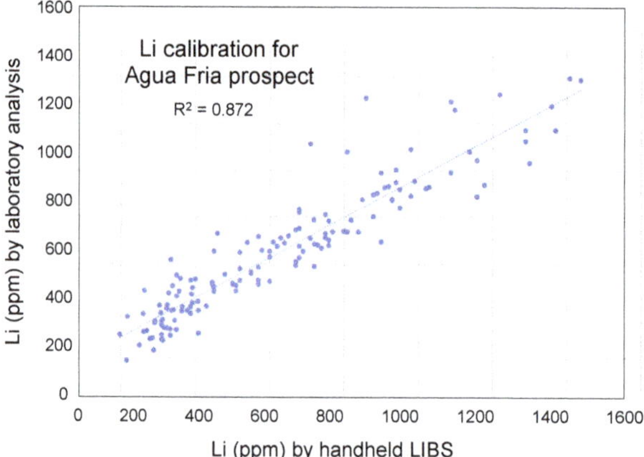

Figure 13. Comparison of Li concentrations for the Agua Fria Li prospect drilling samples by handheld LIBS and certified laboratory assay.

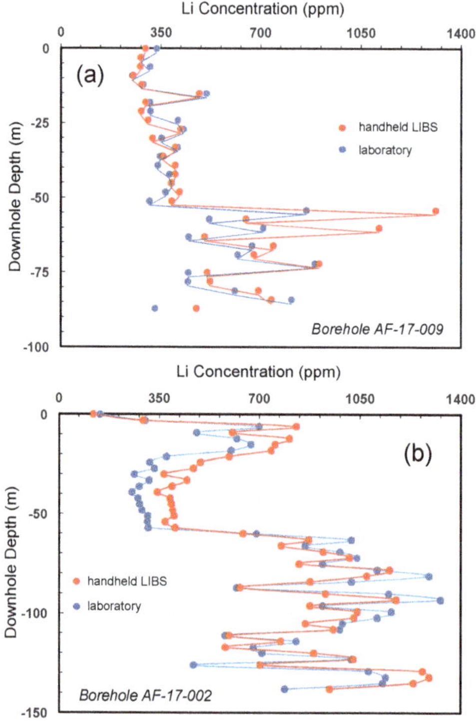

Figure 14. Comparison of Li concentrations for the Agua Fria prospect drill core by handheld LIBS and laboratory analysis for (**a**) drill hole AF-17-004, where the entire hole represents the calibrated clay matrix showing concordance between the laboratory and LIBS analyses, and (**b**) drill hole 17-002, where the upper 60 m of the hole intersects a non-calibrated basalt matrix; although the handheld LIBS still shows a similar overall pattern of downhole Li variation, a significant difference between the laboratory and LIBS analyses is evident. Analytical performance improves once the calibrated clay matrix is encountered at approximately 60 m depth.

Composite samples were collected at 3-m intervals and homogenized by riffle splitting prior to sampling. Aliquots from homogenized 3–5-kg samples were submitted for quantitative, multi-element ($n = 48$) assays using a four-acid digestion with an ICP-MS finish (Australian Laboratory Services (ALS) geochemical procedure ME-MS61). A smaller sample aliquot was selected for further analysis using a SciAps Z-300 handheld LIBS analyzer. Sample pellets were pressed on-site using a portable Reflex hydraulic press. LIBS measurements taking 3 s each to acquire a 3 × 4 raster pattern at 12 locations were averaged to produce a single composite LIBS spectrum for each pressed pellet. Prepared samples from the initial Li calibration set were tested routinely. Handheld LIBS Li analyses agreed well with Li laboratory results, as illustrated in Figure 13, with an observed R^2 of 0.86 for the suite of core samples analyzed. More scattering was observed at higher Li concentrations, but this did not result in incorrect classification of samples based on cut-off grades (i.e., higher-concentration samples still showed elevated levels of Li and, as such, were classified correctly as Li-rich samples). Some loss of accuracy was observed in cases when significantly different matrix types were encountered (Figure 14) or at the bottom of holes where moisture content was markedly higher [98].

4.3. Grain Size Analysis

4.3.1. Introduction

Successful mineral exploration involves not just the discovery of a sufficient concentration of the minerals of interest, but also a demonstration that these minerals can be extracted feasibly and economically. Numerous case studies demonstrated the important relationship between grain size and recovery [99,100]. Whilst many factors affect the recovery of economic minerals (including mineral texture and paragenesis), the size of the grains is a key parameter. Successful grain liberation is vitally important in extraction by techniques such as gravity separation, flotation, leaching, and magnetic and electrical separation, and it dictates the overall recovery [101]. If grain liberation is unsuccessful, recovery will be poor. Since grain size dictates crushing and grinding, which in turn affect overall liberation and recovery, grain size directly affects recovery. Understanding the processing behavior of ore minerals within a deposit is important when assessing the economic viability of an exploration project.

Grain size assessment is traditionally completed by reflected light microscopy or scanning electron microscopy (SEM) in conjunction with advanced mineralogical interpretation software packages [102–104]. Automated SEM systems equipped with advanced mineral identification software, such as the Mineral Liberation Analyzer (FEI Company, Hillsboro, OR, USA) and the quantitative evaluation of mineral by scanning electron microscopy (QEMSCAN, Hillsboro, OR, USA), are now commonly implemented in grain size assessment [102,104]. These systems identify minerals of interest, define the boundaries of the interested minerals, and then calculate the grain size of each mineral. These systems scan the entire sample surface, and typically cost hundreds of dollars per sample. This unit cost restricts the rollout of this type of analysis to a few tens to hundreds of analyses for a given rock volume.

The LIBS technique for in situ analysis captures an optical emission spectrum that reflects the relative proportions of elements at a particular spatial location on a sample. The proportions of copper in an LIBS broadband spectrum are linked to the size of the Cu-bearing minerals present at the spot of analysis. Therefore, LIBS data can be used to develop proxies for determining grain sizes from the LIBS spectral signature. A combination of spectral matching and peak integration techniques was applied to identify copper minerals and calculate grain size proxies from 26 samples. These proxy results were then compared to the traditional MLA grain size measurements collected from the same samples. To determine the minimum number of LIBS analyses required (and, therefore, most rapid analysis time) to produce a representative grain size proxy, a series of bootstrapping experiments was completed.

Quantification of elemental abundances from LIBS spectra is difficult due to a number of factors described above and discussed by Harmon et al. [69] and Senesi [105]. In particular, the interactions between the laser and the sample material are complex and difficult to quantify, there can be multiple atomic emission peak overlaps resulting in signal interferences, and the likelihood of encountering multiple minerals and, therefore, multiple elements in a geological sample is high. Therefore, instead of attempting to quantify exact copper concentrations for grain size assessment, a combination of spectral matching and peak integration was used to identify copper minerals and develop grain size proxies based on relative abundances of each mineral.

4.3.2. Methodology

Data Collection

LIBS spectra were obtained from 26 rock tiles from a porphyry Au–Cu deposit. Compositional grids measuring 3 × 3 cm were collected using an Applied Spectra RT100-HP laboratory LIBS system at Juniata College in Huntingdon, Pennsylvania (USA). Spectra were acquired in air with a 100-μm beam size at 65% laser power operating at a repetition rate of 20 Hz. Typical detection limits reported in the literature for such LIBS copper analyses are in the range of tens of ppm [84]. Prior to LIBS analysis of the entire sample set, a series of LIBS measurements was collected on standard reference materials and an unknown sample to determine the effects of the laser on the sample surface and identify an appropriate analysis step size. Although the laser beam size used was 100 μm, the resulting ablation craters were measured to be approximately 150 μm in diameter. A circular damage zone containing deposits of ejected material was observed 20–50 μm outside of the ablation craters (Figure 15). Thus, a step size between analyses of 300 μm was used to avoid reanalyzing material ejected from the spot of the previous analysis. The analysis pattern consisted of 89 lines across the sample surface with 61 analysis spots per line, for a total of 5429 spot analyses per sample. As a consequence, some 60% of the areas of each 3 × 3 cm sample was covered in LIBS analysis spots (Figure 16).

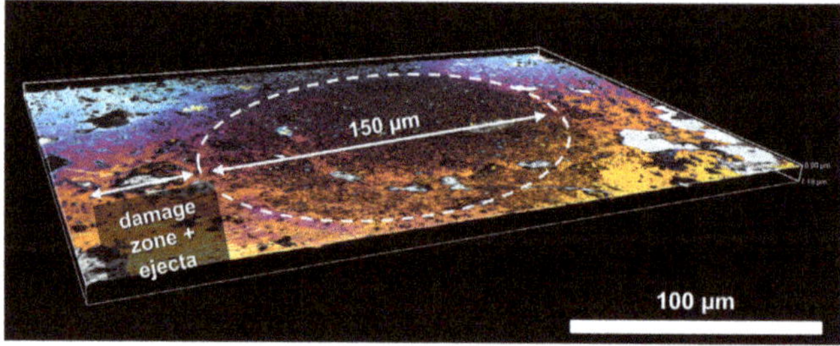

Figure 15. Reflected light three-dimensional (3D) step image of a LIBS crater in chalcopyrite (outlined in white). The ablation depth in copper sulfide minerals is 1–5 μm.

Selection of Spectral Lines and Background

Pure Cu-sulfide minerals were analyzed to (i) select diagnostic Cu emission lines, (ii) determine background thresholds for data reduction, and (iii) identify spectra that contained Cu-sulfide using spectral matching. Specimens of single-phase chalcopyrite ($CuFeS_2$) and bornite (Cu_5FeS_4) were analyzed under the same conditions used for the test sample analysis to serve as standards for these minerals. Cu emission lines were selected from the reference spectra to optimize peak intensity and minimize the effects of interferences from other element emission lines. The emission lines at 324.75 nm and 327.40 nm were selected as the main Cu peaks for the grain size analysis. These emission lines are commonly used to determine Cu concentrations from LIBS spectra, as they typically exhibit high intensities relative to background and have minimal interference from nearby emission lines [106]. Figure 17 shows example spectra collected for the bornite and chalcopyrite reference samples.

Figure 16. Configuration of LIBS analysis spots on each test sample. The upper 60% of each analysis line was used, for a total of 5429 LIBS analyses per sample.

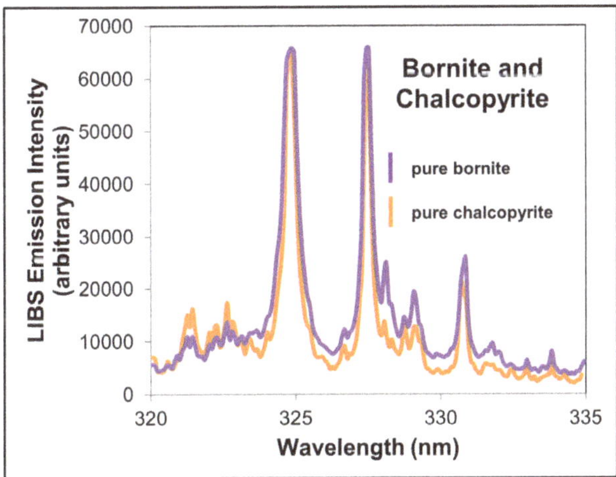

Figure 17. LIBS spectra of the chalcopyrite and bornite standards over the wavelength interval of 335–370 nm. The Cu spectral lines at 324.75 nm and Cu 327.40 nm were selected, and a background value was taken at 329.90 nm.

The background must be removed from the emission spectrum to ensure that only the LIBS signal related to the chemistry of a sample is used for the grain size proxy calculations. The chalcopyrite and bornite standards were used to assess background levels in the vicinity of the diagnostic Cu emission lines at 324.75 nm and 327.40 nm, which was measured at 329.90 nm.

Grain Size Proxy Calculations

Elemental quantification from LIBS data can be challenging; thus, basic techniques were used here to determine if LIBS is a viable option for rapid grain size assessment. Since most Cu-bearing grains are much smaller than the 150-μm spot size of the laser ablation crater, a single LIBS analysis could have sampled a number of different mineral grains. No attempt was made to distinguish between chalcopyrite and bornite grains or the number of grains at a single LIBS analysis spot. Instead, the grain size calculations were undertaken assuming that a single grain occurs in each spot, reporting an appropriate Cu threshold value, and that Cu was contained in combined Cu-sulfide composites consisting of 50% chalcopyrite and 50% bornite.

As a first pass, the spectral analysis software TSG HotCore (Australia Commonwealth Scientific Industrial Research Organisation, Canberra, Australia) was used to compare each unknown LIBS spectrum to the single mineral reference spectra. Copper was identified using the wavelength range of 320–335 nm. A Pearson correlation value between the unknown and the standard reference spectra was used to identify those spectra that were likely to contain Cu-sulfides. Pearson correlation cut-offs were selected by visually assessing numerous unknown spectra over a range of spectral correlation values to limit false positives (Figure 18). A cut-off of 0.3 Pearson correlation spectral match value was used, and spectra with a lower value were discarded from the Cu-sulfide grain size calculations.

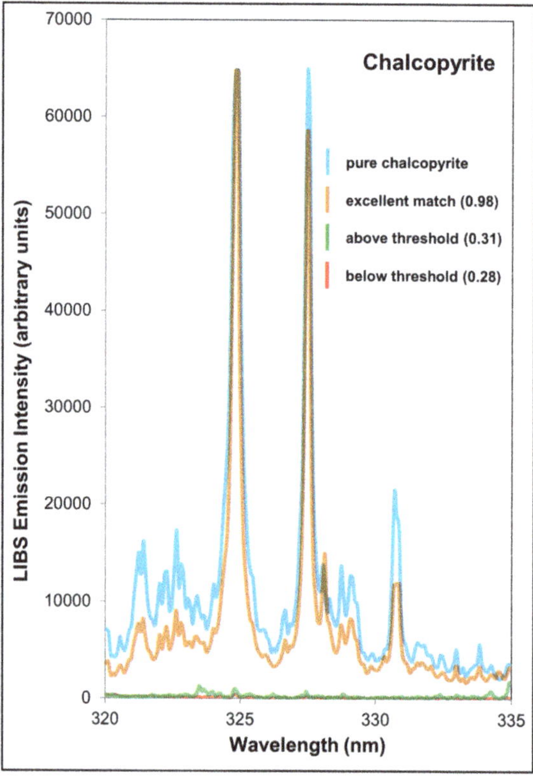

Figure 18. Example of a Pearson spectral match value for a series of unknown spectra with spectral matches just above the 0.3 cutoff threshold (green), just below that threshold (red), and well above the threshold (orange) compared to a pure chalcopyrite mineral spectrum (blue). The Pearson spectral match threshold values were selected by analyzing this relationship for numerous unknown spectra.

Peak integration of the key emission lines was used to determine the proportion of Cu-sulfide present in spots exceeding the Pearson match criteria. To estimate the area under the emission lines, the area of multiple trapezoids drawn between the spectral channels to fit under the curve were summed by calculating an integration value as follows:

$$\text{integration value} = \sum_{i=0}^{n} \frac{\text{intensity}_i + \text{intensity}_{i+1}}{2 \times (\text{wavelength}_{i+1} - \text{wavelength}_i)}. \quad (1)$$

The proportion of combined Cu-sulfide contained in an individual LIBS analysis spot was calculated by firstly background-correcting the integration value (IV) for the sum of both Cu peaks as follows:

$$IV_{\text{Cu Sulf}} = (IV_{\text{Cu 324.75}} + IV_{\text{Cu 327.40}}) - (2 \times IV_{\text{BG 329.9}}). \quad (2)$$

The integration intervals for Cu peaks and background, as well as the expected values for the chalcopyrite and bornite standards, are given in Table 2.

Table 2. Integration intervals used to calculate the proportion of copper minerals present in each spot interrogated by laser-induced breakdown spectroscopy (LIBS) analysis.

Integration Intervals and Values	Chalcopyrite	Bornite	Combined Cu-Sulfides
Mineral integration interval 1 (nm)	324.62–325.13	324.62–325.13	324.62–325.13
Mineral integration interval 2 (nm)	327.19–327.70	327.19–327.70	327.19–327.70
Background integration interval (nm)	329.6–330.18	329.67–330.18	329.67–330.18
Pure mineral integration value 1	26,239.92	29,919.83	-
Pure mineral integration value 2	21,116.59	24,962.61	-
Pure mineral background integration value (same interval used for 1 and 2)	2156.21	3159.25	-

Since the background signal was variable from spectrum to spectrum, an individual "limit of detectable grain size" for each spectrum was determined by taking three times the square root of the integrated value background. This limit was then used to ensure that Cu could be qualitatively distinguished from the background signal [107]. Any spectrum with an integration value less than the limit of detectable grain size was discarded from the grain size calculations.

For LIBS analysis of spots that were not discarded at this point, the calculated background-corrected integration value was then compared to the expected value for combined Cu-sulfide from the reference spectra as follows:

$$\text{Cu Sulf}_{\text{prop}} = \frac{IV_{\text{Cu Sulfide}}\ (\text{measured})}{IV_{\text{Cu Sulfide}}\ (\text{standard})} \quad (3)$$

LIBS craters in chalcopyrite and bornite were measured to be less than 2 μm deep; thus, it was assumed that the LIBS spot is an analysis area rather than an analysis volume. As such, the proportion by area of Cu-sulfide present in a spot can be easily converted into a grain size as follows:

$$\text{Grain Size} = \text{Mineral}_{\text{prop}} \times 150\ \mu\text{m}. \quad (4)$$

Assuming that the reported grain size values represented the length of a square grain, this length was then used to calculate the boundary length and area for each of the grains encountered in each sample. These values were used to determine the diameter by phase-specific surface area (D_{PSSL}) for all of the grains detected in each sample. It should be noted that these proxy methods assume that the Cu-sulfide detected occurs in a single grain completely within a single LIBS analysis spot. It may be the case that an analysis spot contains numerous sub-spots or partial Cu-sulfide grains rather than a single large grain, but this cannot be resolved for grains less than the LIBS analysis spot size, which was 150 μm in this study.

4.3.3. Results

The grain size proxy results from LIBS data assuming that the copper is contained in combined Cu-sulfides were compared to the MLA results (Figure 19). These results had an R^2 value of 0.57 and an r_s correlation value of 0.69. Using all 5429 LIBS analysis spots provided a reasonable estimate of grain size. A series of bootstrapping experiments was completed to assess the minimum number of LIBS spots required to adequately estimate a grain size proxy. These experiments were designed to randomly select groups of 75, 50, 25, 10, and 1 line of LIBS analysis spots from each sample, and calculate the corresponding D_{PSSL} grain size proxy. Fifty iterations of each bootstrapping experiment were completed.

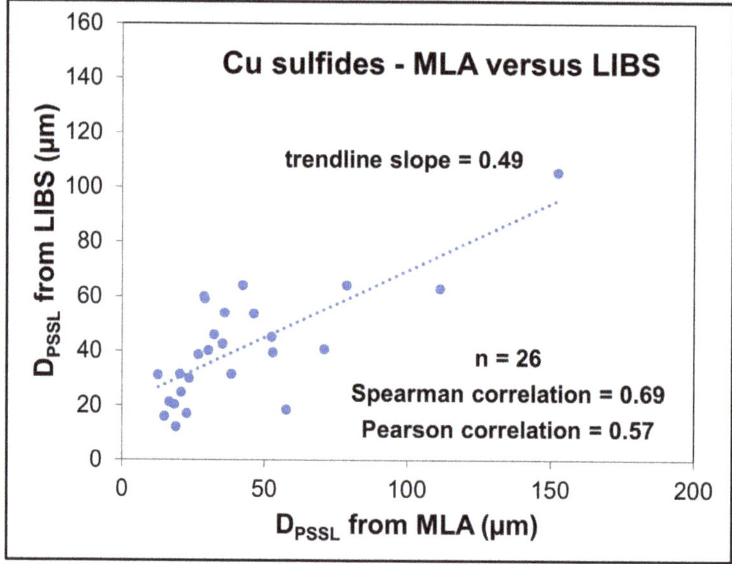

Figure 19. Comparison of the D_{PSSL} (diameter by phase-specific surface area) for combined copper sulfide measured from mineral liberation analysis (MLA) data and the grain size calculated using LIBS data. These calculations assumed that chalcopyrite and bornite equally occur as combined copper sulfides.

To compare the performance of the calculated grain size estimates as the number of analysis lines decreases, a series of box-and-whisker plots containing the bootstrapping results for combined Cu-sulfides were plotted (Figure 20). In general, as the number of lines used to calculate the grain size proxy from LIBS decreased, the mean bootstrapped D_{PSSL} values changed and the second and third quartile range values increased. Grain size proxy values for combined Cu-sulfide calculated from LIBS data had an average standard error of less than 1.5 in 50 experiments, when at least 620 of the total 5429 analysis spots were used for grain size assessment (Figure 21). A standard error value of 1.5 indicates that, with 95% confidence, using 620 random LIBS analyses would provide D_{PSSL} values within 1.5 µm in 50 repeated experiments.

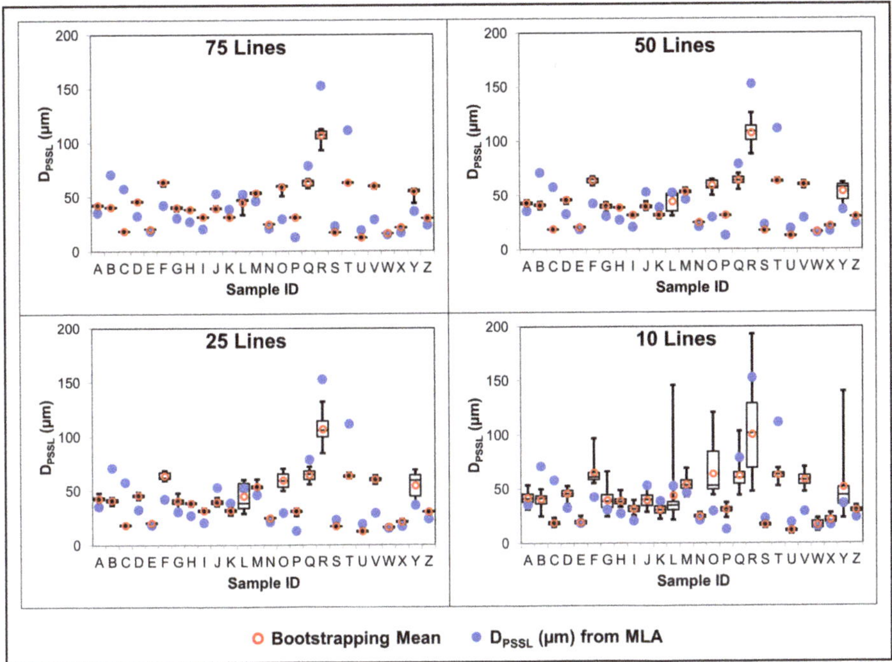

Figure 20. Box-and-whisker plot for the 26 tiles analyzed showing the bootstrapping results for combined Cu-sulfides using 75, 50, 25, and 10 random LIBS lines per sample. The boxes outline the upper and lower quartiles, and the whiskers represent the minimum and maximum values. These are compared with the mineral liberation analysis (MLA) results (blue dots) on the same tiles.

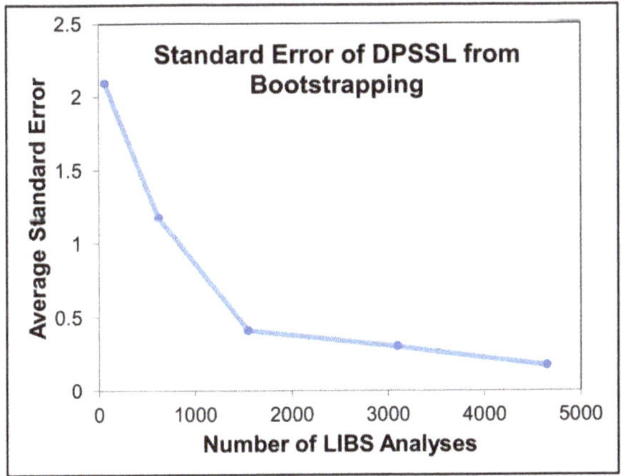

Figure 21. Comparison of the number of LIBS analysis spots used and the average range of grain size proxy values calculated for all 26 test samples. This trend shows that about 620 total 150 μm LIBS analysis are required to rank samples in terms of Cu-sulfide mineral grain size proxies to a satisfactory precision.

4.3.4. Discussion

A combination of spectral matching and peak integration was used to identify combined Cu-sulfides from LIBS spectra. LIBS analysis was successful in detecting and measuring combined

Cu-sulfide grains. When including all 89 LIBS analysis lines, combined Cu-sulfides exhibited a Pearson correlation coefficient of R^2 = 0.57 and a Spearman correlation coefficient of r_s = 0.69 when compared to the D_{PSSL} from MLA. These correlation coefficients show that the LIBS data can be used to rank samples by grain size. The slope of the correlation trendline, when compared to MLA, was below 0.5. Such a low slope value indicates that the LIBS grain size proxy underestimates the grain size by approximately 50%.

There are numerous complexities of LIBS analysis that should be investigated before implementing the approach described here as a grain size assessment tool. LIBS ablation craters are not uniform or symmetrical (Figure 22), which makes an estimation of the surface area or ablation volume challenging. While the ablation depth in quartz is more than 75 µm, the ablation depth in Cu-sulfide minerals is 1–5 µm. Such variation in LIBS ablation crater character indicates that different minerals ablate at different rates as a consequence of distinct laser–material interaction behaviors. Figure 23 shows an SEM image of an LIBS ablation crater in calcite and a Cu-sulfide, both of which display brittle (Figure 23A) and melting behavior (Figure 23B). Additionally, the depth of ablation varies depending on the mineral present at the spot of laser sampling. As illustrated in Figure 24, minerals such as quartz and feldspar produce very deep ablation craters (>100 µm) whilst Cu-sulfide minerals produce very shallow LIBS ablation craters (<5 µm). These changes in mineral behavior and ablation crater depth impact the ablation volume at a given spot. If Cu-sulfide grains are smaller than the spot size and hosted in other minerals, the ablation volume becomes higher than it would be if only Cu-sulfide was ablated.

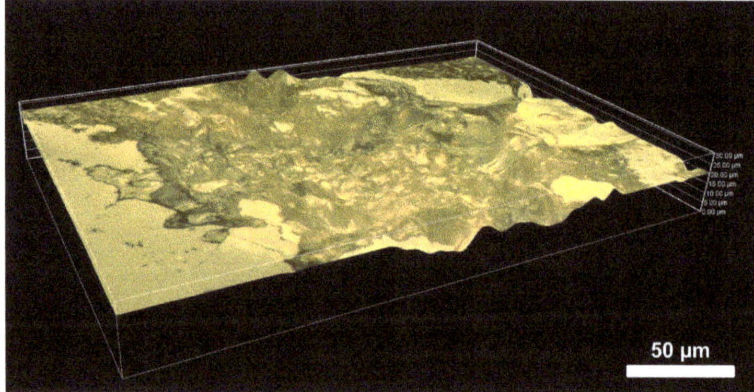

Figure 22. Example of a reflected light 3D step image of an LIBS crater in feldspar. The crater is not uniform or symmetrical, making an estimation of area or ablation volume challenging.

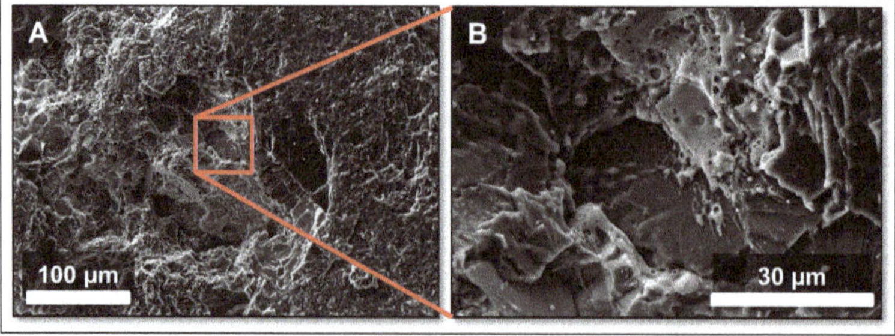

Figure 23. Scanning electron microscope images of an LIBS ablation crater in calcite (darker gray) and a Cu-sulfide (brighter gray). The ablation craters generally show brittle behavior (**A**); however, up close, both calcite and copper sulfide also display melting behavior (**B**).

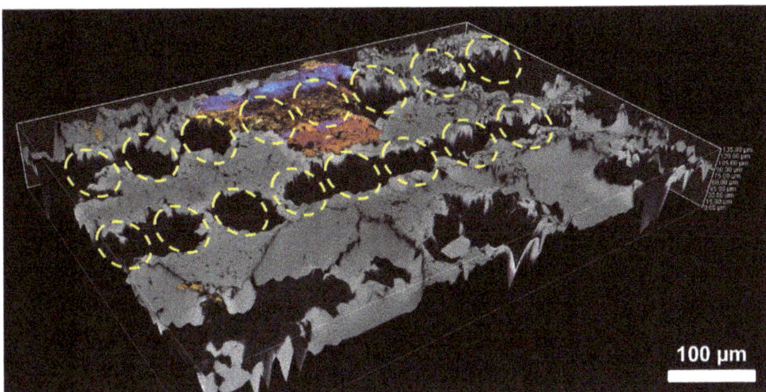

Figure 24. Reflected light 3D step image of LIBS craters (outlined in yellow circles) in quartz (gray) and chalcopyrite (colored). While the ablation depth in quartz is more than 50 μm, the ablation depth in Cu-sulfide minerals is 1–5 μm.

For this study, it was assumed that all ablation craters are shallow, and that the LIBS spectra represent a surface area as opposed to a volume. If a Cu-sulfide grain smaller than the 150 μm laser spot size was hosted in quartz, then the ablation would be a volume not an area. By assuming this volume as an area, the grain size proxy calculation is underestimated. However, much of the ablated material is likely in the surrounding ejecta where the volume is large; thus, it is not possible to estimate the actual volume of material that enters the plasma. Therefore, further research to account for variable ablation behavior is required if the LIBS method for grain size proxies explored here is to be utilized in the future. A visible-light laser (e.g., wavelength-doubled Nd:YAG laser) or an ultraviolet (UV) laser (such as that used by References [50,51] or in laser ablation ICP-MS (LA-ICP-MS) analysis) would reduce the magnitude of the cratering, but these lasers have lower output power and may, therefore, produce smaller spot sizes [108].

4.3.5. Conclusions

Whilst the test dataset used in this study is small, the results show that Cu-sulfide grain size proxies developed from LIBS data can be used to adequately assess grain size. Calculated grain size proxy values are generally within 20 μm of the measured D_{PSSL} values from MLA data, although the slope of the correlation trendline (R^2 = 0.49) indicates that the LIBS analysis underestimates the MLA grain size by approximately 50%. Bootstrapping experiments show that this method requires only 620 LIBS analysis pixels to achieve repeatable grain size results for Cu-sulfides with an average standard error of less than 1.5 μm. In the case of Cu, at least 620 LIBS analyses at 150 μm are required to adequately estimate combined Cu-sulfide grain sizes using proxy calculations developed from LIBS data. The RT100-HP laboratory LIBS system used in this study can analyze the required 620 spots in 31 s. Therefore, currently available LIBS technology shows great promise to acquire early grain size information rapidly (under one minute) and with minimal damage to the samples being analyzed. This information can be used at an early exploration stage to assess project viability and economics.

4.4. Geochemical Imaging and Microscale Mapping

4.4.1. Introduction

Whole-rock lithogeochemistry and micro-analytical spot analysis still represent the two primary tools for mapping the geochemical footprint of ore systems. However, unraveling the complex and multiply overprinting hydrothermal alteration history of most ore systems from whole-rock analysis represents a major challenge. Spot microanalyses also often exclude, or ignore, grain boundaries, fractures, and/or complex mineral intergrowths. Unfortunately, these features are, in fact, important micro-textural sites for transporting and/or depositing ore-forming elements at the microscale. To address these analytical challenges, there is a growing interest in two-dimensional (2D) element mapping, which provides geochemical information in a complete geologic context with all of the adjoining mineral phases. With careful calibration using standards of known composition, these geochemical images can provide fully quantitative results for the targeted area of interest [109,110]. Interpreting geochemical data in a geologic context provides critical constraints on the relative timing relationships between microscale features, which can, in turn, be used to unravel overprinting hydrothermal histories. The interpretive power of geochemical data in a spatial context represents the main reason that 2D, and more recently 3D, element mapping is rapidly becoming the standard mode of data acquisition rather than conventional spot microanalyses. However, geochemical imaging until recently required large and expensive analytical equipment, which mostly restricted element mapping studies to research applications. Advances in handheld LIBS provide an opportunity to acquire geochemical images in the field.

4.4.2. Methodology

The SciAps Z-300 can generate 2D element maps by rastering its laser over the sample surface to create a 16 × 16 grid of spots. Each LIBS spot has a diameter of approximately 50–100 μm, which corresponds to a mapping area of roughly 2 × 2 mm for the standard raster grid pattern. LIBS spectra for each of the 256 spots can be integrated using the SciAps Geochem Pro software to generate qualitative 2D maps of wavelength intensity. The spectral signature of each spot is then matched to the on-board LIBS library before the relative abundance map for each of the identified elements is projected onto the Z-300's results screen. Connors et al. [65] described how geochemical imaging can be applied to identify mineralogy and/or chemical zonation within individual minerals. Both of these applications have significant potential to assist with mapping the geochemical footprints of ore systems in the field.

We extended the capabilities of the SciAps Z-300 to accomplish the mapping of larger areas of interest (cm scale) than permitted by the pre-programmed raster domain of 16 × 16 grids or 2 × 2 mm, by stitching multiple spectral maps of this size together. To integrate data across multiple LIBS mapping experiments, new code was written in the free and open-source R software environment [111]. Semi-automated data processing following this new open-source workflow [112] includes wavelength binning (0.1–0.3 nm), baseline correction (i.e., using a local regression fitted to minimum intensities across the spectral range), local background estimation (i.e., running median to wavelength intensities), and peak finding (i.e., moving window). The most important wavelength intensity peaks are differentiated from the estimated background using a user-selected threshold (i.e., 1.5 × median absolute deviation on running median). Finally, LIBS spectral results are normalized to either the sum of the filtered wavelength peaks intensities or the maximum peak intensity for each pixel. All data processing is completed on a pixel-by-pixel basis, which is important for map areas that include multiple minerals. Here, mapping results are reported as a percentage of the maximum peak wavelength intensity. Wavelength intensities below the local background are reported as not available (NA).

4.4.3. Results

We applied this new LIBS mapping method to a suite of gold-bearing veins from the MacLellan gold deposit, Lynn Lake, Manitoba, Canada. LIBS spectra for the mapping were acquired in Ar on the unprepared surface of the cut drill core. Detailed field and laboratory studies identified multiple vein generations and hydrothermal alteration mineral assemblages at the MacLellan deposit [113–115]. As illustrated in Figure 2, many of the minerals associated with gold in these veins yield characteristic LIBS spectra. Handheld LIBS can, thus, be used to define the spectral fingerprint of the different vein suites and their hydrothermally altered halo in the field and directly on sawed core surfaces (Figures 25–28).

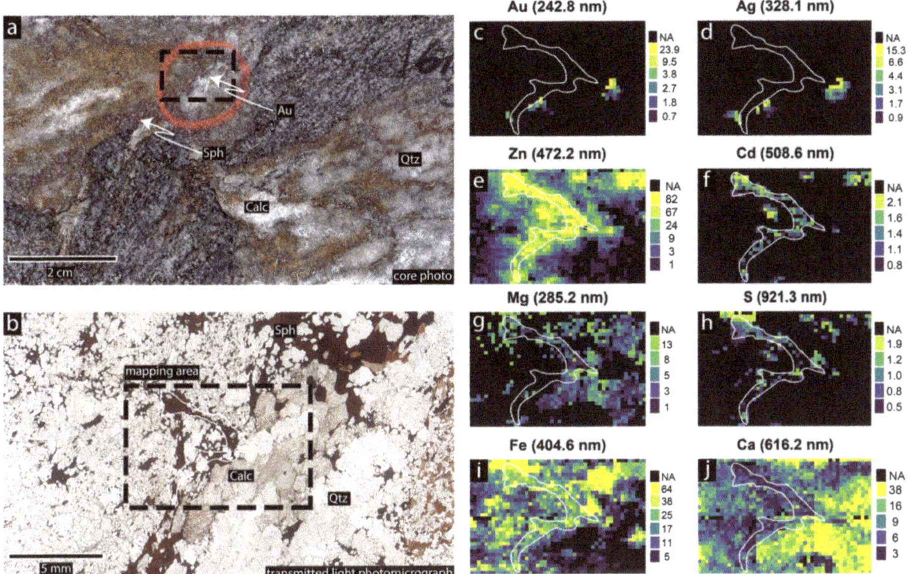

Figure 25. (**a**) Gold- and base metal-rich quartz vein from the MacLellan deposit, Lynn Lake, Manitoba. Mapping area comprises quartz (Qtz), calcite (Calc), arsenopyrite (Asp), pyrrhotite (Po), and sphalerite (Sph); (**b**) transmitted light photomicrograph of LIBS mapping area; (**c**–**j**) LIBS results for Au (**c**), Ag (**d**), Zn (**e**), Cd (**f**), Mg (**g**), S (**h**), Fe (**i**), and Ca (**j**) were processed using new R code and are displayed as relative wavelength intensities (%). Each mineral phase identified within the mapping area can identified based on its LIBS spectral signature and some prior knowledge of the main mineral-forming elements. Minor to trace elements associated with each mineral can be used to support mineral identification (**g**). Geochemical imaging is also effective at mapping the distribution of minute minerals of significant economic interest, including native Au inclusions (**d**,**e**). Maps were acquired in an Ar environment on the unprepared surface of the cut drill core.

The mineralogy of quartz veins is an important visual indicator of gold ore zones at the MacLellan deposit. Gold-bearing vein mineralogy can be complex (Figure 25a), including quartz (SiO_2), calcite ($CaCO_3$), arsenopyrite (FeAsS), pyrrhotite ($Fe_{1-x}S$), sphalerite ([Zn,Fe]S), and rare galena (PbS). Geochemical imaging of the core surface clearly demonstrates how qualitative LIBS spectra can be used to differentiate the complex mineralogy of the vein at the microscale (Figure 25c–j). Sulfide-rich vein domains yield the typical spectral wavelengths for mineral-forming elements and/or minor (wt.%) to trace elements (µg/g or ppm) that readily substitute into sulfide minerals (Figures 25 and 26), including Fe (404.6 nm), S (921.3 nm), As (228.8 nm), Zn (472.2 nm), and lesser Cd (508.6 nm). As expected, minerals that are devoid of these elements, such as the quartz and/or calcite vein matrix, yield relative intensities for these emission wavelengths that are below the estimated background (Figure 25h). The ability to detect minor to trace element signatures of sulfide phases is significant, because the relative abundance of these elements can be used to assist in mineral identification. For example, arsenian pyrite and arsenopyrite, which are the two main As-bearing phases at MacLellan, are readily distinguished from other Fe-bearing minerals in the sample matrix (Figures 25 and 26). The minor to trace element composition of minerals can also be used to differentiate multiple generations of hydrothermal alteration phases and/or gain insight into the composition of the hydrothermal fluid that transported the ore-forming elements. For example, Cd^{2+} occurs at up to wt.% concentrations in sphalerite, which, due to its similar charge, readily substitutes for Zn^{2+} (Figure 25g). This substitution reaction is dependent, in part, on temperature, which makes the Cd concentration of sphalerite a potential geochemical fingerprint of ore-forming processes at different deposits [116].

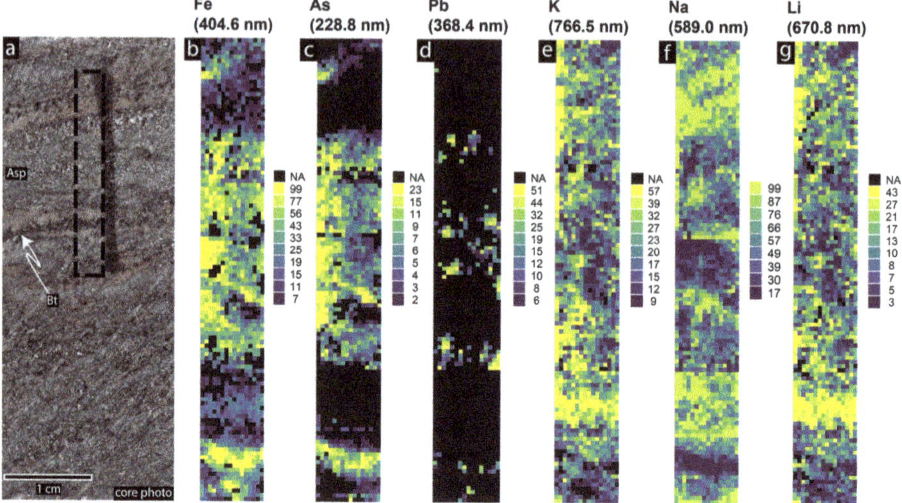

Figure 26. (**a**) Photo of gold- and base metal-rich quartz vein from the MacLellan deposit, Lynn Lake, Manitoba. Mapping area comprises quartz (Qtz), arsenopyrite (Asp), biotite (Bt), muscovite (Mus), and lesser galena (Gl); LIBS mapping results for (**b**) Fe, (**c**) As, (**d**) Pb, (**e**) K, (**f**) Na, and (**g**) Li processed using new R code and displayed as relative wavelength intensities (%). Geochemical imaging is also effective at mapping the distribution of rare galena and the enrichment of light elements along the margins of and within the arsenopyrite-rich vein. Maps were acquired in an Ar environment on the unprepared surface of the cut drill core.

Perhaps most importantly, LIBS mapping is also an effective tool for identifying minute minerals of significant economic interest. For example, geochemical images in Figure 25 map the distribution of at least three precious metal-rich grains and/or clusters of grains across the sample surface. Although the exact mineralogy remains somewhat unclear from the LIBS spectra alone, the main emission wavelengths for Au at 242.8 nm (and 267.5 nm) and Ag at 328.1 nm, together with their reproducible wavelength map patterns for these pixels, tend to support the presence of fine native gold ($\leq$50 μm). Two of these native Au grains were observed prior to LIBS mapping (Figure 25a). Based on these LIBS map results, suspected native gold grains in Figure 25 are intergrown with sphalerite and an Mg-bearing carbonate (Figure 25g). Geochemical imaging, coupled with chemometric methods, can also be used to isolate gold-bearing pixels from the rest of the LIBS mapping results (Figure 27). Although the spatial resolution of the geochemical image in Figure 27 is relatively crude ($n = 256$ spots), the LIBS map results clearly show that native gold is enveloped by a thin quartz rim ($\leq$100 μm) at the margin of the amphibole. Such detailed microscale paragenetic relationships are normally restricted to research applications, but, in this case, were acquired in the field on a non-prepared cut drill core with a handheld LIBS device.

Rare, micrometric galena grains ($\leq$100 μm) spatially associated with arsenopyrite (Figure 26b) are also inferred from elevated wavelength intensities that correspond to Pb (368.4 nm). Some of these accessory, base metal-rich sulfide minerals represent important visual indicators of the high-grade ore zones at many gold deposits and districts (e.g., galena, tellurides, bismuthides), including at the MacLellan deposit [113]. The results indicate the potential of handheld LIBS for accurate mineral identification to distinguish base metal-rich vein types, even in cases where these accessory phases may be very fine grained (Figures 26 and 27).

Geochemical imaging with LIBS is also effective at differentiating other vein minerals (Figures 25–28), including quartz using the spectral emission line for Si at 288.6 nm, for calcite using the spectral emission line for Ca at 616.2 nm, and for biotite/muscovite using the spectral emission line for K at 766.49 nm. The association between K, Na, and Li and biotite–muscovite at the margins of the arsenopyrite-rich band in Figure 26 suggests that auriferous fluids were relatively enriched in light elements. The Li-bearing signature of the veins provides a hitherto unrecognized geochemical vector for mapping the geochemical footprint of the MacLellan deposit.

At many deposits and districts, Au is also often associated with particular composition of carbonate minerals, typically Fe-, Mn-, and/or Mg-bearing carbonate (e.g., ankerite). As discussed above, differentiating the different carbonate minerals and their minor element associations is often difficult in the field. Geochemical imaging in Figure 25 distinguishes at least two generations of carbonate. Coarse grained, Mg-bearing calcite is closely intergrown with sphalerite and gold, whereas finer Mg-poor carbonate within the vein matrix is closely intergrown with quartz. The fine grain size of the latter carbonate generation would have been difficult to identify in the absence of geochemical imaging.

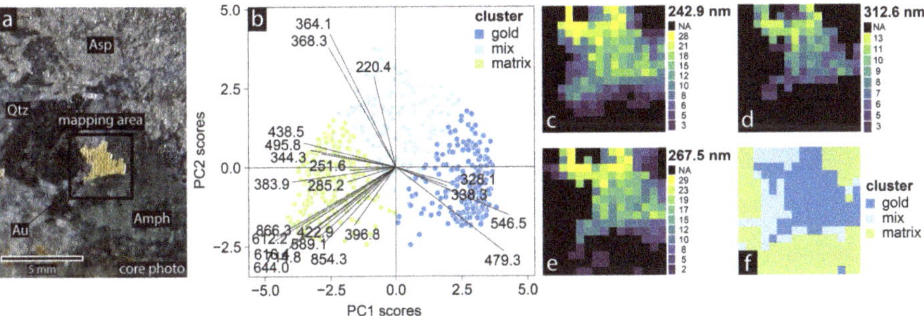

Figure 27. (**a**) Photo of coarse visible gold intergrown with quartz (Qtz) and amphibole (Amph) within an arsenopyrite (Asp)-rich quartz vein; (**b**) principal component analysis (PCA) biplot showing LIBS results and resultant wavelength loadings for one 16 × 16 grid map from 256 laser shots. Wavelengths are reported in nm units. Clusters of PCA scores are based on the k-means algorithm implemented in R [111]; (**c**–**e**) LIBS mapping results for the Au spectral lines at 242.9, 267.5, and 312.6 nm processed using new R code displayed as relative wavelength intensities (%); (**f**) LIBS map color coded to k-means clustering of PCA scores (same as **b**). These maps highlight how chemometrics can be used to re-classify qualitative LIBS results to assist with mineral identification.

The hydrothermally altered halos of veins represent another important component of the hydrothermal footprint of most ore systems. Map transects across different vein generations provide a geochemical fingerprint of the ore-forming fluid. When these hydrothermal alteration assemblages are overprinted, as in Figure 28, the relative timing of the different vein generations provides clues to the geochemical evolution of the ore system. For example, it is clear from LIBS results that the Ca-altered mafic volcanic rocks in Figure 28 were overprinted by younger K-, Na-, and Ba-rich fluids. Both veins types comprise part of the geochemical footprint at the MacLellan deposit; however, based on the whole-rock assay for this particular sample, the younger generation does not appear to be related to auriferous fluids. Repeated transects across each of the identified vein types could be used to build a more complete model for the geochemical evolution at the MacLellan deposit.

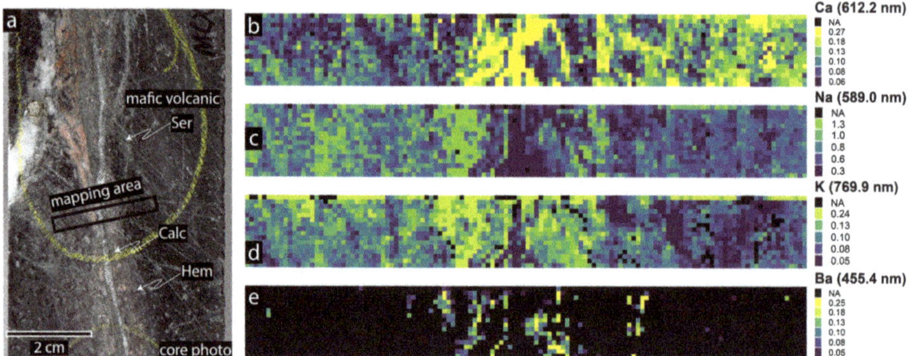

Figure 28. (**a**) Core photo of quartz (Qtz)–calcite (Calc) veinlet with sericite (Ser) and hematite (Hem) halo overprinting carbonate-altered mafic volcanic rock; LIBS mapping results for (**b**) Ca, (**c**) Na, (**d**) K, and (**e**) Ba processed using new R code and displayed as relative wavelength intensities (%). Maps highlight Na- and K-rich halo overprinting Ca-altered mafic volcanic host rock. The most K-rich veinlet domains are associated with Ba, reflecting minor element substitution within hydrothermal K-feldspar. Maps were acquired in an Ar environment on the unprepared surface of the cut drill core.

5. Summary and Future Prospects

The mineral exploration industry faces the enormous challenge of improving discovery rates and lowering discovery costs at a time when future mineral reserves will likely be sourced from deeper, covered, and/or lower-grade deposits. To address that challenge and meet society's demand for critical commodities, new methodologies and technologies are required. Recent advances in both laboratory and field-portable LIBS have an unmatched potential to support mineral exploration efforts through rapid, qualitative to quantitative geochemical analyses conducted in situ under ambient environmental conditions with little to no sample preparation. As described above, qualitative LIBS broadband spectra, appropriately coupled with chemometric methods, can be used to achieve the following:

(i) Define geochemical fingerprints for classification- and/or discrimination-type problems. Provided that a spectral reference library is available, such geochemical fingerprints are being applied to identify and successfully predict the provenance of economic minerals and for tracing their source to ore pathways;

(ii) Raster the laser over the surface of a sample, to convert LIBS spectra to geochemical images of elemental distribution. This element mapping approach represents the preferred mode of data acquisition for micro-analytical geochemistry, which until now was mostly restricted to research investigations;

(iii) Obtain near-complete geochemical characterization for individual spots and mapping analysis at wt.% to ppm concentrations within the constraints of elemental limits of detection because all elements of the periodic table are associated with one or more emission wavelengths between 190 and 900 nm;

(iv) Analyze the light elements (e.g., B, Be, Li, Na, and Mg) to which LIBS is particularly sensitive, which represents one of the primary advantages of LIBS over other handheld technologies;

(v) Acquire semi- to fully quantitative geochemical analyses, based upon proper calibration. Quantitative LIBS calibration strategies were already described, and they represent an emerging tool for mapping cryptic geochemical footprints for minor to trace elements that are invisible to geophysical methods and/or impossible to observe directly.

The specific attributes of LIBS over other analytical techniques make it an important stand-alone tool for mineral explorationists. However, the future of LIBS may, in part, rest with so-called tandem or hyphenated technologies that combine multiple analytical techniques into one sensor. For example, integrated LIBS and laser ablation ICP-MS laboratory systems combine the excellent sensitivity of mass spectrometry methods for element detection down to ppb or ng/g concentrations, with the added benefit of light element analysis [117]. Combining both methods compensates for the limitations of each technique and promises near-complete coverage of the periodic table for characterizing multi-element geochemical footprints. Other types of small, portable tandem instruments are already commercially available (e.g., Olympus TERA portable XRD/XRF analyzer) or are being developed for research applications, including space exploration. More blended handheld technologies should be expected in the future (e.g., Raman/LIBS; XRF/LIBS), which will benefit mineral exploration through improved sample compositional characterization.

In addition to more sensitive spectrometers and other hardware improvements (e.g., double-pulse LIBS, femtosecond LIBS, standoff LIBS), the most promising areas of LIBS development in the short to medium term will likely center around upgraded software and the further application of advanced data analytics. Machine learning tools have the potential to automate all or parts of spectra deconvolution, which represents one of the most complex and time-consuming aspects of the LIBS method. Because of the rich information contained within broadband spectra, chemometric methods are particularly well suited for extracting the most element information possible from the raw LIBS data. Automated or semi-automated data processing will be required for future applications of LIBS mapping or real-time spot analyses deployed at scale. Many of these machine learning tools already exist, but need to be designed for this specific application and better integrated with the software pre-loaded on handheld units. Robust spectral libraries also need to be created, particularly for elements and/or geomaterials relevant to mineral exploration. These spectral libraries will play an important role for quantitative LIBS results and for generating predictive models to convert spectra to element concentrations. In the short term, spectral library development should focus on the ore minerals that host light elements, some of which represent critical commodities (e.g., Li). These spectral libraries will need to be pre-loaded onto handheld units, or, more likely, accessed wirelessly through internet-connected LIBS devices. New data management strategies and methods will also be required to accommodate the storing, accessing, and interpreting of the vast amounts of new information that will be acquired. Wherever possible, these data will need to be labeled so that new results can be integrated into dynamic spectral libraries that improve as more LIBS analyses are acquired.

Going forward, we anticipate that advances in LIBS capability are likely to track and leverage other hardware and software advances in a variety of research fields. Together, these advances will improve LIBS instrument sensitivity for trace element analysis, provide more complete geochemical and/or molecular sample characterization, and improve usability. The mineral exploration industry will benefit from these advanced technologies, which promise the rapid, low-cost, and on-site mapping of geochemical footprints and mineral textures.

Author Contributions: All authors contributed to the analytical work discussed here; R.S.H. and R.R.H. are responsible for Section 4.1, A.M.S. and R.S.H. are responsible for Section 4.2, C.L.H. and R.R.H. are responsible for Section 4.3, and C.J.M.L. and J.W. are responsible for Section 4.4. The original draft was written by R.S.H., C.J.M.L., and C.L.H., supported by A.M.S. and J.W., with all authors contributing to the editing.

Funding: The grain size assessment work of C.L.H. was funded by Transforming the Mining Value Chain (TMVC), an Australian Research Council (ARC) Industrial Transformation Research Hub (project #IH130200004) at the University of Tasmania. C.J.M.L. acknowledges support from the Targeted Geoscience Initiative. J.W. acknowledges support from Alamos Gold and a Society of Economic Geologists student research grant.

Acknowledgments: C.J.M.L. and J.W. would like to acknowledge the support of Alamos Gold over the course of this study. C.L.H. graciously thanks Ron Berry at the University of Tasmania and Mary Harris at Newcrest Mining for their supervision and contributions to this work.

Conflicts of Interest: The authors declare no conflicts of interest.

References

1. Meinert, L.D.; Robinson, G.; Nassar, N.T. Mineral resources: Reserves, peak production and the future. *Resources* **2016**, *5*, 14. [CrossRef]

2. Lusty, P.; Gunn, A. Challenges to global mineral resource security and options for future supply. In *Ore Deposits in An Evolving Earth*; Special Publications; Jenkin, G., Lusty, P., McDonald, I., Smith, M., Boyce, A., Wilkinson, J., Eds.; Geological Society: London, UK, 2019; Volume 393, pp. 265–276.
3. Ali, S.; Giurco, D.; Arndt, N.; Nickless, E.; Brown, G.; Demetriades, A.; Durrheim, R.; Enriquez, M.; Kinnaird, J.; Littleboy, A.; et al. Mineral supply for sustainable development requires resource governance. *Nature* **2017**, *543*, 367–372. [CrossRef] [PubMed]
4. Arndt, N.; Fontboté, L.; Hedenquist, J.; Kesler, S.; Thompson, J.; Wood, D. Future global mineral resources. *Geochem. Perspect.* **2017**, *6*, 171. [CrossRef]
5. Skirrow, R.; Huston, D.; Mernagh, T.; Thorne, J.; Dulfer, H.; Senior, A. *Critical Commodities for A High-Tech World: Australia's Potential to Supply Global Demand*; Geoscience Australia: Canberra, Australia, 2013; p. 126.
6. Mudd, G.M.; Jowitt, S.M.; Werner, T.T. The world's by-product and critical metal resources part I: Uncertainties, current reporting practices, implications and grounds for optimism. *Ore Geol. Rev.* **2017**, *86*, 924–938. [CrossRef]
7. Sillitoe, R. Grassroots Exploration: Between a Major Rock and a Junior Hard Place. *Soc. Econ. Geol. Newsl.* **2010**, *83*, 11–13.
8. Beaty, R. The Declining Discovery Trend: People, Science or Scarcity? *Soc. Econ. Geol. Newsl.* **2010**, *81*, 14–18.
9. Duke, J. *Government Geoscience to Support Mineral Exploration: Public Policy Rationale and Impact*; Prospectors and Developers Association of Canada, PDAC: Toronto, ON, Canada, 2010; p. 72.
10. Blain, C. Fifty-year trends in minerals discovery—Commodity and ore-type targets. *Explor. Min. Geol.* **2001**, *9*, 1–11. [CrossRef]
11. Wood, D. Mineral Resource Discovery—Science, Art & Business. *Soc. Econ. Geol. Newsl.* **2010**, *80*, 12–17.
12. Hagemann, S.G.; Lisitsin, V.A.; Huston, D.L. Mineral system analysis: Quo vadis. *Ore Geol. Rev.* **2016**, *76*, 504–522. [CrossRef]
13. McCuaig, T.C.; Beresford, S.; Hronsky, J. Translating the mineral systems approach into an effective exploration targeting system. *Ore Geol. Rev.* **2010**, *38*, 128–138. [CrossRef]
14. Huston, D.L.; Mernagh, T.P.; Hagemann, S.G.; Doublier, M.P.; Fiorentini, M.; Champion, D.C.; Jaques, A.L.; Czarnota, K.; Cayley, R.; Skirrow, R.; et al. Tectono-metallogenic systems—The place of mineral systems within tectonic evolution, with an emphasis on Australian examples. *Ore Geol. Rev.* **2016**, *76*, 168–210. [CrossRef]
15. Rogers, N. Targeted Geoscience Initiative: 2018 report of activities. *Geol. Surv. Can. Open File 8549* **2018**, *2*, 448.
16. Wyborn, L.; Heinrich, C.; Jaques, A. Australian Proterozoic mineral systems: Essential ingredients and mappable criteria. In *The AusIMM Annual Conference Proceedings*; AusIMM Darwin: Northern Territory, Australia, 1994; pp. 109–115.
17. Gaillard, N.; Williams-Jones, A.E.; Clark, J.R.; Lypaczewski, P.; Salvi, S.; Perrouty, S.; Piette-Lauzière, N.; Guilmette, C.; Linnen, R.L. Mica composition as a vector to gold mineralization: Deciphering hydrothermal and metamorphic effects in the Malartic district, Quebec. *Ore Geol. Rev.* **2018**, *95*, 789–820. [CrossRef]
18. Noble, R.; Christie, A. Update on exploration geochemistry initiatives in Australia and New Zealand. *Explore* **2015**, *166*, 13–17.
19. Lesher, C.; Hannington, M.; Galley, A. Integrated multi-parameter exploration footprints of the Canadian Malartic disseminated Au, McArthur River-Millennium unconformity U, and Highland Valley porphyry Cu Deposits: Preliminary results from the NSERC-CMIC Mineral Exploration Footprints Research. *Proc. Explor.* **2017**, *17*, 23.
20. De Caritat, P.; Main, P.T.; Grunsky, E.C.; Mann, A.W. Recognition of geochemical footprints of mineral systems in the regolith at regional to continental scales. *Aust. J. Earth Sci.* **2017**, *64*, 1033–1043. [CrossRef]
21. Vallée, M.A.; Morris, W.A.; Perrouty, S.; Lee, R.G.; Wasyliuk, K.; King, J.J.; Ansdell, K.; Mir, R.; Shamsipour, P.; Farquharson, C.G.; et al. Geophysical inversion contributions to mineral exploration: Lessons from the Footprints project. *Can. J. Earth Sci.* **2019**, *543*, 525–543. [CrossRef]
22. Halley, S.; Dilles, J.; Tosdal, R. Footprints: Hydrothermal alteration and geochemical dispersion around porphyry copper deposits. *Soc. Econ. Geol. Newsl.* **2015**, *100*, 12–17.
23. Winterburn, P.A.; Noble, R.R.P.; Lawie, D. Advances in exploration geochemistry, 2007 to 2017 and beyond. *Geochem. Explor. Environ. Anal.* **2019**, *30*, 10. [CrossRef]
24. Kyser, K.; Barr, J.; Ihlenfeld, C. Applied geochemistry in mineral exploration and mining. *Elements* **2015**, *11*, 241–246. [CrossRef]

25. Noll, R. *Laser-Induced Breakdown Spectroscopy: Fundamentals and Applications*; Springer: Berlin, Germany, 2012.
26. Cremers, D.; Radziemski, L. *Handbook of Laser-Induced Breakdown Spectroscopy*; Wiley: Hoboken, NJ, USA, 2013.
27. Lee, Y.; Song, K.; Sneddon, J. *Laser-induced Breakdown Spectrometry*; Nova Publishers: Hauppauge, NY, USA, 2000.
28. Miziolek, A.; Palleschi, V.; Schechter, I. (Eds.) *Laser Induced Breakdown Spectroscopy*; Cambridge University Press: Cambridge, UK, 2006.
29. Singh, J.; Thakur, S. (Eds.) *Laser-Induced Breakdown Spectroscopy*; Elsevier: Amsterdam, The Netherlands, 2007.
30. Musazzi, S.; Perini, U. (Eds.) *Laser-Induced Breakdown Spectroscopy, Theory and Applications*; Springer: Berlin, Germany, 2014.
31. Rusak, D.; Castle, B.; Smith, B.; Winefordner, J. Fundamentals and applications of laser-induced breakdown spectroscopy. *Crit. Rev. Anal. Chem.* **1997**, *27*, 257–290. [CrossRef]
32. Harmon, R.; Russo, R.; Hark, R. Applications of laser-induced breakdown spectroscopy for geochemical and environmental analysis: A comprehensive review. *Spectrochim. Acta Part B* **2013**, *87*, 11–26. [CrossRef]
33. Fabre, C.; Boiron, M.; Dubessy, I.; Cathelineau, M.; Banks, D. Palaeofluid chemistry of a single fluid event: A bulk and in-situ multi-technique analysis (LIBS, Raman Spectroscopy) of an Alpine fluid (Mont-Blanc). *Chem. Geol.* **2002**, *182*, 249. [CrossRef]
34. Stipe, C.; Miller, A.; Brown, J.; Guevara, E.; Cauda, E. Evaluation of laser-induced breakdown spectroscopy (LIBS) for measurement of silica on filter samples of coal dust. *Appl. Spectrosc.* **2012**, *66*, 1286–1293. [CrossRef]
35. Tognoni, E.; Cristoforetti, G.; Legnaiolia, S.; Palleschi, V. Calibration-free laser-induced breakdown spectroscopy: State of the art. *Spectrochim. Acta Part B* **2010**, *65*, 1–14. [CrossRef]
36. Praher, B.; Palleschi, V.; Viskup, P.; Heitz, J.; Pedarnig, J. Calibration free laser induced breakdown spectroscopy of oxide materials. *Spectrochim. Acta Part B* **2010**, *65*, 671–679. [CrossRef]
37. Winefordner, J.; Gornushkin, I.; Correll, T.; Gibb, E.; Smith, B.; Omenetto, N. Comparing several atomic spectrometric methods to the super stars: Special emphasis on laser induced breakdown spectrometry, LIBS, a future super star. *J. Anal. At. Spectrom.* **2004**, *19*, 1061–1083. [CrossRef]
38. Hahn, D.; Flower, W.; Hencken, K. Discrete particle detection and metal emissions monitoring using laser-induced breakdown spectroscopy. *Appl. Spectrosc.* **1997**, *51*, 1836–1844. [CrossRef]
39. Derome, D.; Cathelineau, M.; Cuney, M.; Fabre, C.; Lhomme, T.; Banks, D. Mixing of sodic and calcic brines and uranium deposition at McArthur River, Saskatchewan, Canada: A Raman and laser-induced breakdown spectroscopic study of fluid inclusions. *Econ. Geol.* **2005**, *100*, 1529–1545. [CrossRef]
40. Lebedev, V.; Makarchuk, P.; Stepanov, D. Real-time qualitative study of forsterite crystal–Melt lithium distribution by laser-induced breakdown spectroscopy. *Spectrochim. Acta Part B* **2017**, *137*, 23–27. [CrossRef]
41. Fabre, C.; Devismes, D.; Moncayo, S.; Pelascini, F.; Trichard, F.; Leocmte, A.; Bousquet, B.; Cauzid, J.; Motto-Ros, V. Elemental imaging by laser-induced breakdown spectroscopy for the geological characterization of minerals. *J. Anal. At. Spectrom.* **2018**, *33*, 1345–1353. [CrossRef]
42. Kim, T.; Lin, C.; Yoon, Y. Compositional mapping by laser-induced breakdown spectroscopy. *J. Phys. Chem. B* **1998**, *102*, 4284–4287. [CrossRef]
43. Cabalín, L.; Mateo, M.; Laserna, J. Chemical maps of patterned samples by microline-imaging laser-induced plasma spectrometry. *Surf. Interface Anal.* **2003**, *35*, 263–267. [CrossRef]
44. Novotný, K.; Kaiser, J.; Galiová, M.; Konečna, V.; Novotný, J.; Malina, R.; Liška, M.; Kanický, V. Mapping of different structures on large area of granite sample using laser-ablation based analytical techniques, an exploratory study. *Spectrochim. Acta Part B* **2008**, *63*, 1139–1144. [CrossRef]
45. Sweetapple, M.; Tassios, S. Laser-induced breakdown spectroscopy (LIBS) as a tool for in situ mapping and textural interpretation of lithium in pegmatite minerals. *Am. Mineral.* **2015**, *100*, 2141–2151. [CrossRef]
46. Kuhn, K.; Meima, J.; Rammlmair, D.; Ohlendorf, C. Chemical mapping of mine waste drill cores with laser-induced breakdown spectroscopy (LIBS) and energy dispersive X-ray fluorescence (EDXRF) for mineral resource exploration. *J. Geochem. Explor.* **2016**, *161*, 72–84. [CrossRef]
47. Rifai, K.; Laflamme, M.; Constantin, M.; Vidal, F.; Sabsabi, M.; Blouin, A.; Bouchard, P.; Fytas, K.; Castello, M.; Kamwa, N. Analysis of gold in rock samples using laser-induced breakdown spectroscopy: Matrix and heterogeneity effects. *Spectrochim. Acta Part B* **2017**, *134*, 33–41. [CrossRef]
48. Maravlekai, P.; Zafiropulos, V.; Kilikoglou, V.; Kalaitzaki, M.; Fotakis, C. Laser-induced breakdown spectroscopy as a diagnostic technique for the laser cleaning of marble. *Spectrochim. Acta Part B* **1997**, *52*, 41–53. [CrossRef]

49. Ortiz, P.; Vázquez, M.; Ortiz, R.; Martin, J.; Ctvrnickova, T.; Mateo, M.; Nicolas, G. Investigation of environmental pollution effects on stone monuments in the case of Santa Maria La Blanca, Seville (Spain). *Appl. Phys. A* **2010**, *100*, 965–973. [CrossRef]
50. Kaski, S.; Häkkänen, H.; Korppi-Tommola, J. Sulfide mineral identification using laser-induced plasma spectroscopy. *Miner. Eng.* **2003**, *16*, 1239–1243. [CrossRef]
51. Haavisto, O.; Kauppinen, T.; Häkkänen, H. Laser-induced breakdown spectroscopy for rapid elemental analysis of drillcore. *IFAC Proc. Vol.* **2013**, *46*, 87–91. [CrossRef]
52. Popov, A.; Labutin, T.; Zaytsev, S.; Seliverstova, I.; Zorov, N.; Kal'ko, I.; Sidorina, Y.; Bugaev, I.; Nikolaev, Y. Determination of Ag, Cu, Mo and Pb in soils and ores by laser-induced breakdown spectrometry. *J. Anal. At. Spectrom.* **2014**, *29*, 1925–1933. [CrossRef]
53. Díaz, D.; Hahn, D.; Molina, A. Quantification of gold and silver in minerals by laser-induced breakdown spectroscopy. *Spectrochim. Acta Part B* **2017**, *136*, 106–115. [CrossRef]
54. Rifai, K.; Doucet, F.; Özcan, L.; Vidal, F. LIBS core imaging at kHz speed: Paving the way for real-time geochemical applications. *Spectrochim. Acta Part B* **2018**, *150*, 43–48. [CrossRef]
55. Zorba, V.; Mao, X.; Russo, R. Ultrafast laser induced breakdown spectroscopy for high spatial resolution chemical analysis. *Spectrochim. Acta Part B* **2011**, *66*, 189–192. [CrossRef]
56. Čtvrtníčková, T.; Fortes, F.; Cabalin, L.; Laserna, J. Optical restriction of plasma emission light for nanometric sampling depth and depth profiling of multilayered metal samples. *Appl. Spectrosc.* **2007**, *61*, 719–724. [CrossRef] [PubMed]
57. Čtvrtníčková, T.; Fortes, F.; Cabalin, L.; Kanický, V.; Laserna, J. Depth profiles of ceramic tiles by using orthogonal double-pulse laser induced breakdown spectrometry. *Surf. Interface Anal.* **2009**, *41*, 714–719. [CrossRef]
58. Yamamoto, K.; Cremers, D.; Ferris, M.; Foster, L. Detection of metals in the environment using a portable laser-induced breakdown spectroscopy instrument. *Appl. Spectrosc.* **1996**, *50*, 222–233. [CrossRef]
59. Wainner, R.; Harmon, R.; Miziolek, A.; McNesby, K.; French, P. Analysis of environmental lead contamination: Comparison of LIBS field and laboratory instruments. *Spectrochim. Acta Part B* **2001**, *56*, 777–793. [CrossRef]
60. Palanco, S.; Alises, A.; Cunat, J.; Baena, J.; Laserna, J. Development of a portable laser-induced plasma spectrometer with fully-automated operation and quantitative analysis capabilities. *J. Anal. At. Spectrom.* **2003**, *18*, 933–938. [CrossRef]
61. Goujon, J.; Giakoumaki, A.; Piñon, V.; Musset, O.; Georgiou, E.; Boquillon, J. A compact and portable laser-induced breakdown spectroscopy instrument for single and double pulse applications. *Spectrochim. Acta Part B* **2008**, *63*, 1091–1096. [CrossRef]
62. Cuñat, J.; Fortes, F.; Cabalín, L.; Carrasco, F.; Simón, M.; Laserna, J. Man-portable laser-induced breakdown spectroscopy system for in situ characterization of karstic formations. *Appl. Spectrosc.* **2008**, *62*, 1250–1255. [CrossRef]
63. Rakovský, J.; Musset, O.; Buoncristiani, J.; Bichet, V.; Monna, F.; Neige, P.; Veis, P. Testing a portable laser-induced breakdown spectroscopy system on geological samples. *Spectrochim. Acta Part B* **2012**, *74*, 57–65. [CrossRef]
64. Bertolini, A.; Carelli, G.; Francesconi, F.; Francesconi, M.; Marchesini, L.; Marsili, P.; Sorrentino, F.; Cristoforetti, G.; Legnaioli, S.; Palleschi, V.; et al. Modì: A new mobile instrument for in situ double-pulse LIBS analysis. *Anal. Bioanal. Chem.* **2006**, *385*, 240–247. [CrossRef]
65. Connors, B.; Somers, A.; Day, D. Application of handheld laser-induced breakdown spectroscopy (LIBS) to geochemical analysis. *Appl. Spectrosc.* **2016**, *70*, 810–815. [CrossRef] [PubMed]
66. Radziemski, L. From LASER to LIBS, the path of technology development. *Spectrochim. Acta Part B* **2002**, *57*, 1109–1113. [CrossRef]
67. Bogue, R. Boom time for LIBS technology. *Sens. Rev.* **2004**, *24*, 353–357. [CrossRef]
68. McMillan, N.; McManus, C.; Harmon, R.; DeLucia, F.; Miziolek, A. Laser-induced breakdown spectroscopy analysis of complex silicate minerals—Beryl. *Anal. Bioanal. Chem.* **2006**, *385*, 263–271. [CrossRef] [PubMed]
69. Harmon, R.; Remus, J.; McMillan, N.; McManus, C.; Collins, L.; JL, G.; DeLucia, F.; Miziolek, A. LIBS analysis of geomaterials: Geochemical fingerprinting for the rapid analysis and discrimination of minerals. *Appl. Geochem.* **2009**, *24*, 1125–1141. [CrossRef]
70. Grant, K.; Paul, G.; O'Neill, J. Time-resolved laser-induced breakdown spectroscopy of iron ore. *Appl. Spectrosc.* **1990**, *44*, 1711–1714. [CrossRef]

71. Grant, K.; Paul, G.; O'Neill, J. Quantitative elemental analysis of iron ore by laser induced breakdown spectroscopy. *Appl. Spectrosc.* **1991**, *45*, 701–705. [CrossRef]
72. Sun, Q.; Tran, M.; Smith, B.; Winefordner, J. Determination of Mn and Si in iron ore by laser induced plasma spectroscopy. *Anal. Chim. Acta* **2000**, *413*, 187–195. [CrossRef]
73. Michaud, D.; Leclerc, R.; Proulx, É. Influence of particle size and mineral phase in the analysis of iron ore slurries by Laser-Induced Breakdown Spectroscopy. *Spectrochim. Acta Part B* **2007**, *62*, 1575–1581. [CrossRef]
74. Death, D.; Cunningham, A.; Pollard, L. Multi-element analysis of iron ore pellets by laser induced breakdown spectroscopy and principal components regression. *Spectrochim. Acta Part B* **2008**, *63*, 763–769. [CrossRef]
75. Gaft, M.; Sapir-Sofer, I.; Modiano, H.; Stana, R. Laser induced breakdown spectroscopy for bulk minerals online analyses. *Spectrochim. Acta Part B* **2007**, *62*, 1496–1503. [CrossRef]
76. Rosenwasser, S.; Asimellis, G.; Bromley, B.; Hazlett, R.; Martin, J.; Pearce, T.; Zigler, A. Development of a method for automated quantitative analysis of ores using LIBS. *Spectrochim. Acta Part B* **2001**, *56*, 707–714. [CrossRef]
77. Harhira, A.; Bouchard, P.; Rifai, K.; El Haddad, J.; Sabsabi, M.; Blouin, A.; Laflamme, M. Advanced Laser-induced breakdown spectroscopy (LIBS) sensor for gold mining. In Proceedings of the Conference of Metallurgists (COM), Vancouver, BC, Canada, 27–30 August 2017; p. 11.
78. Díaz, D.; Molina, A.; Hahn, D. Effect of laser irradiance and wavelength on the analysis of gold-and silver-bearing minerals with laser-induced breakdown spectroscopy. *Spectrochim. Acta Part B* **2018**, *145*, 86–95. [CrossRef]
79. Harmon, R.S.; Hark, R.R.; Throckmorton, C.S.; Rankey, E.C.; Wise, M.A.; Somers, A.M.; Collins, L.M. Geochemical fingerprinting by handheld laser-induced breakdown spectroscopy. *Geostand. Geoanal. Res.* **2017**, *41*, 563–584. [CrossRef]
80. Somers, A. Application of hand-held laser induced breakdown spectroscopy to drilling samples: New technology providing new in-field analytical capabilities. In Proceedings of the Drilling for Geology II, Brisbane, Australia, 26–28 July 2017; pp. 89–96.
81. Hoefs, J. Geochemical fingerprints: A critical appraisal. *Eur. J. Mineral.* **2010**, *22*, 3–15. [CrossRef]
82. Gottfried, J.; Harmon, R.; De Lucia, F.; Miziolek, A. Multivariate analysis of laser-induced breakdown spectroscopy chemical signatures for geomaterial classification. *Spectrochim. Acta Part B* **2009**, *64*, 1009–1019. [CrossRef]
83. Alvey, D.; Morton, K.; Harmon, R.; Gottfried, J.; Remus, J.; Collins, L.; Wise, M. Laser-induced breakdown spectroscopy-based geochemical fingerprinting for the rapid analysis and discrimination of minerals: The example of garnet. *Appl. Opt.* **2010**, *49*, C168–C180. [CrossRef]
84. Hark, R.; Harmon, R. Geochemical fingerprinting using LIBS. In *Laser-Induced Breakdown Spectroscopy–Theory & Applications*; Musazzi, S., Perini, U., Eds.; Springer: Berlin, Germany, 2014; pp. 309–348.
85. Harmon, R.; Throckmorton, C.; Hark, R.; Gottfried, J.; Wörner, G.; Harpp, K.; Collins, L. Discriminating volcanic centers with handheld laser-induced breakdown spectroscopy (LIBS). *J. Archaeol. Sci.* **2018**, *98*, 112–127. [CrossRef]
86. Death, D.L.; Cunningham, A.P.; Pollard, L. Multi-element and mineralogical analysis of mineral ores using laser induced breakdown spectroscopy and chemometric analysis. *Spectrochim. Acta Part B* **2009**, *64*, 1048–1058. [CrossRef]
87. Young, S.; Dias, G. LCM of metals supply to electronics: Tracking and tracing "conflict minerals". In Proceedings of the LCM 2011—Towards Life Cycle Sustainability Management, Berlin, Germany, 28–31 August 2011; p. 12.
88. Harmon, R.; Shughrue, K.; Remus, J.; Wise, M.; East, L.; Hark, R. Can the provenance of the conflict minerals columbite and tantalite be ascertained by laser-induced breakdown spectroscopy? *Anal. Bioanal. Chem.* **2011**, *400*, 3377–3382. [CrossRef]
89. Černý, P.; Ercit, T. Mineralogy of niobium and tantalum: Crystal chemical relationships, paragenetic aspects and their economic implications. In *Lanthanides, Tantalum and Niobium*; Springer: Berlin, Germany, 1989; pp. 27–79.
90. Robert, F.; Brommecker, R.; Bourne, B.; Dobak, P.; McEwan, C.; Rowe, R.; Zhou, X. Models and exploration methods for major gold deposit types. In Proceedings of the Exploration 07: Fifth Decennial International Conference on Mineral Exploration, Toronto, ON, Canada, 9–12 September 2007; Volume 48, pp. 691–711.

91. Mortensen, J.; Chapman, R.; LeBarge, W.; Jackson, L. Application of placer and lode gold geochemistry to gold exploration in western Yukon. In *Yukon Exploration and Geology 2004*; Emond, D., Lewis, L., Bradshaw, G., Eds.; Yukon Geological Surve: Whitehorse, YT, Canada, 2004; pp. 205–212.
92. Chapman, R.; Mortensen, J.; LeBarge, W. Styles of lode gold mineralization contributing to the placers of the Indian River and Black Hills Creek, Yukon Territory, Canada as deduced from microchemical characterization of placer gold grains. *Miner. Depos.* **2011**, *46*, 881–903. [CrossRef]
93. McCandless, T.; Baker, M.; Ruiz, J. Trace element analysis of natural gold by laser ablation ICP-MS: A combined external/internal standardization approach. *J. Geostand. Geoanal.* **1997**, *21*, 271–278. [CrossRef]
94. McInnes, M.; Greenough, J.; Fryer, B.; Wells, R. Trace elements in native gold by solution ICP-MS and their use in mineral exploration. A British Columbia example. *Appl. Geochemi.* **2008**, *23*, 1076–1085. [CrossRef]
95. Liu, H.; Mao, X.; Mao, S.; Greif, R.; Russo, R. Particle size dependence chemistry from laser ablation of brass. *Anal. Chem.* **2005**, *77*, 6687–6691. [CrossRef] [PubMed]
96. Sillitoe, R.; Hedenquist, J. Linkages between volcanotectonic settings, ore-fluid compositions, and epithermal precious metal deposits. *Soc. Econ. Geol. Spec. Publ.* **2003**, *10*, 315–343.
97. Afgan, M.; Hou, Z.; Wang, Z. Quantitative analysis of common elements in steel using a handheld μ-LIBS instrument. *J. Anal. At. Spectrom.* **2017**, *32*, 1905–1915. [CrossRef]
98. Griffin, A. Lithium Australia Success with Innovative Lithium Analyser at Agua Fria, Mexico. Available online: http://www.asx.com.au/asxpdf/20170502/pdf/43hz02zvjm8f0y.pdf (accessed on 2 May 2017).
99. Sutherland, D. Estimation of mineral grain size using automated mineralogy. *Miner. Eng.* **2007**, *20*, 452–460. [CrossRef]
100. Berry, R.; Hunt, J. *Proxy Methods for Domaining ore Deposits for Au Grain Size: Geometallurgical Mapping and Mine Modelling*; Technical Report, AMIRA Project P843; University of Tasmania: Hobart, Australia, 2013.
101. Wills, B. *Wills' Mineral Processing Technology: An Introduction to the Practical Aspects of Ore Treatment and Mineral Recovery*; Elsevier: Amsterdam, The Netherlands, 2011.
102. Fandrich, R.; Gu, Y.; Burrows, D.; Moeller, K. Modern SEM-based mineral liberation analysis. *Int. J. Miner. Process.* **2007**, *84*, 310–320. [CrossRef]
103. Goodall, W.; Butcher, A. The use of QEMSCAN in practical gold deportment studies. *Miner. Process. Extr. Metall.* **2012**, *121*, 199–204. [CrossRef]
104. Gu, T. Automated scanning electron microscope based mineral liberation analysis an introduction to JKMRC/FEI mineral liberation analyser. *J. Miner. Mater. Charact. Eng.* **2003**, *2*, 33. [CrossRef]
105. Senesi, G.S. Laser-Induced Breakdown Spectroscopy (LIBS) applied to terrestrial and extraterrestrial analogue geomaterials with emphasis to minerals and rocks. *Earth Sci. Rev.* **2014**, *139*, 231–267. [CrossRef]
106. Bolger, J. Semi-quantitative laser-induced breakdown spectroscopy for analysis of mineral drill core. *Appl. Spectrosc.* **2000**, *54*, 181–189. [CrossRef]
107. Potts, P. Laboratory methods of analysis. In *Analysis of Geological Materials*; Riddle, C., Ed.; Marcel Dekker: New York, NY, USA, 1993; pp. 123–220.
108. Tamura, S.; Horisawa, H.; Yamaguchi, S.; Yasunaga, N. Laser ablation of sapphire with a pulsed ultra-violet laser beam. In *Initiatives of Precision Engineering at the Beginning of a Millennium*; Springer: Berlin, Germany, 2002; pp. 224–228.
109. Lawley, C.; Creaser, R.A.; Jackson, S.E.; Yang, Z.; Davis, B.; Pehrsson, S.; Dubé, B.; Mercier-Langevin, P.; Vaillancourt, D. Unravelling the western Churchill Province Paleoproterozoic gold metallotect: Constraints from Re-Os arsenopyrite and U-Pb xenotime geochronology and LA-ICPMS arsenopyrite trace element chemistry at the BIF-hosted Meliadine Gold District, Nunavut, Canada. *Econ. Geol.* **2015**, *110*, 1425–1454. [CrossRef]
110. Lawley, C.; Kjarsgaard, B.; Jackson, S.; Yang, Z.; Petts, D.; Roots, E. Trace metal and isotopic depth profiles through the Abitibi cratonic mantle. *Lithos* **2018**, *314–315*, 520–533. [CrossRef]
111. R Development Core Team. R: A Language and Environment for Statistical Computing, Vienna. 2018. Available online: https://www.r-project.org (accessed on 12 February 2018).
112. Wehrens, R. *Chemometrics with R*; Springer: Berlin/Heidelberg, Germany, 2011.
113. Samson, I.; Gagnon, J. Episodic fluid infiltration and genesis of the Proterozoic MacLellan Au-Ag deposit, Lynn Lake greenstone belt. *Explor. Min. Geol.* **1995**, *4*, 33–50.
114. Samson, I.M.; Blackburn, W.H.; Gagnon, J.E. Paragenesis and composition of amphibole and biotite in the MacLellan gold deposit, Lynn Lake greenstone belt, Manitoba, Canada. *Can. Mineral.* **1999**, *37*, 1405–1421.

115. Hastie, E.C.G.; Gagnon, J.E.; Samson, I.M.; Lake, L. The Paleoproterozoic MacLellan deposit and related Au-Ag occurrences, Lynn Lake greenstone belt, Manitoba: An emerging, structurally-controlled gold camp. *Ore Geol. Rev.* **2018**, *94*, 24–45. [CrossRef]
116. Cook, N.J.; Ciobanu, C.L.; Pring, A.; Skinner, W.; Shimizu, M.; Danyushevsky, L.; Saini-eidukat, B.; Melcher, F. Trace and minor elements in sphalerite: A LA-ICPMS study. *Geochim. Cosmochim. Acta* **2009**, *73*, 4761–4791. [CrossRef]
117. Laser-Induced Breakdown Spectroscopy—An Emerging Analytical Tool for Mineral Exploration. Available online: https://appliedspectra.com/products.html# (accessed on 10 November 2019).

Article

3D Mineral Prospectivity Modeling for the Low-Sulfidation Epithermal Gold Deposit: A Case Study of the Axi Gold Deposit, Western Tianshan, NW China

Xiancheng Mao [1], Wei Zhang [1,2], Zhankun Liu [1,3,*], Jia Ren [1], Richard C. Bayless [1] and Hao Deng [1]

1 Key Laboratory of Metallogenic Prediction of Nonferrous Metals and Geological Environment Monitoring (Ministry of Education), School of Geosciences and Info-Physics, Central South University, Changsha 410083, China; mxc@csu.edu.cn (X.M.); viviengis@csu.edu.cn (W.Z.); renjia10281230@163.com (J.R.); richardbayless02@gmail.com (R.C.B.); haodeng@csu.edu.cn (H.D.)

2 School of Municipal and Mapping Engineering, Hunan City University, Yiyang 413000, China

3 Department of Geology, Lakehead University, 955 Oliver Road, Thunder Bay, ON P7B 5E1, Canada

* Correspondence: zkliu0322@csu.edu.cn; Tel.: +86-7318-887-7571

Received: 20 January 2020; Accepted: 1 March 2020; Published: 4 March 2020

Abstract: The Axi low-sulfidation (LS) epithermal deposit in northwestern China is the result of geological controls on hydrothermal fluid flow through strike-slip faults. Such controls occur commonly in LS epithermal deposits worldwide, but unfortunately, these have not been quantitatively analyzed to determine their spatial relationships with gold distribution and further guide mineral prospecting. In this study, we conduct a 3D mineral prospectivity modeling approach for the Axi deposit involving 3D geological modeling, 3D spatial analysis, and prospectivity modeling. The spatial analysis of geometric features revealed the gold mineralization trends in convex segments (0–20 m) with a specific distance from fault 2, the lower interface of late volcanic phase, and the upper interface of phyllic alteration with steep slopes (>65°), implying that gold deposition was significantly controlled by the morphological characteristics and distance fields of geologic features. The present alteration–mineralization zone at Axi has a larger width in bending sites (sections No. 35–15 and No. 40–56) than elsewhere, indicating the location of two fluid conduits extending to depth. The prediction-area plots and receiver operating characteristic curves demonstrated that (genetic algorithm optimized support vector regression (GA-SVR)) outperformed multiple nonlinear regression and fuzzy weights-of-evidence, which was proposed as a robust method to solve complicated nonlinear and high-dimensional issues in prospectivity modeling. Our study manifests spatial controls of structure, host rock, and alteration on LS epithermal gold deposition, and highlights the capability of GA-SVR for identifying deposit-scale potential epithermal gold mineralization.

Keywords: 3D mineral prospectivity modeling; spatial analysis; GA-SVR; epithermal gold deposits; Axi deposit

1. Introduction

Low-sulfidation (LS) epithermal deposits with major Ag and Au production are formed in shallow parts (<1 km) of magmatic-hydrothermal systems at moderate-low temperature conditions of 100 °C to 300 °C [1–5]. The ore-forming fluids of LS epithermal deposits have been fully investigated and identified as dominantly meteoric water with minor but crucial additions of magmatic water, and their mixing, boiling, and interaction with country rocks are proposed as principal mineralization mechanisms [5–8]. In spite of the consensus about metallogenic aspects, the spatial pattern of

LS epithermal mineralization at the deposit scale remains poorly understood. One fundamental understanding of the localization of LS epithermal mineralization is the control on metal transport and deposition during epizonal fluid flow [1,9–13]. It, to some extent, requires a special tool to quantitatively analyze the spatial association between the metal distribution and geologic features plausibly related to mineralization [14–16]. However, research regarding geological controls on LS epithermal mineralization at the deposit-scale is rarely conducted, hindering the further understanding of the LS epithermal ore-forming processes and mineral exploration.

3D spatial analysis has been demonstrated as an effective approach for identifying the spatial features of geological objects in 3D space and extracting diverse potential ore-controlling features for 3D prospectivity modeling [17–22]. However, the spatial association between ore and ore-controlling features would be significantly more complicated with the expansion of geological features in space [23–25]. In terms of the non-linearity and high-dimensionality of spatial data, traditional 3D prospectivity methods remain inadequate, thereby resulting in larger error and bias in prospectivity modeling [19,20]. Recent studies employ machine learning algorithms in prospectivity modeling, for they have good spatial dataset processing performance unrelated to statistical distribution and automatically generate high-level representations from complex data [19,23–28]. Genetic algorithm optimized support vector regression (GA-SVR), a supervised learning regression method extended from support vector machine (SVM) with parameters optimized by a genetic algorithm, is a machine learning algorithms with better stability and generalization abilities in trouble-shooting complicated nonlinear and high dimensional issue and may be a promising tool in 3D prospectivity modeling [29–32].

The Axi gold deposit, as the largest LS epithermal deposit in the West Tianshan, has been the subject of extensive geological studies and drill exploration, that provide important insights into ore genesis and abundant spatial information of geological bodies. In this study, the Axi deposit was chosen as a representative research case of LS epithermal deposits to implement improved 3D prospectivity modeling, including 3D geological modeling, multiple spatial analysis of potential ore-controlling geological features, and GA-SVR prospectivity modeling. The work presented herein focuses on the spatial pattern of LS epithermal mineralization and its controls from geological features, as well as reducing bias and uncertainty in prospectivity modeling by a machine learning method. Its implementation would help us (1) to better understand the gold spatial deposition during fluid flow, (2) to guide deep mineral exploration at Axi, (3) to verify the availability of GA-SVR in prospectivity modeling, and finally, (4) to provide the reference for mineral prospecting in other LS epithermal gold deposits.

2. Geological Background

2.1. Regional and Deposit Geology

The Tulasu Basin in Western Tianshan, NW China, hosting several Late Paleozoic porphyry, skarn, and epithermal deposits, is sandwiched between the NWW-striking North Yili Basin Fault and South Keguqinshan Mountain Fault (Figure 1) [33–35]. The basement of the Tulasu basin is mainly subdivided into the lower part with mesoproterozoic and neoproterozoic sedimentary clastic rocks and the upper part with ordovician shallow marine clastic rocks with interlayered carbonates [36,37]. It is unconformably covered by paleozoic strata of the lower carboniferous Dahalajunshan formation and overlying the lower carboniferous aqialehe formation. The Dahalajunshan formation, generated in a continental arc setting related to the subduction of the North Tianshan Ocean [33,38,39], consists of tuff, basaltic andesite, andesite, dacite, and rhyolite, with minor volcanoclastic breccia. The aqialehe formation is mainly comprised of conglomerate, sandstone, and limestone [38]. The granitic intrusions (e.g., diorite, granodiorite, and granitic porphyry) exposed in the Tulasu basin commonly intrude into the Dahalajunshan andesite, with the early carboniferous age of ca. 357–348 Ma (Figure 1a) [40].

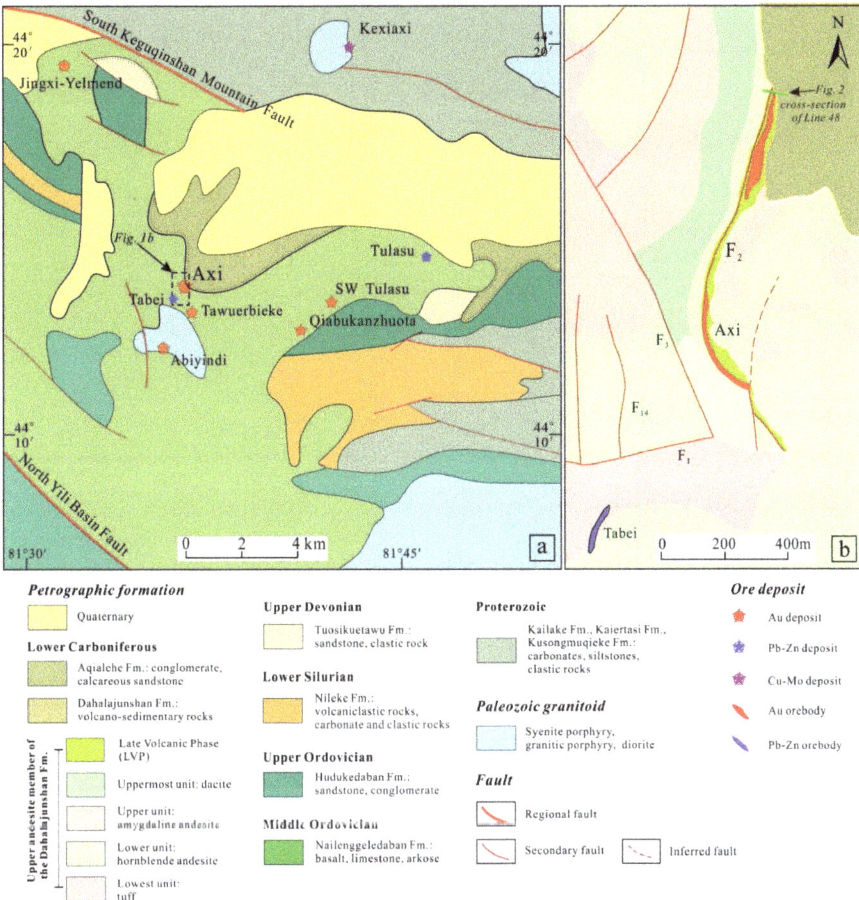

Figure 1. (**a**) Geological map of the Tulasu basin showing the locations of porphyry and epithermal deposits. (**b**) Geological map of the Axi district. Both subfigures are based on Liu et al. [35].

The Axi gold deposit, located in the central part of the Tulasu Basin, is hosted in andesitic and dacitic volcanic rocks and breccia of the uppermost member of the Dahalajunshan formation (Figure 1b) [34,35,37]. The strike-slip fault (F_2), which is N-trending and east-dipping, controls nearly all economic orebodies in the Axi gold deposit. The radial fault series developed in the western segment rarely exhibit alteration and mineralization, rather, these likely reflect a structural overprinting on the late evolution of a diatreme at Axi. The fault F_2 displays variable strikes from south to north due to multiple phases of deformation associated with pre-ore strike-slip extension, syn-ore normal-strike slip extension, and post-ore compression and overprinting by secondary reverse faults (Figure 1b) [41]. The well-developed alteration-mineralization zoning pattern in the Axi epithermal deposit, from the core outward, mainly consists of silicification, quartz-carbonate, pyrite-sericite-quartz, and propylitic alteration (Figure 2) [33–36]. In general, pyrite-sericite-quartz and silicification close to gold mineralization occur in the hanging wall of fault F_2, and the pervasive propylitic alteration affects all lithologies in the Axi district.

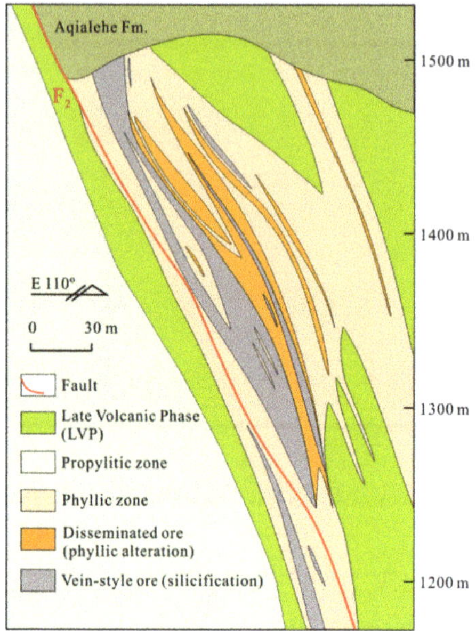

Figure 2. Representative cross-section (Line 48 on Figure 1b) illustrating the spatial relationships among orebodies, alteration, and fault in the Axi deposit. After Liu et al. [37].

Gold orebodies in the controlled exploration engineering area extend approximately 1300 m from north to south (Figure 1b), with dipping angles of 45° to 80°, an average thickness of 20 m and the maximum depth of about 425 m. Nine orebodies that generally parallel fault F_2 in the hanging wall have been identified. These gold orebodies exhibit a similar spatial variation to fault F_2, both in vertical and horizontal directions (Figure 2). Based on the ore textures, compositions, and crosscutting relationships, two subtypes of primary gold ore were identified: disseminated ore (20% of documented gold resources) and quartz-vein ore (80% of documented gold resources) [37]. The disseminated ore is hosted in rocks with strong phyllic alteration that typically has a gray to yellowish-brown color and a relatively lower gold grade compared to vein ore (2 g/t vs. 5.6 g/t). Quartz-vein ore with a width of 0.1 m to 20 m consists mostly of gray, smoky gray quartz-sulfide veins, stockworks, and breccias. The major orebodies commonly contain phyllic altered fragments that were replaced and cemented by quartz-sulfide veins. The absolute ages via auriferous pyrite Re-Os dating for the two types of ore are ca. 356–348 Ma (disseminated) and ca. 332 Ma (veins) [37]. Gold-bearing minerals in disseminated ore are described as coarse-grained euhedral pyrite, whereas the gold in quartz-vein ore is typically related to electrum, native gold, zoned pyrite, arsenopyrite, and marcasite [33–36]. Gangue minerals in both types of ore are mainly quartz, chalcedony, illite, chlorite, adularia, calcite, and barite.

2.2. Metallogenic Model

The formation of the Axi epithermal gold deposit involved two major stages; the disseminated ore was mainly formed in the early ore-stage whereas the vein-type ore was generated in the main ore-stage [35,37]. In the early ore-stage, ore-forming fluids were predominantly magmatic water and had high-temperature and a high concentration of Cu, Ni, Co, Mn, HS^-, Cl^- [15]. Relatively hot hydrothermal fluids ascended from depth along faults or cracks to shallower levels. These structures, as fluid conduits, were beneficial for gold deposition and hydrothermal alteration when fluid infiltrated and diffused along fracture zones and interplayed with country rocks [35]. During the later ore-stage, ore-forming fluids were progressively diluted and cooled by significant inputs of meteoric water [33,35,36]. Gold was transported as Au-bearing chloride- and sulfur- complexes in ore-forming fluids and the mechanism of gold deposition was interpreted as fluid boiling caused by hydraulic fracturing and/or seismic activity in fluid conduits [33,35,36].

The following empirical exploration markers are recognized in the Axi deposit:

(1) Gold occurrence is structurally controlled by the development of fault F_2, generally occurring in the hanging wall of F_2. The spatial distribution of gold mineralization is closely associated with the geometry of F_2. The dilational sites along F_2 with changing dip angle always hold high-grade orebodies.

(2) The late volcanic phase rock (LVP) is the major host rock for gold orebodies. Ore-forming fluids preferentially interacted with LVP that had undergone brecciation and structural deformation. The gold deposition is possibly related to the location and shape of LVP.

(3) Orebodies are commonly within regions that have undergone phyllic alteration and silicification, and high-economic lodes are located in the center of alteration zonation, i.e., silicification. Typically, the scale of the gold mineralization zone is positively correlated to the thickness of alteration zone.

3. Methodology

3.1. Datasets

In this study, data employed to conduct 3D prospectivity modeling of the Axi gold deposit covered several surface geological maps, 29 cross-sections with 40 m intervals, 211 drill holes with a maximum depth of 653.7 m, as well as the gold assay data of 13,628 samples. These datasets were provided by Western Region Gold Co., Ltd. (Urumqi, Xinjiang Uygur Autonomous Region, China). Databases may be requested by contacting the primary author.

3.2. 3D Geological Modeling

It is notable that 3D geological modeling was the foundation for spatial analysis and prospecting models, and its quality is closely related to the complexity of geological features, geological interpretation, and exploration data density [20,42]. Two categories of 3D geological modeling methods exist, explicit and implicit. Explicit modeling is more applicable to 3D reconstruction of deposits with abundant geological data with low-uncertainty and can be made more accurate by skilled geological interpretation [20,43,44]. Conversely, implicit modeling is suitable for 3D geological modeling at the district-scale, where geological data is sparse and the geology is poorly understood [43,44].

Since an extensive geological database with comprehensive geological data and many scholarly geological interpretations exist for the Axi deposit, we preferred using a data- and knowledge-driven explicit modeling method. We constructed a series of 3D geological models with the SKUA-GOCAD™ software program (version 2018, Emerson Paradigm Holding LLC, MO, USA), consisting of orebodies, fault F_2, late volcanic phase (LVP), and phyllic alteration zone (Figure 3).

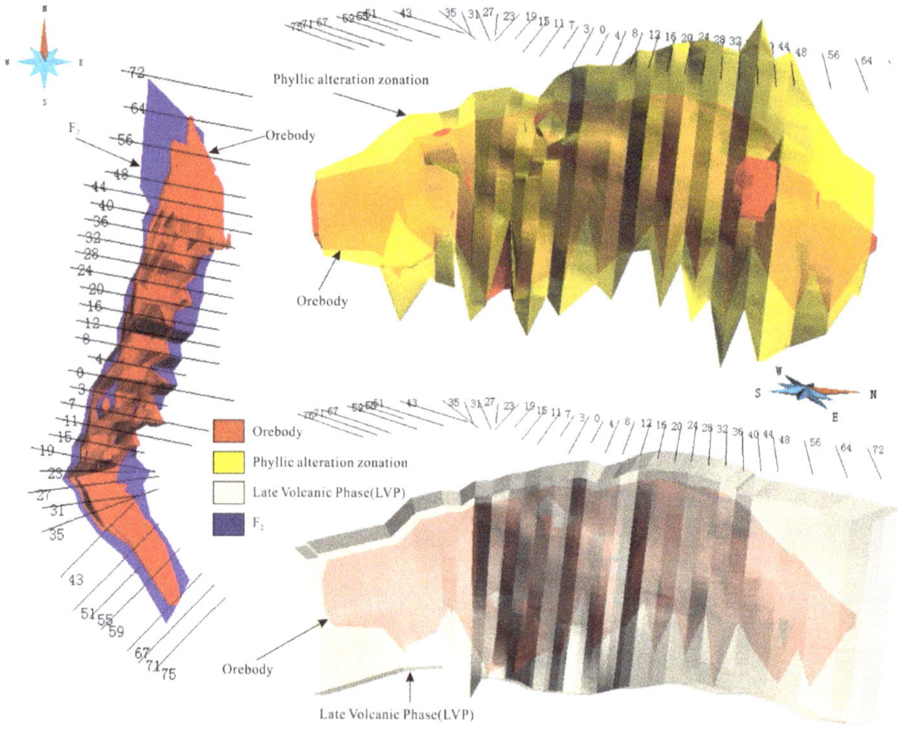

Figure 3. Above; 3D geological models of orebodies, phyllic alteration zone, late volcanic phase (LVP), and fault F_2 in the Axi gold deposit.

3.3. *Spatial Analysis of Geological Features*

Many highly regarded studies have determined that the spatial distribution and morphology of geological features control orebody occurrence in epithermal deposits [3,43,45]. Thus, the spatial analysis of geological features is crucial for providing effective mineral exploration criteria [17,18,46–48]. In this study, multi-spatial analysis methods were integrated to quantify their relative spatial control on gold mineralization and further generate a series of predictive maps by spatial distance analysis, slope analysis, and morphological undulation analysis.

The location and enrichment of gold orebodies in the Axi deposit are spatially associated with geological features of F_2, LVP, and phyllic alteration zone. The spatial distance analysis method was performed to quantitatively represent the influence of ore-controlling factors [19,20]. We defined distance value (d) as the minimum Euclidean distance between each voxel in 3D prospecting space and the ore-controlling factors and used a sign on the Euclidean distance to depict orientation relationships. For a given voxel, $d < 0$ indicates the voxel is located in the hanging wall of the geological body, inversely $d > 0$ illustrated it situated in the footwall.

Gold deposition from ore fluid linked to the destabilization of Au-bearing complexes was affected by geometric features [1,3,47,48]. Slope, as a significant indicator for quantitatively representing geometry, was computed. In this study, 3D geological models of F_2, LVP and the phyllic alteration zone were expressed as TIN models with enormous triangular patches, and the plane equation for each patch can be formulated as $z = ax + by + c$. The slope was defined as the angle between the triangular patch and horizontal surface, calculated as $g = \arccos\frac{1}{\sqrt{a^2+b^2+1}}$.

Changes in the slope of geological surfaces, such as the changes in the slope of a fault plane, commonly determine the formation and location of orebodies [18–20]. We conducted a morphological undulation analysis method to compute undulation based on trend analysis [18,20,48]. It was subdivided into two steps: (1) extracting trend surface $T(x, y, z)$ from discrete smooth interpolation (DSI) to original geological interface $Z(x, y, z)$; (2) for each point in $Z(x, y, z)$, calculating residual value $R(x, y, z)$, defined as the minimum Euclidean distance from $T(x, y, z)$ to $Z(x, y, z)$ in elevation. $R(x, y, z) > 0$ indicated a uplift shape, oppositely, $R(x, y, z) < 0$ illustrated a depression shape.

3.4. Prospectivity Modeling and Assessing

Prospectivity modeling entails the analysis and synthesis of various predictive maps derived from the spatial analysis of geological models. Diverse predictive maps can be treated as n-Dimension feature vectors with a nonlinear relationship, hence prospectivity modeling actually is a solution for the complex nonlinear and high dimensional problem [24,49]. This problem is always treated as a classification process that divides the study area into favorable and unfavorable parts [50–52], but essentially, it is a multivariate nonlinear regression problem to quantify the spatial association between predictive maps and gold occurrence [24,53]. Support vector regression (SVR), as a supervised learning regression algorithm branching from SVM, has been widely applied for solving the nonlinear regressing estimation and time series forecasting issues [29,31,54,55] and has also been used in geological engineering and the mineral industry [56,57]. Additionally, it can also be a preferable solution for performing complicated nonlinear prospectivity modeling. In this study, a hybrid GA-SVR (genetic algorithm optimized support vector regression) was adopted for resource evaluation and prediction analysis. Simultaneously, both multiple nonlinear regression (MNR) and the fuzzy weights-of-evidence (fuzzy WofE) method were chosen for comparative assessment through receiver operating characteristic (ROC) and prediction-area (PA) analysis [58,59].

3.4.1. Support Vector Regression Model

SVM, proposed on the basis of statistical learning theory, is a supervised learning algorithm for pattern classification [50,51,60]. It implements the global optimum based on the structural risk minimization principle [60]. Extended from SVM for trouble-shooting of nonlinear regression issues, SVR attempts to fit an optimal approximating hyperplane to the training set, through mapping nonlinear input datasets into a multi-dimensional or hyper-dimensional feature space [56,60,61]. Thus, the optimal hyperplane quantitatively describes the relationship between features and target values in the training dataset, to predict the future target values with input features. This method has been shown to be superior in minimizing the expected error and reducing the problem of over-fitting [56,60,62]. ε-SVR, as a popular SVR model, locates the hyperplane with an epsilon-insensitive (ε-insensitive) loss function [60,61]. The ε-SVR function can be obtained as formula (1),

$$\mathrm{f}(\mathrm{x}) = \sum_{i=1}^{l} (a_i - \hat{a}_i) k(x, x_i) + b \tag{1}$$

where $\mathrm{f}(\mathrm{x})$ denotes regression function, x is the training sample, a_i and $\hat{a}_i$ is Lagrange multiplier, $k(x, x_i)$ is the kernel function, x_i is the i-th eigenvector of x (the vector of the i-th predictive map), and b is the offset.

The kernel function $k(x, x_i)$ is the key for mapping nonlinear input data into a high dimensional space. Diverse kernel functions, such as the polynomial kernel function, the Sigmoid kernel function and the Gaussian radial basis function (RBF), have been used in previous research. Among them, RBF is the most conventionally devoted to ε-SVR modeling, for its high efficiency and lower deviation in sample training [63]. It is defined as follows:

$$k(x,x') = exp\left(\frac{-\|X_j - X_i\|^2}{2\sigma^2}\right) \quad (2)$$

where σ stands for the width of the RBF, controlling the solution complexity.

In the aforementioned ε-SVR model with RBF, three parameters need to be optimized: the penalty parameter (c), the insensitive loss function parameter (ε), and the width parameter of the RBF (σ), respectively representing the trade-off between training deviation and model complexity, the sparseness of solution, and the complexity of solution [61,62].

The traditional parameter optimization methods (e.g., grid search, gradient descend, and cross-validation) are insufficient in search efficiency and accuracy [31,54]. Genetic algorithm (GA), an adaptive heuristic algorithm mimicking biological evolution, is one type of global optimization algorithm that has proved to be superior in parameter selection, because of its simple structure, good robustness and intelligence searching [30–32,64]. GA treats parameters to be optimized as chromosomes, and determines the optimal model via a computational iterative process based on the survival of the fittest in a population, realizing intelligent exploitation of a random search in a predefined searching space [30,32]. With a good performance in finding the global solution for the optimization problem, GA has been gradually applied to synchronously optimize parameters of Machine Learning, and then hybrid algorithms (e.g., GA-SVR and genetic regulatory networks) have been proposed for predication [64–67]. Particularly, plenty of studies demonstrate that GA-SVR improves the generalization ability and forecasting accuracy, which has been extensively employed in various research fields [64–66].

The hybrid GA-SVR (genetic algorithm optimized support vector regression) that was adopted for prospectivity modeling involved several processes (Figure 4): (1) normalizing all predictive maps proposed in Section 3.3 into values with zero mean, to eliminate the differences of units and magnitudes between these selected parameters; (2) dividing known data into two datasets, namely a training dataset and a test dataset. The former is applied to train the prospectivity model, whereas the latter is used for model assessment; (3) encoding parameters (c, ε, σ) to be optimized, and determining initial population and maximum iterations; (4) training initial population with SVR, and then calculating the fitness function; (5) if stopping criteria have been met, parameters optimization would be finished with maximum fitness function defined as preferential parameters. Otherwise, go to the next step; (6) generating a new population of chromosomes by genetic operations, including selection, crossover, and mutation, and then return steps (4) and (5).

The GA was implemented by using GATBX toolbox of MATLAB (developed by The University of Sheffield, https://www.shef.ac.uk/polopoly_fs/1.60188!/file/manual.pdf). We adopted a binary-coded genetic algorithm with stochastic universal sampling selection, single point crossover, and discrete mutation. Selective probability, crossover probability, and mutation probability were 0.9, 0.7 and 0.035 respectively. Initial population size was chosen as 20, and ranges of (c, ε, σ) were defined as [0, 100], [0.01, 1] and [0, 100] respectively. Ultimately, c = 1.17, ε = 0.01, and σ = 16.86 were fixed as optimum parameters for the construction of SVR. SVR was performed by using LIBSVM toolbox of MATLAB [68]. Furthermore, we compared GA-SVR and GS-SVR (grid search optimized SVR) via RMSE (root mean squared error), MAE (mean absolute error), R^2 (squared correlation coefficient) with known data (Table 1). The results of this comparison showed a more robust ability of GA-SVR relative to GS-SVR for the mineral prospectivity data.

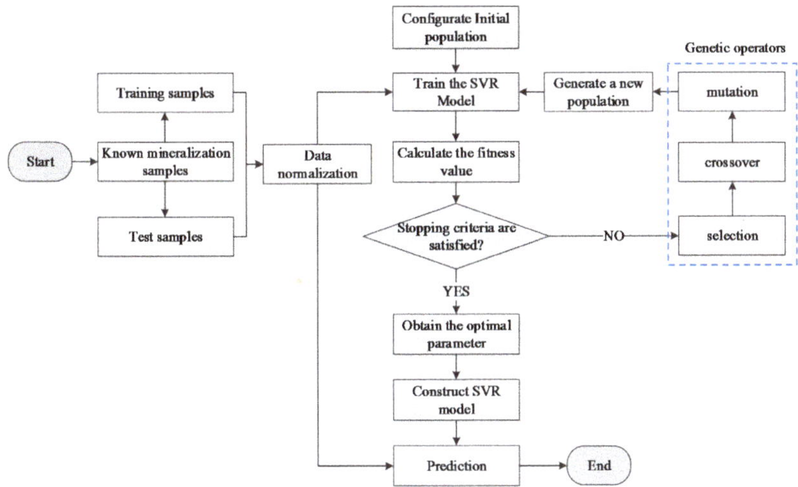

Figure 4. Genetic algorithm optimized support vector regression (GA-SVR) flowchart involved in prospectivity modeling for the Axi epithermal gold deposit.

Table 1. Performance comparison of GA-SVR and grid search optimized support vector regression (GS-SVR).

Training Set	RMSE	MAE	R^2
GA-SVR	0.49	0.41	0.89
GS-SVR	1.12	0.58	0.71

3.4.2. Comparison and Assessment

Known mineralization samples in the prospectivity space are thought to be an effective verification for model validity assessment [19,20,59]. To test the performance of SVR, another two typical methods, MNR and fuzzy WofE [69,70], a statistics evaluation model and a Bayesian probabilistic model were chosen as comparison methods.

The aforementioned methods were compared in terms of PA plot and ROC curve. In the P-A plot, two curves, namely the prediction-rate curve and the occupied area curve, are mapped for estimating the ability of each to predict mineralization, mineral occurrences and the sizes of predicted target areas [58]. Additionally, the point where curves intersect straightway reflects the probability of mineral deposit occurrence. When the intersection point shows a higher prediction rate and smaller occupied area, it indicates high mineralization probability. ROC curve, a validation technique widely used in machine learning, has been introduced into prospectivity models assessment [24,59,71]. It takes both the true positive rate and false positive rate into consideration. Additionally, the area under the ROC curve (AUC), with values varying from 0 to 1, can be used to measure the performance of prospectivity models. An AUC value of 1 indicates the result has perfect accuracy, whereas an AUC value below 0.5 determines the result is a random model [59,71].

4. Results

4.1. Gold Distribution

Based on the gold assay data and the 3D model of orebodies, gold grade (Au) was assigned to every voxel of orebodies via the 3D Kriging interpolation method [20,72]. The 3D assay model displays a discontinuous gold spatial distribution and an obvious gold-enrichment in the footwall of the orebody in the north domain (Figure 5). Gold mineralization with Au > 3 g/t in the north domain is mainly confined into sections of No. 48 and No. 12, where mineralization extends about 300 m from south to north. This area is a gently undulating (overall trend NNE) with a maximum thickness of orebody (about 40 m). In the northernmost orebody changes to NNW-striking, roughly paralleling with F_2 (Figure 2) and maintains a relatively high grade proximal to the fault. Industrial-grade mineralization occurs in the domain between sections No. 35 and No. 15 (Figure 5). This area is characterized by an arcuate zone, and high-grade mineralization occurs as a lens. Overall, the Axi orebodies expand in sections of No. 48–12 and No. 27–35 and bend in the vicinity of sections No. 44 and No. 23, emerging, enlarging and narrowing with undulations along the dip and strike (Figure 5).

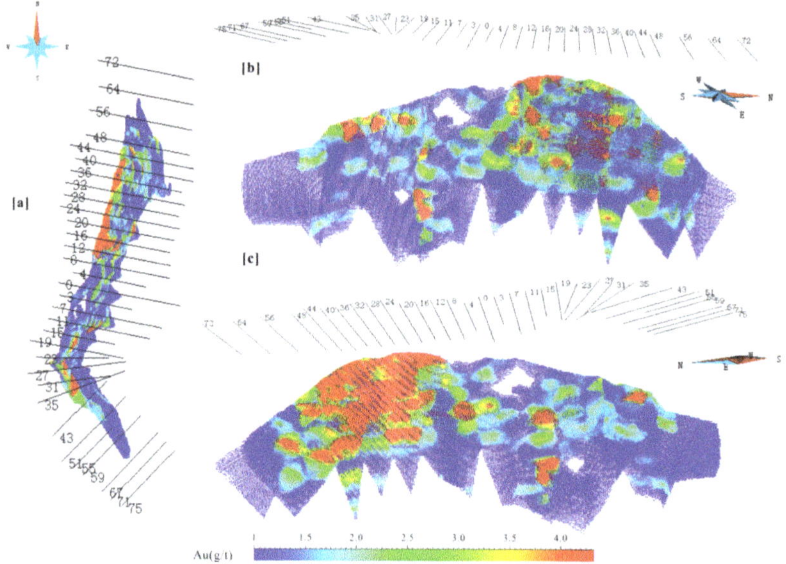

Figure 5. Above; 3D assay models showing the spatial distribution of gold grade of each voxel in the horizontal (**a**) and vertical (**b**,**c**) directions in the Axi deposit.

4.2. Spatial Analysis of Geological Features

The spatial distance analysis results demonstrate that the higher values (yellow to red) of distance to F_2, LVP, and the upper interface of the phyllic alteration zone are similarly confined in the areas between sections No. 56 and No. 16, and No. 35 and No. 15. The *d*F, *d*GV and *d*Alt show overall lower values in the south and higher values in the north (Figure 6a–c). Gold occurrences are principally located in the hanging wall of F_2 and LVP lower interface with a distance of 0 to 50 m (Figure 6a,b), and in the footwall of phyllic alteration upper interface zone with a distance of 0 to −80 m (Figure 6c). Distinct bimodal distribution has been found between the gold grade and distances (*d*F, *d*GV, and *d*Alt) (Figure 7(a1–c1)), showing multiple-peak relationships with gold grade. For instance, the Au-*d*Alt has two peaks, the flat peak is in the range from 40 to 60 with a grade peak of 3.4 g/t, whereas the sharp peak centralizes closely to 95 with a grade peak of 3 g/t (Figure 7(c1)). However, the relationship between AuMet and distances (*d*F, *d*GV, and *d*Alt) is similar to left-skewed Gaussian distribution with peaks respectively in 30, 40 and 50 (Figure 7(a2–c2)).

The slope extraction results show that the highest values of *g*F are concentrated near-surface between sections No. 36 and No. 4, but at different depths between sections No. 11 and No. 27 (Figure 6d). The reverse is true for *g*Alt, which is lower at shallow depth (Figure 6f). Gold is concentrated in relatively sloped areas (>65°; Figure 7(d1–f1)). Analogously, there is a bimodal distribution between the slopes (*g*F, *g*GV, and *g*Alt) and grade (Figure 7(d1–f1)), such as Au vs. *g*F which has two peaks, 65°

and 85° (Figure 7(d1)). A right-skewed distribution between the slope and AuMet demonstrates the segments with steeper slopes host a higher percentage of gold than other intervals (Figure 7(d2–f2)).

The morphological undulation analysis displays there are several distinct high-value centers of *wr*F, *wr*GV, and *wr*Alt in the areas between sections No. 48 to 32, No. 12 to 3, and No. 19 to 43, respectively (Figure 6g–i). Concentrated gold distribution areas of undulation (*wr*F, *wr*GV, and *wr*Alt) are in the interval of 0 to 20 m, −10 to 10 m, and 0 to 20 m, respectively (Figure 7(g1–i1)). The relationships between the undulation (*wr*F, *wr*GV, and *wr*Alt) and grade show a multimodal distribution (Figure 7(g1–i1)). Differing from the normal distribution of the other AuMet undulations, the relationship between AuMet and *wr*F is a bimodal distribution with two peaks of −2 and 16 (Figure 7(i2)).

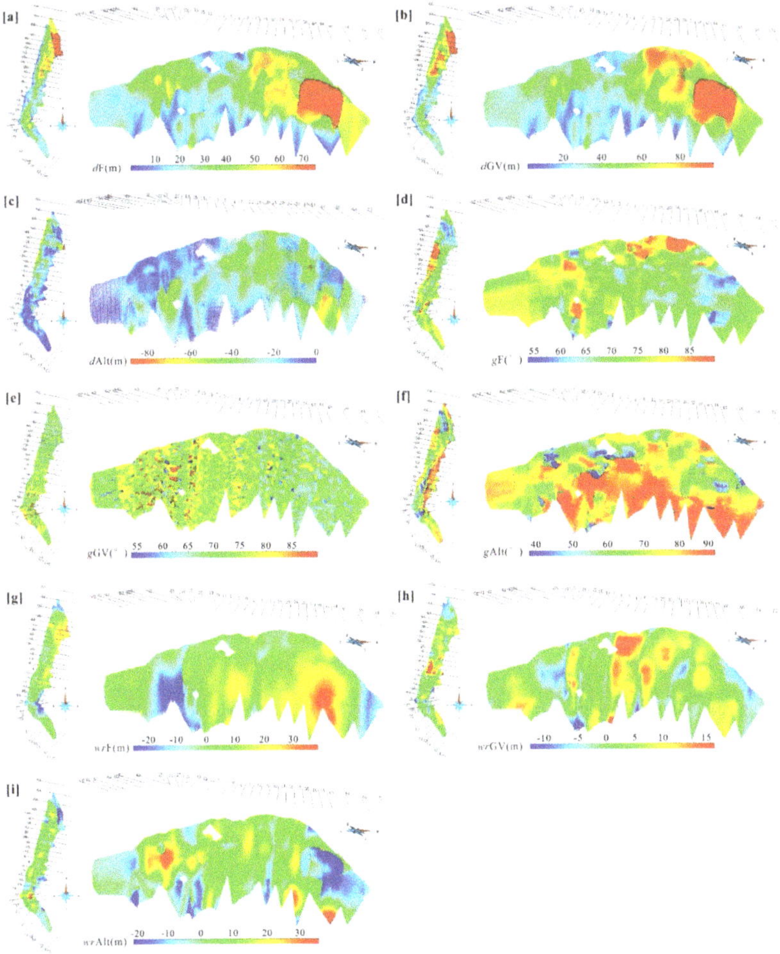

Figure 6. Above; 3D raster models showing the distribution of geological features of the Axi deposit. (**a**) *d*F (distance value to F_2); (**b**) *d*GV (distance value to the lower interface of LVP); (**c**) *d*Alt (distance value to the upper interface of the phyllic alteration zone); (**d**) *g*F (slope of F_2); (**e**) *g*GV (slope of the lower interface of LVP); (**f**) *g*Alt (slope of the upper interface of the phyllic alteration zone); (**g**) *wr*F (undulation of F_2); (**h**) *wr*GV (undulation of the lower interface of LVP); and (**i**) *wr*Alt (undulation of the upper interface of the phyllic alteration zone).

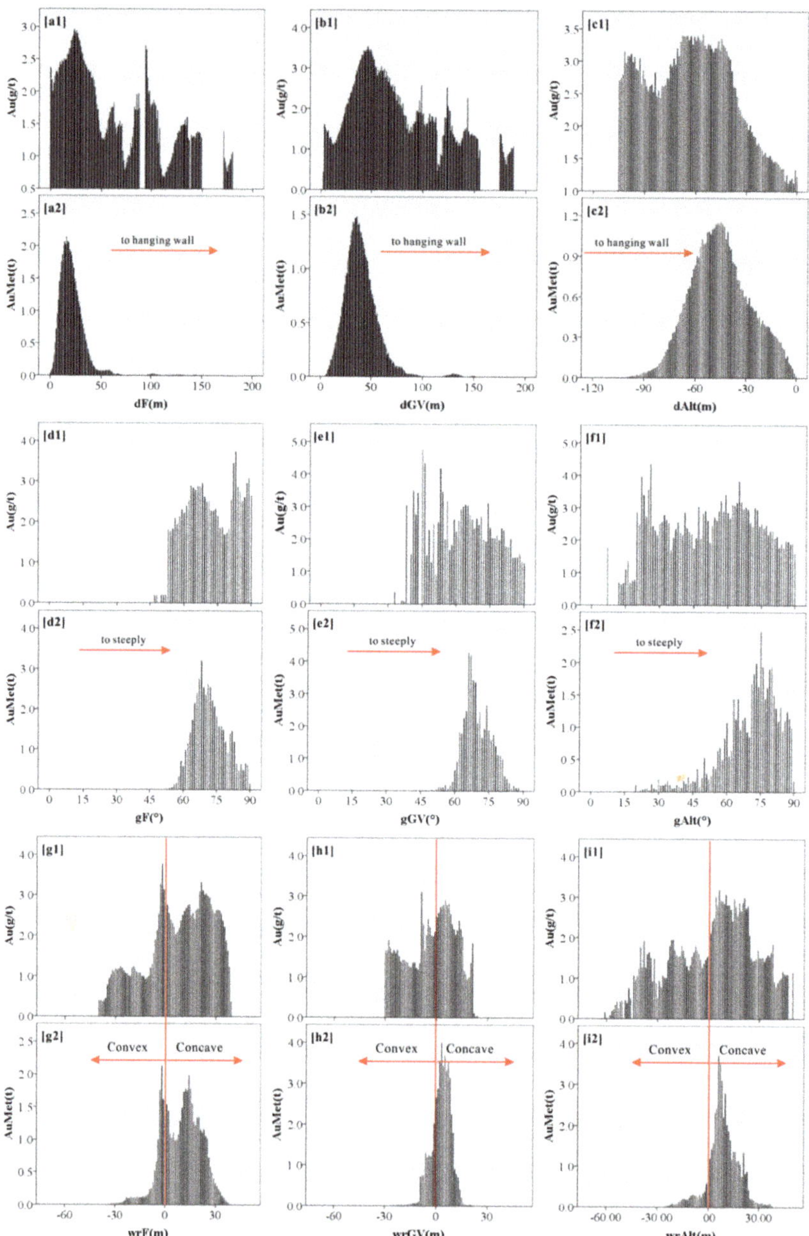

Figure 7. Bar charts of geological features and mineralization indexes. (**a**) *d*F v.s. Au and AuMet; (**b**) *d*GV v.s. Au and AuMet; (**c**) *d*Alt v.s. Au and AuMet; (**d**) *g*F v.s. Au and AuMet; (**e**) *g*GV v.s. Au and AuMet; (**f**) *g*Alt v.s. Au and AuMet; (**g**) *wr*F v.s. Au and AuMet; (**h**) *wr*GV v.s. Au and AuMet; (**i**) *wr*Alt v.s. Au and AuMet. Au stands for the gold grade, and AuMet is the gold resource summation of all ore voxels in the same factor value.

4.3. The Efficiency of Prospectivity Models

The P-A plots of three prospectivity models show that GA-SVR had a prediction rate of 74% in 26% of the study area; 62% of the known ore voxels were predicted in 38% of the study area by MNR, and only 55% of the known occurrences were forecasted in 45% of the study area via fuzzy WofE (Figure 8a–c). In ROC curves, GA-SVR models with the AUC value of 0.95 exhibits a better performance of prospectivity, whereas MNR and fuzzy WofE are less accurate, with the same AUC value of 0.73 (Figure 8d).

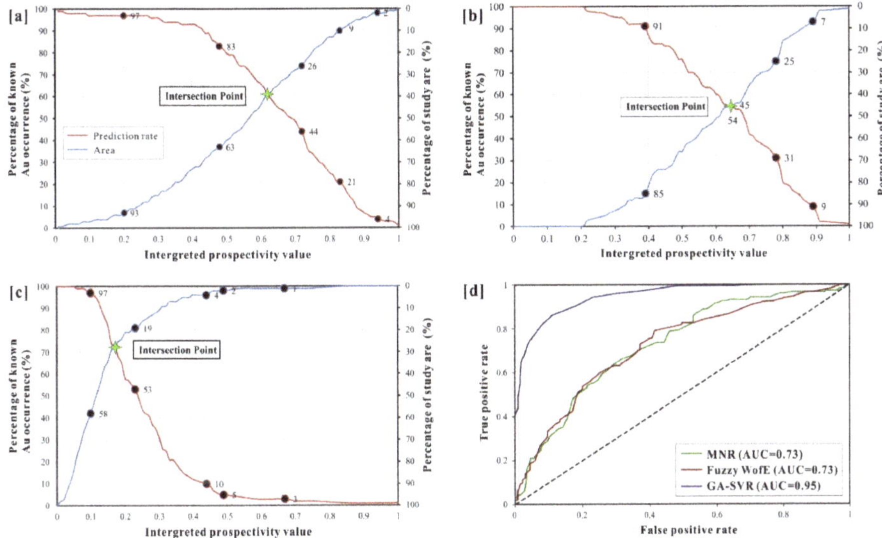

Figure 8. Prediction-area (P-A) plots and receiver operating characteristic (ROC) curves for different prospectivity models. (**a**) PA plot of MNR; (**b**) PA plot of fuzzy weights-of-evidence (WofE); (**c**) PA plot of GA-SVR; (**d**) ROC curves of three prospectivity models.

4.4. Target Appraisal

Based on the criterion of the highest predictive score in the smallest occupied area, three exploration targets (targets *I* to *III*) with high-value mineralization probability have been inferred in the study area using the GA-SVR prospectivity model (Figure 9). Target *I* is located at a depth beneath the known orebody in the north segment and in the hanging wall of the fault F_2, with the burial depth of 440 m to 750 m. Close to the largest and economically productive orebody group, target *I* is most likely a group of deep, ore-forming fluid conduits stretching from the deep magmatic intrusion of the Axi deposit (see below discussion) and is considered as the highest priority target in the study area. Target *II* in the south part of the deposit occurs below the known orebodies and in the hanging wall of the fault F_2 as well, at a depth of >800 m. It is situated in the transitional zone of the fault, orebodies, and alteration zone. Target *III* is located to north stretching farther along the N-trend of known orebodies, with a depth of >1000 m, probably demonstrating that it is the north branch of the proven orebody.

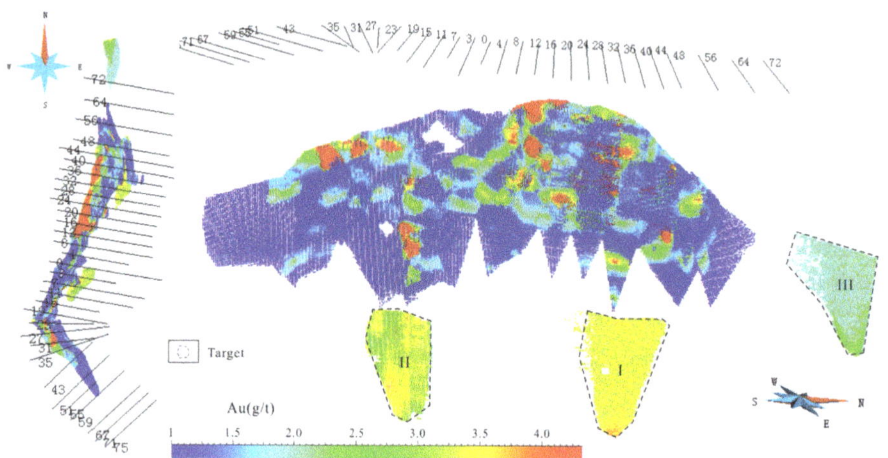

Figure 9. Targets determined by the 3D prospectivity modeling of the Axi gold deposit.

5. Discussion

5.1. Insights from 3D Models and Spatial Analysis

Our spatial analysis shows that there is a bimodal distribution of gold grade (Au) and a single peak of gold resource (AuMet) in the pyrite-sericite-quartz alteration buffer field (Figure 7(c1)). Here, *d*Alt (40, 60) may represent the position of main orebodies, because of the higher grades (avg. up to 3.5 g/t) and approximate distance to width (about 20 m). The intervals of *d*Alt (0, 20) and *d*Alt (20, 40) most likely reflect the location of disseminated ore and the mixing zone of disseminated ore and quartz-sulfide stockwork, respectively, taking account of the value and increasing rate of Au and AuMet. Such vein-alteration distribution is well supported by field investigations (Figure 2). It is very interesting that almost all siliceous veins with gold mineralization are hosted in the phyllic alteration zone, with major vein-type orebodies located in their centers. Apparently, phyllic alteration, to some extent, controls the occurrence of late quartz veins that are products of distributed hydrothermal fluid infiltrating a brecciated zone [33]. However, the crosscutting relationship and 20 M.y. time gap between phyllic alteration and auriferous quartz veins preclude the coeval formation in one hydrothermal event [33,35,37,39]. Previous studies have proposed that the interactions between hydrothermal fluids and rocks could enhance the rock permeability by mass and volume changes linked to mineral dissolution and precipitation, and further control or facilitate the brecciation and formation of new fractures [73–76]. Since the fault F_2 underwent at least three-phases of deformation during strike-slip movement during ore-forming processes [41], the occurrence of quartz veins was mostly likely promoted by late structural deformation on the earlier altered rock.

Two major types of economic ore (i.e., disseminated and vein-type) at Axi are genetically related to early fluid-rock reactions during phyllic alteration and later fluid mixing and boiling in the fault damage zone [33,35,37,39]. The above mineralization mechanism is ubiquitous in LS epithermal gold deposits and is essentially attributed to fluid flow and convergence in the epithermal environment [2,3,6]. The 3D models and spatial analysis indicate that the morphological characteristics and distance field of geological features significantly control the distribution of gold resources in the Axi gold deposit. Specifically, the AuMet is commonly confined to the zones distance to F_2 (0–50 m), the LVP lower interface (10–60 m), and the phyllic upper interface (20–70 m), with high dipping angles of 65–75° and gentle undulation, but with increased frequency on convex areas (0–20 m; Figure 7). These characteristics underline the critical roles of these geological interfaces for driving and gathering ore-forming fluids at Axi. It is notable that all convex areas have high grade and resources (Figure 7(g1,g2)). This implies strain localization (or release) with orientation similar to global strain, accompanied gold deposition, and likely led to the brecciation or caused fluid preferential confluence in more outward bending areas at Axi.

The 3D models and distance field analysis reveal a rule that the alteration-mineralization zone in bending sites is clearly wider than the overall distribution (Figure 3), and notably, also have relatively high gold grades (Figure 7(c1)). Interestingly, these are locations where the main orebodies are comprised of quartz vein-type ores, suggesting at least a rough spatial or genetic correlation between vein-type orebodies and normal-strike slip extension deformation. The fragmentized contact zone ascribed to the overprinting structural deformation on the early altered zone may be the pathway and dilation space for hydrothermal fluid transmission [3,75,76]. In addition, the altered rocks near permeable channels commonly have different mechanical properties from nearby wall rocks, thereby showing distinct deformation during fault propagation, as discussed above. Thus, the quartz vein orebodies in the transition zone reflect the original location of fluid conduits, which is consistent with the empirical mineral assemblage of quartz/chalcedony-carbonates-adularia-sericite as the indicator for fluid conduits in the epithermal system [3,8,77,78]. We, therefore, infer that the major fluid conduits of the Axi epithermal gold deposit are located in the bending areas (No. 35–15 and No. 40–56).

5.2. Application of Support Vector Regression Model

The extracted metallogenic information in 3D prospectivity modeling of the Axi epithermal gold deposit has the following features: (1) non-linearity in that gold mineralization is spatially controlled by a series of geological factors with non-linear relationships [14,49,59]; (2) distribution diversification in that the ore-controlling indicators datasets exhibit different distribution statistics [23–25]; (3) high-dimensionality in that 9 predictor maps construct a nine-dimensional feature space [23–25].

In this study, the MNR method for integrating ore-forming information from the Axi epithermal gold deposit shows weak prospective ability. MNR has been proved fit for prospectivity modeling of deposits with a relatively mild grade and amount of metal variation, distributed normally [23,26]. However, the gold grade of Axi gold deposit changes sharply and yields a log-normal distribution (Figure 6), hence, MNR is a poor fit for prospectivity modeling of this LS epithermal gold deposit, and especially performs worse in fitting samples with erratic high-grade. Although the fuzzy WofE identified more known ore voxels in a smaller occupied space when compared to MNR, it did not produce a robust prospectivity model. In fuzzy WofE, spatially continuous predictor map values are first simplified and discretized into a series of arbitrary classes, and the same weight is assigned to all values in each class. This process is excessively dependent on expert knowledge, resulting in some geological information loss or inconsistencies, which would generate systemic bias in the weighting of spatial evidential layers in the prospectivity model [20,79]. Furthermore, multiple layer integration based on fuzzy WofE is analogous to a nonlinear multiplicative cascade process. Thus, the accuracy and reliability of predictive results may decline, if there is dependence or weak conditional independence between diverse predictive maps [57].

Both the PA plot and ROC curve indicated GA-SVR was superior to MNR and fuzzy WofE in delimiting exploration targets in the Axi gold deposit. These results could be interpreted in several ways:

(1) As previously mentioned, the known Au-grade in Axi gold deposit changes sharply and unevenly distributes in space, resulting in the poor performance of traditional prospectivity methods (e.g., MNR and fuzzy WofE). Nevertheless, GA-SVR is able to overcome the aforesaid disadvantages, because of it is irrespective of data distribution and outliers, making better generalization performance [30,32,65,66].

(2) The spatial analysis results indicate that there is a complicated non-linear spatial association between gold occurrence and geological features, due to the superimposition of various geological events in the ore-forming processes that are difficult to quantify [57]. SVR, characterized by non-linearity, self-learning, and robustness, constructs a higher dimensional feature space from

the lower dimensional input space via non-linear mapping [20,50,51]. Then, the complicated non-linear spatial association in the original low dimensional input space is treated as an oversimplified linear regression solution in high dimensional feature space, and therefore it can be commendably quantified and the prediction ability is improved [54,62].

(3) There are many parameters that have to be determined using expert opinion in traditional prospectivity models (e.g., weight definition in fuzzy WofE), resulting in systemic bias and error [20,79]. By contrast, GA-SVR obtains optimal parameters that could produce the best prediction via GA with less subjectivity [30,32,61].

Consequently, the hybrid model of GA-SVR based on the combination of SVR and GA takes advantage both of GA and SVR in the prospecting modeling of the Axi gold deposit, and therefore is a promising method in 3D mineral prospectivity modeling.

5.3. *Exploration Significanture for LS Epithermal Deposit*

LS epithermal mineralization is typically developed at shallow depths in terms of the high-permeability networks favoring magmatic water ascent, external fluid influx, and the interplay of fluid and rock, thereby resulting in the extensional vein systems and intensive water-rock interactions [1–7]. These characteristics are clear in our 3D models and spatial analysis results and allow understanding of the metallogenic implications in the Axi LS epithermal gold deposit. In addition, the integration of obtained spatial data about F_2, LVP and the phyllic alteration zone controls on mineralization was used to identify the fertility of unknown areas at Axi. One of the targets in the northern Axi district is buried in the Aqialehe sedimentary formation and indicates ore-forming fluid migration along F_2 from conduits to the north in the shallow areas. This suggests a major horizontal fluid pathway controlled by structures in the epizonal segments of this LS epithermal deposit. Thus, in the exploration of epithermal mineralization in a mine, more attention should be given to the (sub-) horizontal extending directions from present orebodies with particular concern paid to the local structural deformation.

Although deep segments of ore fluid conduits host less gold in epithermal deposits, the areas could offer clues to find a new type of mineralization, such as base metal mineralization and porphyry [80,81]. Hence, determining as much as possible about the LS epithermal deposit fluid conduits, even at depth, has the potential to be quite valuable in guiding further mineral exploration. In this study, a comprehensive analysis of geological models and spatial analysis obtained important information about the ore fluid conduits in the Axi epithermal system that include the relative specific spatial location, meaningful geological indicators, and overprinting structural influences on up-flow zones. The above findings highlight the great potential of 3D geological modeling and spatial analysis in understanding the overprinting structural-hydrothermal processes and determining the location of fluid conduits.

In summary, the framework of 3D prospectivity modeling presented here, including 3D geological modeling, spatial analysis, and prospectivity modeling, can improve the understanding of ore genesis and allow for the improvement of exploration strategies in LS epithermal deposits. We encourage all LS mine operators and prospectors to develop similar models.

6. Conclusions

(1) The 3D prospectivity modeling for the Axi gold deposit quantitatively determines the spatial association between orebodies and geological features, and further improves evaluation of the mineralization potential in 3D space at the deposit-scale, and can aid in developing and enhancing mineral exploration strategies for the LS epithermal gold deposit.

(2) The hybrid model GA-SVR is proven to be better at quantifying the complicated non-linear spatial association between gold occurrence and geological features, with less subjectivity when compared to MNR and fuzzy WofE, demonstrating a better prediction capability in prospectivity mapping.

(3) The majority of gold mineralization in the Axi deposit is confined to zones close to fault F_2 (0–50 m), the LVP lower interface (10–60 m), and the phyllic alteration upper interface (20–70 m), with steeply dipping angles (>65°) and gentle undulation with priority on the convex areas (0–20 m). Two fluid conduits of the Axi epithermal gold deposit are speculated in sections No. 35–15 and No. 40–56.

(4) In future work, incorporating various geoscience data into 3D mineral prospectivity modeling, such as geochemical data, geophysical data, and remote sensing data, may be an effective step to provide more reliable exploration decision-making [82–84]. The deep learning (e.g., neuroevolution) with enormous potential in learning complicated functions is encouraged to be further applied for improving prospectivity modeling.

Author Contributions: X.M., W.Z., Z.L., and J.R., conceived and designed the experiments. X.M., and Z.L., conducted the field investigation at Axi. X.M., W.Z., and Z.L., wrote the manuscript with a significant contribution of discussing the results with H.D. Geoscience generalist R.C.B. primarily provided assistance with English. All authors have read and agreed to the published version of the manuscript.

Funding: This work was jointly funded by projects from the National Natural Science Foundation of China (No. 41772349, No. 41972309, and No. 41472301), the National Key R&D Program of China (No. 2017YFC0601503), and the Open Research Fund Program of Key Laboratory of Metallogenic Prediction of Nonferrous Metals and Geological Environment Monitoring (Central South University), Ministry of Education (No. 2019YSJS21).

Acknowledgments: Two anonymous reviewers who provide critical comments to improve this paper are greatly appreciated. We are grateful to Jianmin Han, Hongxi Fan, and Xiujuan Lei of the Western Region Gold Co., Ltd. for fieldwork and data collection. Xiaoxia Liu, Yuting Lin, Min Pan, and Dan Luo are thanked for their efforts in 3D modeling and geology investigation.

Conflicts of Interest: The authors declare no conflict of interest.

References

1. Hedenquist, J.W.; Arribas, A.; Gonzales-Urien, E. Exploration for epithermal gold deposits. *Rev. Econ. Geol.* **2000**, *13*, 245–277.
2. Sillitoe, R.H.; Hedenquist, J.W. Linkages between volcanotectonic settings, ore-fluid compositions, and epithermal precious metal deposits. In *Volcanic, Geothermal, and Ore-Forming Fluids: Rulers and Witnesses of Processes within the Earth*; Simmons, S.F., Graham, I., Eds.; Society of Economic Geologists, Inc.: Littleton, CO, USA, 2003; pp. 315–345.
3. Simmons, S.F.; White, N.C.; John, D.A. Geological characteristics of epithermal precious and base metal deposits. In *Economic Geology One Hundredth Anniversary Volume 1905–2005 Society of Economic Geologists*; Hedenquist, J.W., Thompson, J.F.H., Goldfarb, R.J., Richards, J.P., Eds.; Society of Economic Geologists, Inc.: Littleton, CO, USA, 2005; pp. 485–522.
4. Sillitoe, R.H. Epithermal paleosurfaces. *Miner. Depos.* **2015**, *50*, 767–793. [CrossRef]
5. Cooke, D.R.; Simmons, S.F. Characteristics and genesis of epithermal gold deposits. *Rev. Econ. Geol.* **2000**, *13*, 221–244.
6. John, D.A.; Hofstra, A.H.; Fleck, R.J.; Brummer, J.E.; Saderholm, E.C. Geologic setting and genesis of the Mule Canyon low-sulfidation epithermal gold-silver deposit, north-central Nevada. *Econ. Geol.* **2003**, *98*, 425–463. [CrossRef]
7. Clark, L.V.; Gemmell, J.B. Vein stratigraphy, mineralogy, and metal zonation of the kencana low-sulfidation epithermal Au–Ag deposit, Gosowong goldfield, Halmahera island, Indonesia. *Econ. Geol.* **2018**, *113*, 209–236. [CrossRef]
8. Simmons, S.F.; Browne, P.R. Hydrothermal minerals and precious metals in the Broadlands-Ohaaki geothermal system: Implications for understanding low-sulfidation epithermal environments. *Econ. Geol.* **2000**, *95*, 971–999. [CrossRef]
9. Henley, R.W. The geothermal framework of epithermal deposits. *Rev. Econ. Geol.* **1985**, *2*, 1–24.
10. Rowland, J.V.; Simmons, S.F. Hydrologic, magmatic, and tectonic controls on hydrothermal flow, Taupo Volcanic Zone, New Zealand: Implications for the formation of epithermal vein deposits. *Econ. Geol.* **2012**, *107*, 427–457. [CrossRef]

11. John, D.A.; Vikre, P.G.; du Bray, E.A.; Blakely, R.J.; Fey, D.L.; Rockwell, B.W.; Mauk, J.L.; Anderson, E.D.; Graybeal, F.T. *Descriptive Models for Epithermal Gold-Silver Deposits: Chapter Q in MINERAL Deposit Models for Resource Assessment (No. 2010-5070-Q)*; US Geological Survey: Reston, VA, USA, 2018.
12. Cox, S.F.; Knackstedt, M.A.; Brown, J. Principle of structural control on permeability and fluid flow in hydrothermal systems. In *Structural Control on Ore Genesis*; Richards, J.P., Tosdal, R.M., Eds.; Society of Economic Geologists, Inc.: Littleton, CO, USA, 2001; Volume 14, pp. 1–24.
13. Berger, B.R.; Tingley, J.V.; Drew, L.J. Structural localization and origin of compartmentalized fluid flow, Comstock Lode, Virginia City, Nevada. *Econ. Geol.* **2003**, *98*, 387–408. [CrossRef]
14. Carranza, E.J.M. Controls on mineral deposit occurrence inferred from analysis of their spatial pattern and spatial association with geological features. *Ore Geol. Rev.* **2009**, *35*, 383–400. [CrossRef]
15. Xiao, F.; Chen, J.; Hou, W.; Wang, Z.; Zhou, Y.; Erten, O. A spatially weighted singularity mapping method applied to identify epithermal Ag and Pb-Zn polymetallic mineralization associated geochemical anomaly in Northwest Zhejiang, China. *J. Geochem. Explor.* **2018**, *189*, 122–137. [CrossRef]
16. De Palomera, P.A.; van Ruitenbeek, F.J.; Carranza, E.J.M. Prospectivity for epithermal gold–silver deposits in the Deseado Massif, Argentina. *Ore Geol. Rev.* **2015**, *71*, 484–501. [CrossRef]
17. Li, N.; Bagas, L.; Li, X.H.; Xiao, K.Y.; Li, Y.; Ying, L.J.; Song, X.L. An improved buffer analysis technique for model-based 3D mineral potential mapping and its application. *Ore Geol. Rev.* **2016**, *76*, 94–107. [CrossRef]
18. Mao, X.; Zhang, B.; Deng, H.; Zou, Y.; Chen, J. Three-dimensional morphological analysis method for geologic bodies and its parallel implementation. *Comput. Geosci.* **2016**, *96*, 11–22. [CrossRef]
19. Zhang, M.; Zhou, G.; Shen, L.; Zhao, W.; Liao, B.; Yuan, F.; Li, X.; Hu, X.; Wang, C. Comparison of 3D prospectivity modeling methods for Fe–Cu skarn deposits: A case study of the Zhuchong Fe–Cu deposit in the Yueshan orefield (Anhui), eastern China. *Ore Geol. Rev.* **2019**, *114*, 103126. [CrossRef]
20. Mao, X.; Ren, J.; Liu, Z.; Chen, J.; Tang, L.; Deng, H.; Bayless, R.C.; Yang, B.; Wang, M.; Liu, C. Three-dimensional prospectivity modeling of the Jiaojia-type gold deposit, Jiaodong Peninsula, Eastern China: A case study of the Dayingezhuang deposit. *J. Geochem. Explor.* **2019**, *203*, 27–44. [CrossRef]
21. Qin, Y.Z.; Liu, L.M. Quantitative 3D Association of Geological Factors and Geophysical Fields with Mineralization and Its Significance for Ore Prediction: An Example from Anqing Orefield, China. *Minerals* **2018**, *8*, 300. [CrossRef]
22. Yuan, F.; Li, X.H.; Zhang, M.M.; Jowitt, S.M.; Jia, C.; Zheng, T.K.; Zhou, T.F. Three-dimensional weights of evidence-based prospectivity modeling: A case study of the Baixiangshan mining area, Ningwu Basin, Middle and Lower Yangtze Metallogenic Belt, China. *J. Geochem. Explor.* **2014**, *145*, 82–97. [CrossRef]
23. Carranza, E.J.M.; Laborte, A.G. Data-driven predictive mapping of gold prospectivity, Baguio district, Philippines: Application of Random Forests algorithm. *Ore Geol. Rev.* **2015**, *71*, 777–787. [CrossRef]
24. Rodriguez-Galiano, V.; Sanchez-Castillo, M.; Chica-Olmo, M.; Chica-Rivas, M. Machine learning predictive models for mineral prospectivity: An evaluation of neural networks, random forest, regression trees and support vector machines. *Ore Geol. Rev.* **2015**, *71*, 804–818. [CrossRef]
25. Sun, T.; Chen, F.; Zhong, L.; Liu, W.; Wang, Y. GIS-based mineral prospectivity mapping using machine learning methods: A case study from Tongling ore district, eastern China. *Ore Geol. Rev.* **2019**, *109*, 26–49. [CrossRef]
26. Porwal, A.; Carranza, E.J.M.; Hale, M. Artificial Neural Networks for Mineral-Potential Mapping: A Case Study from Aravalli Province, Western India. *Nat. Resour. Res.* **2003**, *12*, 155–171. [CrossRef]
27. Harris, J.R.; Grunsky, E.; Behnia, P.; Corrigan, D. Data- and knowledge-driven mineral prospectivity maps for Canada's North. *Ore Geol. Rev.* **2015**, *71*, 788–803. [CrossRef]
28. Juliani, C.; Ellefmo, S.L. Prospectivity Mapping of Mineral Deposits in Northern Norway Using Radial Basis Function Neural Networks. *Minerals* **2019**, *9*, 131. [CrossRef]
29. Hong, W.C.; Dong, Y.; Zhang, W.Y.; Chen, L.Y.; Panigrahi, B.K. Cyclic electric load forecasting by seasonal SVR with chaotic genetic algorithm. *Int. J. Electr. Power* **2013**, *44*, 604–614. [CrossRef]
30. Oyehan, T.A.; Alade, I.O.; Bagudu, A.; Sulaiman, K.O.; Olatunji, S.O.; Saleh, T.A. Predicting of the refractive index of haemoglobin using the Hybrid GA-SVR approach. *Comput. Biol. Med.* **2018**, *98*, 85–92. [CrossRef]
31. Tang, X.; Wang, L.; Cheng, J.; Chen, J.; Sheng, V. Forecasting Model Based on Information-Granulated GA-SVR and ARIMA for Producer Price Index. *Tech Sci. Press* **2019**, *58*, 463–491. [CrossRef]
32. Roushangar, K.; Koosheh, A. Evaluation of GA-SVR method for modeling bed load transport in gravel-bed rivers. *J. Hydrol.* **2015**, *527*, 1142–1152. [CrossRef]

33. An, F.; Zhu, Y. Geology and geochemistry of the Early Permian Axi low-sulfidation epithermal gold deposit in North Tianshan (NW China). *Ore Geol. Rev.* **2018**, *100*, 12–30. [CrossRef]
34. Zhang, B.; Li, N.; Shu, S.P.; Wang, W.; Yu, J.; Chen, X.; Ye, T.; Chen, Y.J. Textural and compositional evolution of Au-hosting Fe–S–As minerals at the Axi epithermal gold deposit, Western Tianshan, NW China. *Ore Geol. Rev.* **2018**, *100*, 31–50. [CrossRef]
35. Liu, Z.; Mao, X.; Deng, H.; Li, B.; Zhang, S.; Lai, J.; Bayless, R.C.; Pan, M.; Li, L.; Shang, Q. Hydrothermal processes at the Axi epithermal Au deposit, western Tianshan: Insights from geochemical effects of alteration, mineralization and trace elements in pyrite. *Ore Geol. Rev.* **2018**, *102*, 368–385. [CrossRef]
36. Zhai, W.; Sun, X.; Sun, W.; Su, L.; He, X.; Wu, Y. Geology, geochemistry, and genesis of Axi: A Paleozoic low-sulfidation type epithermal gold deposit in Xinjiang, China. *Ore Geol. Rev.* **2009**, *36*, 265–281. [CrossRef]
37. Liu, Z.; Mao, X.; Ackerman, L.; Li, B.; Dick, J.M.; Yu, M.; Peng, J.; Shahzad, S.M. Two-stage gold mineralization of the Axi epithermal Au deposit, Western Tianshan, NW China: Evidence from Re–Os dating, S isotope, and trace elements of pyrite. *Miner. Depos.* **2019**. [CrossRef]
38. Zhao, X.; Xue, C.; Chi, G.; Wang, H.; Qi, T. Epithermal Au and polymetallic mineralization in the Tulasu Basin, western Tianshan, NW China: Potential for the discovery of porphyry CuAu deposits. *Ore Geol. Rev.* **2014**, *60*, 76–96. [CrossRef]
39. An, F.; Zhu, Y.; Wei, S.; Lai, S. An Early Devonian to Early Carboniferous volcanic arc in North Tianshan, NW China: Geochronological and geochemical evidence from volcanic rocks. *J. Asian Earth Sci.* **2013**, *78*, 100–113. [CrossRef]
40. Tang, G.J.; Wang, Q.; Wyman, D.A.; Sun, M.; Zhao, Z.H.; Jiang, Z.Q. Petrogenesis of gold-mineralized magmatic rocks of the Taerbieke area, northwestern Tianshan (western China), Constraints from geochronology, geochemistry and Sr–Nd–Pb–Hf isotopic compositions. *J. Asian Earth Sci.* **2013**, *74*, 113–128. [CrossRef]
41. Wei, J.L.; Cao, X.Z.; Xu, B.J.; Zhang, W.S.; Rao, D.P.; Huang, L.W. Erosion and post-mineralization change of Axi epithermal gold deposit in Western Tianshan Mountains. *Miner. Depos.* **2014**, *33*, 241–254. (In Chinese)
42. Jessell, M.; Ailleres, L.; De Kemp, E.; Lindsay, M.; Wellmann, F.; Hillier, M.; Laurent, G.; Carmichael, T.; Martin, R. Next Generation Three-Dimensional Geologic Modeling and Inversion. In *Building Exploration Capability for the 21st Century*; Kelley, K.D., Golden, H.C., Eds.; Soc Economic Geologists, Inc.: Littleton, CO, USA, 2014; pp. 261–272.
43. Li, X.; Yuan, F.; Zhang, M.; Jowitt, S.M.; Ord, A.; Zhou, T.; Dai, W. 3D computational simulation based mineral prospectivity modeling for exploration for concealed Fe–Cu skarn-type mineralization within the Yueshan orefield, Anqing district, Anhui Province, China. *Ore Geol. Rev.* **2019**, *105*, 1–17. [CrossRef]
44. Li, N.; Song, X.; Xiao, K.; Li, S.; Li, C.; Wang, K. Part II: A demonstration of integrating multiple-scale 3D modelling into GIS-based prospectivity analysis: A case study of the Huayuan-Malichang district, China. *Ore Geol. Rev.* **2018**, *95*, 292–305. [CrossRef]
45. Sporli, K.B.; Cargill, H. Structural Evolution of a World-Class Epithermal Orebody: The Martha Hill Deposit, Waihi, New Zealand. *Econ. Geol.* **2011**, *106*, 975–998. [CrossRef]
46. Jowitt, S.M.; Cooper, K.; Squire, R.J.; Thebaud, N.; Fisher, L.A.; Cas, R.A.F.; Pegg, I. Geology, mineralogy, and geochemistry of magnetite-associated Au mineralization of the ultramafic-basalt greenstone hosted Crusader Complex, Agnew Gold Camp, Eastern Yilgarn Craton, Western Australia; a Late Archean intrusion-related Au deposit? *Ore Geol. Rev.* **2014**, *56*, 53–72. [CrossRef]
47. Mao, X.; Tang, Y.; Lai, J.; Zou, Y.; Chen, J.; Peng, S.; Shao, Y.J. Three Dimensional Structure of Metallogenic Geologic Bodies in the Fenghuangshan Ore Field and Ore-controlling Geological Factors. *Acta Geol. Sin.* **2011**, *85*, 1507–1518. (In Chinese)
48. Mao, X.; Zhao, Y.; Tang, Y.; Peng, Z.; Chen, J.; Deng, H. Three-dimensional morphological analysis method for geological interfaces based on TIN and its application. *J. Cent. South Univ.* **2013**, *44*, 1493–1499. (In Chinese)
49. Zuo, R. A nonlinear controlling function of geological features on magmatic-hydrothermal mineralization. *Sci. Rep.* **2016**, *6*, 27127. [CrossRef] [PubMed]
50. Abedi, M.; Norouzi, G.-H.; Bahroudi, A. Support vector machine for multi-classification of mineral prospectivity areas. *Comput. Geosci.* **2012**, *46*, 272–283. [CrossRef]
51. Zuo, R.; Carranza, E.J.M. Support vector machine: A tool for mapping mineral prospectivity. *Comput. Geosci.* **2011**, *37*, 1967–1975. [CrossRef]

52. Shabankareh, M.; Hezarkhani, A. Application of support vector machines for copper potential mapping in Kerman region, Iran. *J. Afr. Earth Sci.* **2017**, *128*, 116–126. [CrossRef]
53. Carranza, E.J.M. Geocomputation of mineral exploration targets. *Comput. Geosci.* **2011**, *37*, 1907–1916. [CrossRef]
54. Li, X.; Luo, A.; Li, J.; Li, Y. Air Pollutant Concentration Forecast Based on Support Vector Regression and Quantum-Behaved Particle Swarm Optimization. *Environ. Model. Assess.* **2018**, *24*, 205–222. [CrossRef]
55. Santamaria-Bonfil, G.; Reyes-Ballesteros, A.; Gershenson, C. Wind speed forecasting for wind farms: A method based on support vector regression. *Renew. Energy* **2016**, *85*, 790–809. [CrossRef]
56. Patel, A.K.; Chatterjee, S.; Gorai, A.K. Development of a machine vision system using the support vector machine regression (SVR) algorithm for the online prediction of iron ore grades. *Earth Sci. Inf.* **2019**, *12*, 197–210. [CrossRef]
57. Cheng, Q. Non-Linear Theory and Power-Law Models for Information Integration and Mineral Resources Quantitative Assessments. *Math. Geosci.* **2008**, *40*, 503–532. [CrossRef]
58. Yousefi, M.; Carranza, E.J.M. Prediction–area (P–A) plot and C–A fractal analysis to classify and evaluate evidential maps for mineral prospectivity modeling. *Comput. Geosci.* **2015**, *79*, 69–81. [CrossRef]
59. Nykänen, V.; Lahti, I.; Niiranen, T.; Korhonen, K. Receiver operating characteristics (ROC) as validation tool for prospectivity models—A magmatic Ni–Cu case study from the Central Lapland Greenstone Belt, Northern Finland. *Ore Geol. Rev.* **2015**, *71*, 853–860. [CrossRef]
60. Vapnik, V.N. *The Nature of Statistical Learning Theory*; Springer: New York, NY, USA, 1995. [CrossRef]
61. Wang, L.; Wang, C.; Khoshnevisan, S.; Ge, Y.; Sun, Z. Determination of two-dimensional joint roughness coefficient using support vector regression and factor analysis. *Eng. Geol.* **2017**, *231*, 238–251. [CrossRef]
62. Wang, J.; Li, Y. Short-Term Wind Speed Prediction Using Signal Preprocessing Technique and Evolutionary Support Vector Regression. *Neural Process. Lett.* **2017**, *48*, 1043–1061. [CrossRef]
63. Scholkopf, B.; Sung, K.K.; Burges, C.J.C.; Girosi, F.; Niyogi, P.; Poggio, T. Comparing support vector machines with Gaussian kernels to radial basis function classifiers. *IEEE Trans. Signal Proces.* **1997**, *45*, 2758–2765. [CrossRef]
64. Ghaedi, M.; Dashtian, K.; Ghaedi, A.M.; Dehghanian, N. A hybrid model of support vector regression with genetic algorithm for forecasting adsorption of malachite green onto multi-walled carbon nanotubes: Central composite design optimization. *Phys. Chem. Chem. Phys.* **2016**, *18*, 13310–13321. [CrossRef]
65. Xin, N.; Gu, X.F.; Wu, H.; Hu, Y.Z.; Yang, Z.L. Application of genetic algorithm-support vector regression (GA-SVR) for quantitative analysis of herbal medicines. *J. Chemom.* **2012**, *26*, 353–360. [CrossRef]
66. Liu, S.; Tai, H.; Ding, Q.; Li, D.; Xu, L.; Wei, Y. A hybrid approach of support vector regression with genetic algorithm optimization for aquaculture water quality prediction. *Math. Comput. Model.* **2013**, *58*, 458–465. [CrossRef]
67. Stanley, K.O.; Clune, J.; Lehman, J.; Miikkulainen, R. Designing neural networks through neuroevolution. *Nat. Mach. Intell.* **2019**, *1*, 24–35. [CrossRef]
68. Chang, C.C.; Lin, C.J. LIBSVM: A Library for Support Vector Machines. *ACM Trans. Intell. Syst. Technol.* **2011**, *2*. [CrossRef]
69. Cheng, Q.M.; Agterberg, F.P. Fuzzy Weights of Evidence Method and Its Application in Mineral Potential Mapping. *Nat. Resour. Res.* **1999**, *8*, 27–35. [CrossRef]
70. Mao, X.; Zou, Y.; Lu, X.; Wu, X.; Dai, T. Quantitative analysis of geological ore-controlling factors and stereoscopic quantitative prediction of concealed ore bodies. *J. Cent. South Univ. Technol.* **2009**, *16*, 0987–0993. [CrossRef]
71. Bradley, A.P. The use of the area under the ROC curve in the evaluation of machine learning algorithms. *Pattern Recogn.* **1997**, *30*, 1145–1159. [CrossRef]
72. Xiao, K.Y.; Li, N.; Alok, P.; Holden, E.J.; Leon, B.; Lu, Y.J. GIS-based 3D prospectivity mapping: A case study of Jiama copper-polymetallic deposit in Tibet, China. *Ore Geol. Rev.* **2015**, *71*, 611–632. [CrossRef]
73. Sibson, R.H. Structural permeability of fluid-driven fault-fracture meshes. *J. Struct. Geol.* **1996**, *18*, 1031–1042. [CrossRef]
74. Micklethwaite, S. Mechanisms of faulting and permeability enhancement during epithermal mineralisation: Cracow goldfield, Australia. *J. Struct. Geol.* **2009**, *31*, 288–300. [CrossRef]
75. Williams, J.N.; Toy, V.G.; Smith, S.A.F.; Boulton, C. Fracturing, fluid-rock interaction and mineralisation during the seismic cycle along the Alpine Fault. *J. Struct. Geol.* **2017**, *103*, 151–166. [CrossRef]

76. Yang, L.; Zhao, R.; Wang, Q.F.; Liu, X.F.; Carranza, E.J.M. Fault geometry and fluid-rock reaction: Combined controls on mineralization in the Xinli gold deposit, Jiaodong Peninsula, China. *J. Struct. Geol.* **2018**, *111*, 14–26. [CrossRef]
77. De Palomera, P.A.; van Ruitenbeek, F.J.A.; van der Meer, F.D.; Fernandez, R. Geochemical indicators of gold-rich zones in the La Josefina epithermal deposit, Deseado Massif, Argentina. *Ore Geol. Rev.* **2012**, *45*, 61–80. [CrossRef]
78. Chang, Z.S.; Hedenquist, J.W.; White, N.C.; Cooke, D.R.; Roach, M.; Deyell, C.L.; Garcia, J.; Gemmell, J.B.; McKnight, S.; Cuison, A.L. Exploration Tools for Linked Porphyry and Epithermal Deposits: Example from the Mankayan Intrusion-Centered Cu–Au District, Luzon, Philippines. *Econ. Geol.* **2011**, *106*, 1365–1398. [CrossRef]
79. Yousefi, M.; Carranza, E.J.M. Fuzzification of continuous-value spatial evidence for mineral prospectivity mapping. *Comput. Geosci.* **2015**, *74*, 97–109. [CrossRef]
80. Spry, P.G.; Paredes, M.M.; Foster, F.; Truckle, J.S.; Chadwick, T.H. Evidence for a genetic link between gold-silver telluride and porphyry molybdenum mineralization at the Golden Sunlight Deposit, Whitehall, Montana; fluid inclusion and stable isotope studies. *Econ. Geol.* **1996**, *91*, 507–526. [CrossRef]
81. Sillitoe, R.H. Some metallogenic features of gold and copper deposits related to alkaline rocks and consequences for exploration. *Miner. Depos.* **2002**, *37*, 4–13. [CrossRef]
82. Farahbakhsh, E.; Chandra, R.; Eslamkish, T.; Müller, R.D. Modeling geochemical anomalies of stream sediment data through a weighted drainage catchment basin method for detecting porphyry Cu–Au mineralization. *J. Geochem. Explor.* **2019**, *204*, 12–32. [CrossRef]
83. Scalzo, R.; Kohn, D.; Olierook, H.; Houseman, G.; Chandra, R.; Girolami, M.; Cripps, S. Efficiency and robustness in Monte Carlo sampling for 3-D geophysical inversions with Obsidian v0.1.2: Setting up for success. *Geosci. Model. Dev.* **2019**, *12*, 2941–2960. [CrossRef]
84. Farahbakhsh, E.; Chandra, R.; Olierook, H.K.H.; Scalzo, R.; Clark, C.; Reddy, S.M.; Mueller, R.D. Computer vision-based framework for extracting tectonic lineaments from optical remote sensing data. *Int. J. Remote Sens.* **2020**, *41*, 1760–1787. [CrossRef]

Article

Stochastic Modeling of Chemical Compounds in a Limestone Deposit by Unlocking the Complexity in Bivariate Relationships

Nurassyl Battalgazy and Nasser Madani *

School of Mining and Geosciences, Nazarbayev University, Nur-Sultan 010000, Kazakhstan; nurassyl.battalgazy@nu.edu.kz

* Correspondence: nasser.madani@nu.edu.kz; Tel.: +7-747-1755327

Received: 12 October 2019; Accepted: 2 November 2019; Published: 4 November 2019

Abstract: Modeling multivariate variables with complexity in a cross-correlation structure is always applicable to mineral resource evaluation and exploration in multi-element deposits. However, the geostatistical algorithm for such modeling is usually challenging. In this respect, projection pursuit multivariate transform (PPMT), which can successfully handle the complexity of interest in bivariate relationships, may be particularly useful. This work presents an algorithm for combining projection pursuit multivariate transform (PPMT) with a conventional (co)-simulation technique where spatial dependency among variables can be defined by a linear model of co-regionalization (LMC). This algorithm is examined by one real case study in a limestone deposit in the south of Kazakhstan, in which four chemical compounds (CaO, Al_2O_3, Fe_2O_3, and SiO_2) with complexity in bivariate relationships are analyzed and 100 realizations are produced for each variable. To show the effectiveness of the proposed algorithm, the outputs (realizations) are statistically examined and the results show that this methodology is legitimate for reproduction of original mean, variance, and complex cross-correlation among the variables and can be employed for further processes. Then, the applicability of the concept is demonstrated on a workflow to classify this limestone deposit as measured, indicated, or inferred based on Joint Ore Reserves Committee (JORC) code. The categorization is carried out based on two zone definitions, geological, and mining units.

Keywords: mineral resource classification; JORC code; limestone deposit; project pursuit multivariate transform; (co)-simulation

1. Introduction

Accurate evaluation of viable mineral resources is fundamentally important in optimal sustainable development and mine planning procedures since it has an enormous impact on the value of produced metals and formation of technical plans, from extraction to closure of the mine [1–3]. In the world of emerging new technologies and increasing complexity for modeling the grades and chemical compounds in deposits, precise estimation and classification of resources is becoming more crucial, as it can vary depending on the technological modifications that change the notion of mine planning [4]. Nowadays, the "best practice" in resource classification takes into account assumptions made in grades and tonnages of orebodies, as well as in the methods applied, future expenses, and commodity costs. One of the main ways to solve these issues is to consider uncertainty and risk assessment [5–7]. Consequently, integrating enhanced geostatistical techniques in order to quantify uncertainty in the process of evaluating resources allow practitioners to effectively unlock the complexity in the modeling of mineral deposits [8,9].

In every mining project, the discrepancy between the estimated resource model and actual production value is the main concern, which leads to low probability of meeting production targets in respect to quality and quantity of ore content, which can be stated as unsuccessful operation of the mine [10]. In this respect, resource estimation basically depends on the block model, which can be constructed by different deterministic and stochastic approaches. Commonly, the procedure of resource estimation and classification is done on the basis of a block model of a deposit, in which it can be constructed by conventional (deterministic) methods of geostatistics such as kriging. This type of linear interpolation method provides only one unique scenario from a deposit [11] that, in fact, is inadequate to represent the variable under study, with a direct impact on further steps of a mining project, including determining optimum pit shell for scheduling, identifying the sequence of the block extraction intended to generate the best Net Present Value (NPV), and projecting the pushbacks (phases of the mine) [1]. Hence, the main problems in conventional methods not only affect the poor reproduction of original variability of grade (e.g., smoothing effect), but also lead to other issues such as the inability to quantify the uncertainty within each block. Ignoring this type of valuable information, such as risk and uncertainty, in mineral resource evaluation may result in unrealistic outcomes of planned production and cash flow of mine projects [10,12]. The existence of uncertainties in the block model and other parameters makes consideration of a single deterministic model doubtful for resource classification [13]. In contrast, the application of stochastic models to the estimation of resources based on international standards is important in terms of both reproducing original variability of grade and quantifying uncertainty [14]. In this regard, there are two commonly used simulation methods, turning bands (co)-simulation [15] and sequential Gaussian (co)-simulation [16,17]. However, these approaches are restricted to the linear bivariate characteristics among the variables and are insufficient to properly handle the orebodies wherever complexities such as heteroscedasticity, nonlinearity, and geological constraints in cross-correlation structures inherently exist [18,19]. To address these complexities, another geostatistical approach, such as stepwise conditioning transformation (SCT) [20], flow anamorphosis [21], or projection pursuit multivariate transform [22,23], can be an option for unlocking the complexity among the variables. The latter honors the complex multivariate normality after transformation and allows back-transformation of the simulated variables to the original dataset as well as restoring the multivariate characteristics [24,25]. Applications of PPMT in 3D multivariate geosciences mapping include petroleum reservoirs [23], environmental studies [26], grade control [27], and mineral resource classification [24]. However, PPMT, in some circumstances, does not completely remove the cross-dependency among the variables, and PPMT forward transformation may show some slight dependency [21] in bivariate relationships. In order to circumvent this impediment, one more step of factorization with a technique such as minimum/maximum autocorrelation factor (MAF) can be applied to make sure the correlation is removed not only in lag 0, but also in other arbitrary lag. However, MAF may not be enough to eliminate the correlation at lags other than these two lag separations. In this study, we propose an approach that is a combination of PPMT and (co)-simulation techniques. This takes into account the linear model of co-regionalization for defining cross-spatial dependency in order to establish the (co)-kriging system in the (co)-simulation algorithm right after transforming multivariate variables to PPMT factors. In the proposed algorithm, it is not necessary to employ MAF or any other factorization technique for further decorrelation.

The discussion of this paper is thus relevant for multivariate analysis of any complex bivariate features and can be applied to many cases.

The objectives of this paper are fourfold: (1) to briefly present the concept of (co)-simulation and PPMT and the proposed algorithm; (2) to apply the proposed algorithm in a real case study of a limestone deposit in Kazakhstan; (3) to classify resources based on Joint Ore Reserves Committee (JORC) code over the results obtained for objective 2; and (4) to provide a discussion and conclusion.

2. Methodology

2.1. Gaussian (Co)-Simulation

Stochastic simulation requires the transformation of variables into normal score values with zero mean and variance equal to 1. Generally, the transformation can be executed through Gaussian anamorphosis, which transforms variables into standard Gaussian form [28], or a quantile–quantile based approach [29]. With two or more variables, multivariable geostatistical modelling is based on the multivariate Gaussianity assumption [16,30]. In this technique, the variables should be independently transferred to normal Gaussian distribution. Then, (co)-simulation can be performed over normal score data, taking into account the cross-dependency functions that are defined by, for instance, a linear model of co-regionalization. The obtained realizations can subsequently be back-transformed to original scale in order to approximately reproduce the original linear multivariate relationships among the variables. Despite the fact that independent normal score transform ensures that each item of normal score data is multi-Gaussian, the multivariate Gaussianity assumption is violated when there is complexity in bivariate relations between pairs of variables. For example, with features such as geological constraints, heteroscedasticity, and nonlinearity, as illustrated in Figure 1 [31], that employ the typical (co)-simulation algorithm, using the multi-Gaussianity assumption is problematic. In other words, in a typical (co)-simulation paradigm, in order to implement the algorithms effectively, the normal score transformed data should follow the elliptical (Figure 1a). However, in the case of nonlinearity, heteroscedasticity, or linearity constraints (Figure 1), the multi-Gaussianity assumption cannot be respected and the outputs of typical (co)-simulation algorithms are unsatisfactory. To cope with this difficulty, one idea is to use another transformation technique such as projection pursuit multivariate transformation (PPMT) [23,24] rather than the regular normal score transformation (e.g., quantile–quantile or Gaussian anamorphosis [28,29]) that is common for univariate transformation.

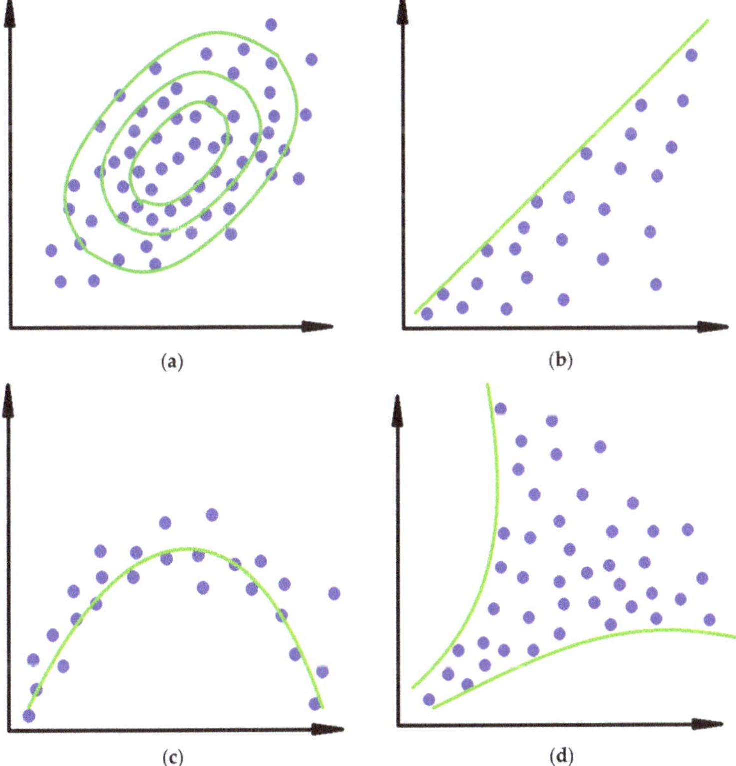

Figure 1. Schematic pattern of (**a**) multi-Gaussianity; multivariate complexities such as (**b**) constraint, (**c**) non-linearity and (**d**) heteroscedasticity.

2.2. Projection Pursuit Multivariate Transform Steps

The steps of projection pursuit multivariate transform are composed of preprocessing and projection pursuit. In the preprocessing steps, variables are transformed to normal score values and linear dependency is removed. In projection pursuit, complexity dependency is removed.

2.2.1. Preprocessing Steps

Normal Score Transformation

Consider data matrix A, which has M variables and N observations such that $A_{M \times N}$ is transformed through Equation (1) into normal score [29]:

$$A' = C^{-1}(E(A)) \tag{1}$$

where E and C^{-1} are original data cumulative distribution function (CDF) standard and normal data CDF, respectively.

Data Sphering

One of the requirements of the projection pursuit algorithm is to center the data with an orthogonal covariance matrix and variance, which can be done by applying the last preprocessing step, data sphering, which uses Equation (2):

$$P = R^{-\frac{1}{2}} B^T (A' - E\{A'\}) \tag{2}$$

where R and B are eigenvector and diagonal eigenvalue matrices, respectively, obtained from the A' covariance matrix's spectral decomposition.

Projection Pursuit

After completing the preprocessing steps, the projection pursuit algorithm can be computed, taking into consideration that projection is $Q = A\beta$, where β is the unit length vector of h × 1 dimension related to projection Q. With multi-Gaussianity of A, every unit length vector β must yield a Q that is univariate Gaussian. In order to measure the univariate non-Gaussianity, the projection index $T(\beta)$ test statistic is designated; when projection Q is finely Gaussian, $T(\beta)$ is equal to zero. The projection pursuit algorithm uses optimized search, which is focused on identifying the β where the projection index $T(\beta)$ is the highest. Finding the highest projection index will also find the maximum non-Gaussian projection Q. Consequently, by using the determined optimum unit length vector β, data A is transformed to A'', in which the projection is standard Gaussian, as $Q' = A''\beta$. In order to achieve this transformation, Equation (3) is used:

$$Z = \left[\beta, \gamma_1, \gamma_2, \ldots, \gamma_{h-1,}\right] \tag{3}$$

where values of γ are unit vectors obtained by applying the Gram–Schmidt algorithm to compute the orthogonal matrix of Z [32]. The next step is the transformation, which can be achieved by multiplying A and Z by each other (Equation (4)):

$$AZ = \left[Q, A\gamma_1, A\gamma_2, \ldots, A\gamma_{h-1,}\right] \tag{4}$$

The next step considers the X transformation, which yields the projection of standard Gaussian Q' while holding the same orthogonal matrix:

$$(AZ) = \left[Q\prime, A\gamma_1, A\gamma_2, \ldots, A\gamma_{h-1,}\right] \tag{5}$$

The last step is multiplication of Z^T by $X(AZ)$, which results in back-transformation to the initial basis (Equation (6)):

$$A'' = X(AZ)Z^T \quad (6)$$

After back-transformation, optimized search can be used again to detect other complexities of A''.

Stopping Criteria

To select the optimum value for projection index $T(\beta)$, considering the optimum h dimensions and number of N observations is critical due to the fact that a high value of h dimensions results in trouble with identifying the complexity, and a small amount of N observations leads to low reliability in detection. To choose the target projection index, a special algorithm is applied [22].

Application

Overall, the implementation of projection pursuit multivariate transformation on multivariate data is based on the forward- and backward-transformation techniques. For the sake of simplicity, Figure 2 shows a schematic illustration to explain the implementation of the PPMT technique [22].

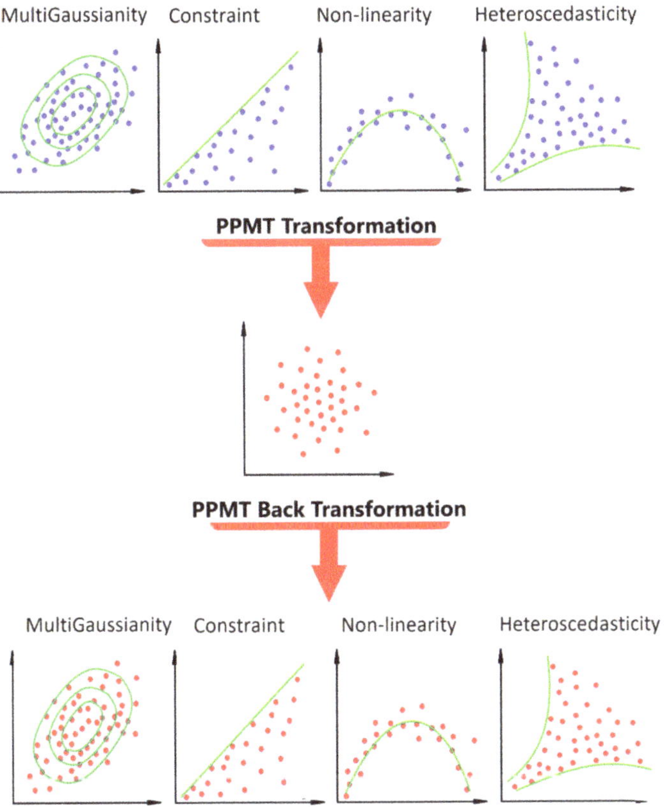

Figure 2. Representation of projection pursuit multivariate transformation (PPMT).

2.3. Proposed Algorithm

As already explained, the (co)-simulation approaches based on multi-Gaussianity assumptions (e.g., turning bands [33] and sequential Gaussian [34] co-stimulation) are not suitable for modelling purposes with such bivariate complexity. On the contrary, PPMT is able to model these characteristics based on factor transformation. In this technique, the dataset is first transferred to PPMT factors, in which the correlation among the factors become almost zero [22,24]. Since the cross-dependency among the factors is zero, the independent simulation can be applied over each factor and the outputs should be back-transformed to original scale. However, one of the main difficulties is related to the decorrelation step, where sometimes the forward transformation is not able to completely remove the inherent correlation between pairs of variables and a mild dependency may remain among the transformed factors that must not be negligible. One solution is to employ one more decorrelation steps, such as minimum/maximum autocorrelation factor (MAF) or principal component analysis (PCA), to make sure the correlation at lag 0 and other arbitrary lag is substantially removed. However, these methods do not guarantee that the correlation will be entirely eliminated through all other lags [35]. A way around this impediment is to co-simulate the PPMT factors even with small correlation that are left after forward transformation of the original variables to factors by inference of cross-dependency functions using the linear model of co-regionalization [25]. The output of the simulation is then back-transformed to original scale in the same way as in the independent simulation. In this paradigm, no MAF or PCA transformation is needed. The steps of the proposed algorithm are as follows:

1. Exploratory data analysis of multivariate data
2. Investigation of the level of complexity in bivariate relation analysis
3. PPMT forward transformation
4. Examination of removing cross-correlations among variables by using cross-correlogram
5. Inference of cross-dependency functions by linear model of co-regionalization (LMC)
6. (Co)-simulation of PPMT transformed factors taking into account the fitted LMC
7. PPMT backward transformation of simulated results (realizations) into original scale
8. Validation of the output by statistical analysis tools

In order to show the capability of the proposed model, a case study is presented and the outputs after validation are taken into account for mineral resource classification, a critical issue in international reporting of deposits.

3. Case Study: Aktas-South Deposit in Kazakhstan

This case study is the Aktas-South deposit, which is located in Reddipalayalam, in the state of Tamil Nadu, south of Kazakhstan. The deposit is limestone being used at a cement plant that belongs to the Aktas Group. For confidentiality reasons, the exact location of the deposit is not disclosed. The regional geology of this deposit falls in the Masanchi and Uzun formations in the group of Sabanbay rocks. The limestone in the deposit refers to Masanchi formation. The whole sedimentary formation of this zone is related to marine origin due to marine transgressions and regressions with corresponding fluctuations. As the limestone beds lack marl intrusions, they are composed of clusters of marine organisms. The limestone in the Aktas South area is divided into four groups according to their outward aspects, as shown in Table 1.

Table 1. Division of limestone in Aktas South region.

Lithology	Physical Appearance	Chemical Characteristics	Comment
Cherty limestone (CL)	Yellow in color with alternating cherty bands	Not ascertained	Likely to be used in cement after removing cherty bands
Pale yellow limestone (PYLS)	Yellow to pale brown in color	>40% CaO and <3.5% Fe_2O_3	Very good for cement manufacturing
Brown cherty limestone (BCLS)	Brown to dark brown	>8% Fe_2O_3	
Ferruginous limestone (FLS)	Brown to dark brown	>40% CaO and Fe_2O_3 > 3.5% to 10	Considered as low grade limestone

3.1. Exploratory Data Analysis in Limestone Deposit

The dataset consists of 4553 samples with homotopic sampling patterns. This configuration means that the data are available through all sample points and all variables have been assayed on equal sets of sample locations [36].

The dataset is composed of four chemical compounds: iron oxide (Fe_2O_3), aluminum oxide (Al_2O_3), silicon oxide (SiO_2), and calcium oxide (CaO), whose values are assayed in percentages. Similar to all geostatistical projects, the first step is exploratory data analysis to identify global statistical characteristics of the underlying variables. First, possible outliers and duplicated data were recognized. The presence of outliers in the dataset makes the inference of statistical parameters problematic and nonrepresentative [37,38]. These aberrant values intentionally influence the variance and result in sharp fluctuations in variogram analysis [39]. In addition, detection and removal of duplicate data is also important prior to any geostatistical analysis. One of the problems is that these repeated values generate singular matrices in kriging systems, leading to unestimated blocks surrounding the duplicate locations [40]. After removing duplicate locations in this study, the variables are plotted in a probability plot. This statistical tool helps to detect and fix extreme and innermost values [41]. Following this, some outlier values are detected for Fe_2O_3 and CaO (Figure 3). In the distribution of Fe_2O_3, values more than about 18% are considered as extreme values, and in the distribution of CaO, values less than about 19% are recognized as innermost values, while the other two distributions (SiO_2 and Al_2O_3) sound reasonable given that no significant outliers were identified.

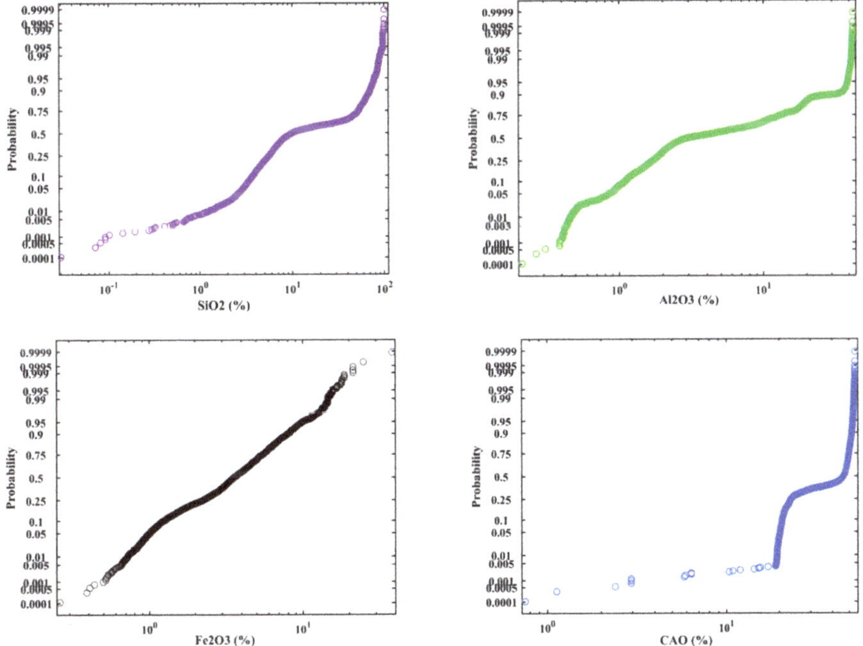

Figure 3. Probability plots of chemical compounds that identify extreme and innermost values.

Once these outlier values are detected, the capping approach [41] is applied to treat them accordingly in order to preserve them in the dataset after examining whether they are valid samples and not erroneous. In this technique, the values in the upper and lower tails of the distribution should be moved back to the previous maximum value and forward to the next minimum value, respectively.

The next step in statistical analysis is related to the declustering process. This step was not necessary, as the sampling pattern is almost regular in the region (Figure 4). The most critical chemical compounds for this deposit are related to the variability of CaO and SiO_2. These two variables introduce high-quality limestone for high amounts of CaO (>10%) and low amounts of SiO_2 (<40%). In addition, Fe_2O_3 plays an important role, yet does not have as significant an influence as SiO_2 in favor of CaO. As can be seen, the majority of drillholes illustrate a high concentration of CaO, which is distributed homogeneously in the region. Interestingly, SiO_2 reveals poor concentration in the same locations, meaning it satisfies the quality of limestone for the entire deposit, corroborating high CaO and low SiO_2. This visual inspection also indicates a strong spatial dependency between CaO and SiO_2, in which there is a negative impact on the local distribution of these two chemical compounds. This shows a high concentration of CaO versus a low concentration of SiO_2, which is important for this deposit. This phenomenon motivates a further investigation into the cross-correlation structures among these chemical variables toward better decision making for the selection of efficient geostatistical algorithms for 3D modelling and mineral resource evaluation.

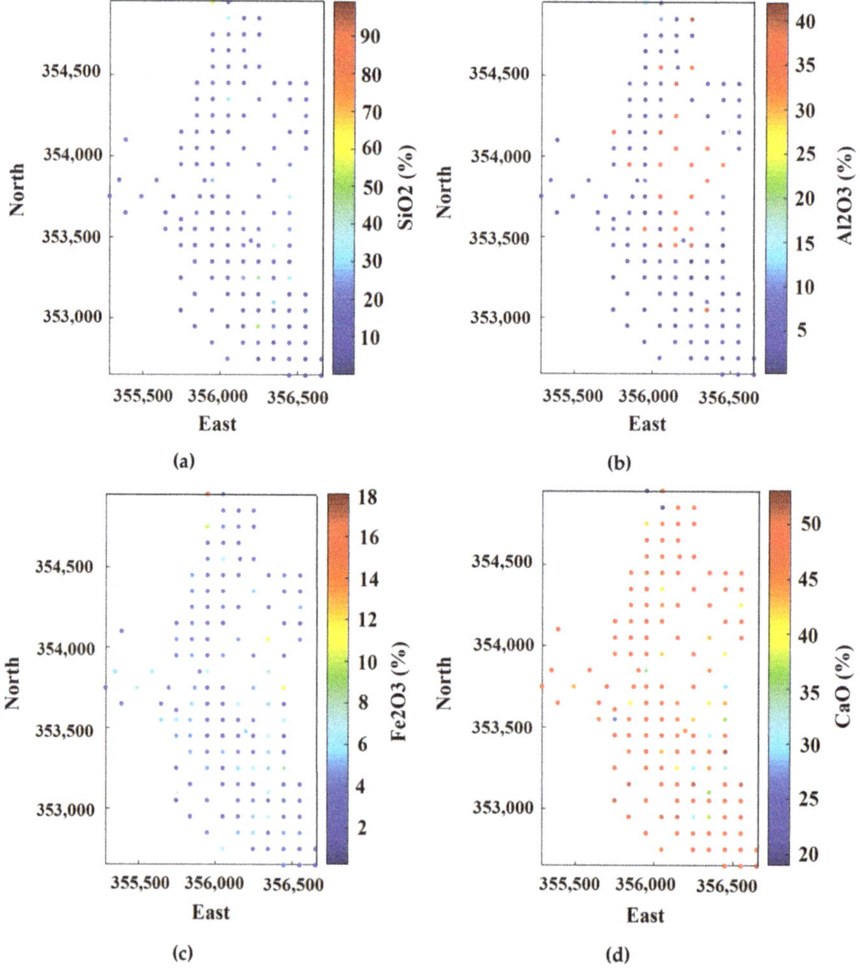

Figure 4. Location map of drilling patterns in limestone deposit.

Then, statistical parameters of the data were computed, as shown in Table 2. The coefficient of variation (COV) for all variables, particularly CaO and SiO_2, is less than 2.0, which indicates that the distribution of data has no significant harsh variability and predictive models can be suitable and meaningful [41]. In this limestone deposit, as previously mentioned, CaO and SiO_2 are two critical variables for ore/waste selection. For this purpose, areas with more than 10% CaO and less than 40% SiO_2 define ore zones, and areas with less than 10% CaO and more than 40% SiO_2 introduce waste zones based on mining excavation destination. Before initiating the modelling process, it might be of interest to calculate two global recovery functions, fraction of recoverable ore above or below the cutoff and mean grade above or below the cutoff [28,42]. These two parameters are calculated by bivariate cumulative distribution functions computed over CaO and SiO_2 as follows:

Fraction of total tonnage in the specified cutoff for ore:

$T(CAO > 10\% \mid SiO2 \leq 40\%) = 66.20\%$

Mean grade in the specified cutoff for CaO in ore:

$m(CAO > 10\% \mid SiO2 \leq 40\%) = 9.19\%$

These two parameters show that about 66.2% and 33.8% of the entire deposit may be ore and waste, respectively, which is an interesting economical characteristic of this mine deposit.

Table 2. Statistical univariate parameters of original Al_2O_3, CaO, Fe_2O_3, and SiO_2 in the dataset of Aktas-South deposit.

Variable (%)	Number of Samples	Minimum	Maximum	Mean	Variance	Coefficient of Variation (COV)
Al_2O_3	4553	0.21	42.32	9.33	132.92	1.23
CaO	4553	0.75	53.94	38.79	164.14	0.33
Fe_2O_3	4553	0.26	38.24	4.29	9.04	0.70
SiO_2	4553	0.03	99.37	27.24	744.72	1.00

In order to investigate the cross-dependency among these four chemical compounds, the global correlation coefficient was calculated, as presented in Table 3. Fairly good negative and positive correlations can be seen between Fe_2O_3 and CaO (−0.64) and between Fe_2O_3 and SiO_2 (approximately 0.53). The highest dependency is seen between CaO and SiO_2, which has negative correlation with almost −0.94 correlation coefficient. This corroborates the visual interpretation already provided in the location map of sampling points (Figure 4). Other correlations, such as those between Fe_2O_3 and Al_2O_3, Al_2O_3 and CaO, and Al_2O_3 and SiO_2, are somewhat low. These values only give a general perspective on the linear dependency that exists among the variables and may not be suitable for examining whether or not complex characteristics such as nonlinearity, heteroscedasticity, and geological constraints may exist. In order to examine the latter characteristics, the bivariate relation in Figure 5 is presented as scatter plot between pairs of the variables Al_2O_3, CaO, Fe_2O_3, and SiO_2. This statistical diagram is suitable to explore relationships such as complexities and linearity features between pairs of variables. This is an interesting illustration of different complexities between co-variables, starting from heteroscedastic characteristics between Fe_2O_3 and Al_2O_3, and CaO and SiO_2; nonlinearity observed between SiO_2 and CaO; and geologic constraints between CaO and Al_2O_3.

Table 3. Correlation coefficients between pairs of variables in Aktas-South limestone deposit.

Variables	Fe_2O_3	Al_2O_3	CaO	SiO_2
Fe_2O_3	1	0.13	−0.64	0.53
Al_2O_3		1	−0.17	0.13
CaO			1	−0.94
SiO_2				1

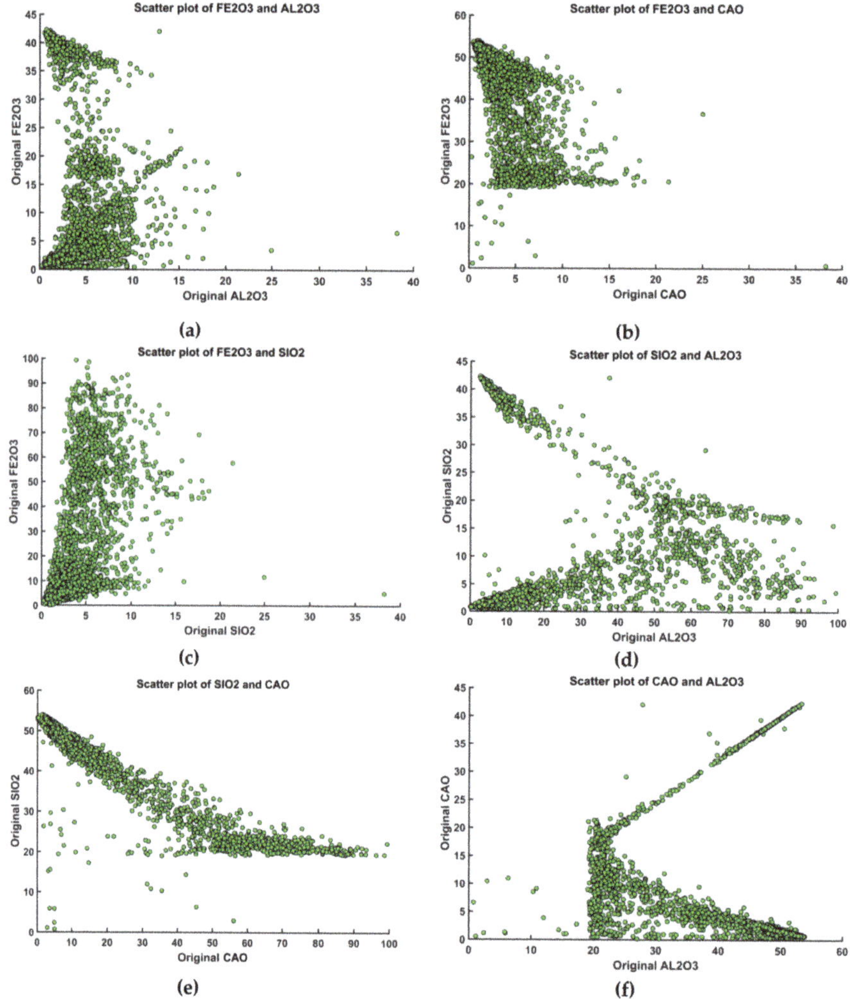

Figure 5. Scatter plots of (**a**) Fe_2O_3 and Al_2O_3, (**b**) Fe_2O_3 and CaO, (**c**) Fe_2O_3 and SiO_2, (**d**) SiO_2 and Al_2O_3, (**e**) SiO_2 and CaO, (**f**) CaO and Al_2O_3.

3.2. PPMT Forward Transformation

As mentioned before, one of the objectives of this paper is to jointly model variables by considering original correlations among variables to reconstruct the shape of complex bivariate relations. It is explained that correlation shapes between pairs of variables are complex (Figure 5). These types of features motivate the use of factorization approaches such as projection pursuit multivariate transform (PPMT) to jointly model the underlying four variables (Al_2O_3, CaO, Fe_2O_3, and SiO_2). The result of this modelling approach is then applied for mineral resource classification. However, prior to any geostatistical modelling, whether it is independent simulation or co-simulation after forward PPMT transformation, the removal of correlations after this first transformation should be assessed. Therefore, in this study, as a common practice in forward PPMT transformation, the variables are first transformed to PPMT factors and then undergo one further transformation by MAF, implemented to completely remove cross-correlation. This can be evaluated by a cross-correlogram (Figure 6).

The results of the cross-correlogram between MAF factors show that a small amount of correlation is still resistant in some lags. For instance, one can recognize that the correlation in the first lag (~100 m) is around 25% between SiO_2 and Al_2O_3 even after this MAF transformation (Figure 6). This signifies that MAF is not able to decorrelate the variables over all the lag separations. In this respect, it is not advocated to use the independent simulation due to the remaining small correlations among factors. In order to cope with the proposed algorithm in this case, it was decided to employ co-simulation right after the initial PPMT forward transformation of underlying variables irrespective of considering any further MAF transformation. For this, once again, the cross-correlation among the transformed PPMT factors is examined through the cross-correlogram and before the MAF step. The results show that the correlation manifests itself through some lags (e.g., ~12% between SiO_2 and CaO at a lag separation of 150 m and ~11% between SiO_2 and Al_2O_3 at a lag separation of 400 m; Figure 7), although at a lag separation of 0, the correlation is significantly removed (Figure 8). Even these small amounts of correlation among the PPMT factors are important and provoke applying the co-simulation algorithms via PPMT transformed factors and considering the linear model of co-regionalization.

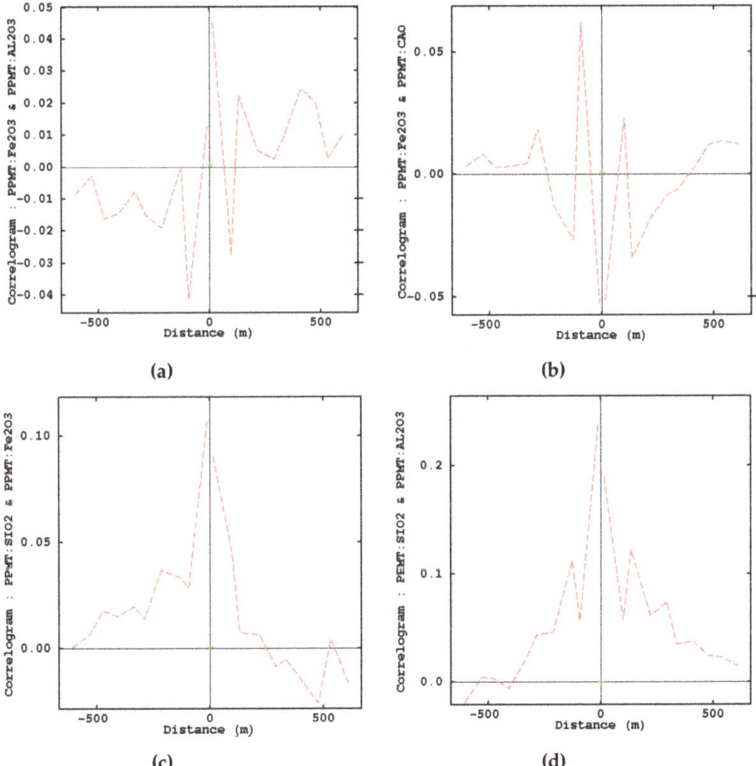

Figure 6. *Cont.*

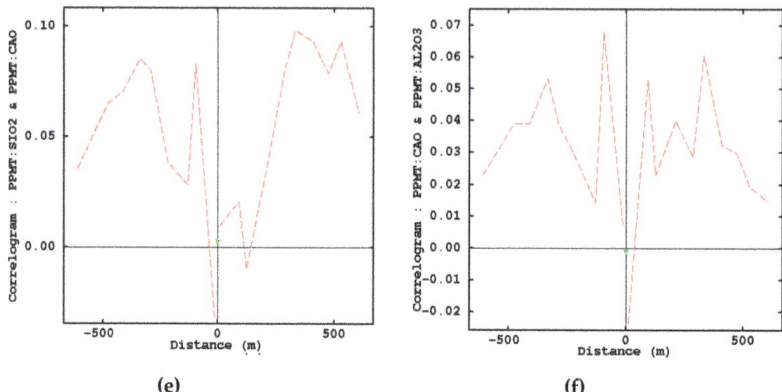

(e) (f)

Figure 6. Correlograms of PPMT factors with minimum/maximum autocorrelation factor (MAF) transformation ((**a**) PPMT Fe_2O_3 and PPMT Al_2O_3, (**b**) PPMT Fe_2O_3 and PPMT CaO, (**c**) PPMT SiO_2 and PPMT Fe_2O_3, (**d**) PPMT SiO_2 and PPMT Al_2O_3, (**e**) PPMT SiO_2 and (**f**) PPMT CaO, PPMT CaO and PPMT Al_2O_3).

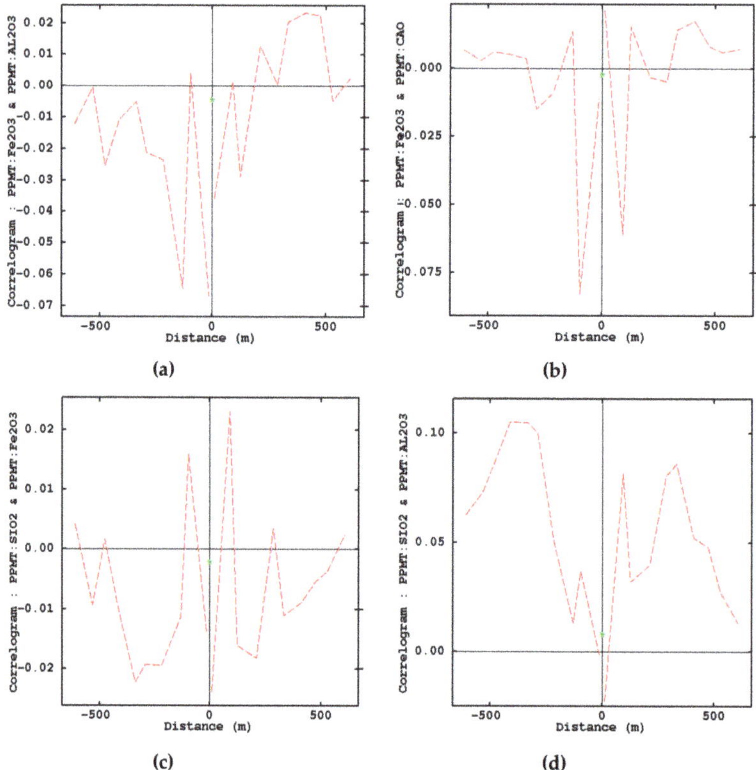

(a) (b)

(c) (d)

Figure 7. *Cont.*

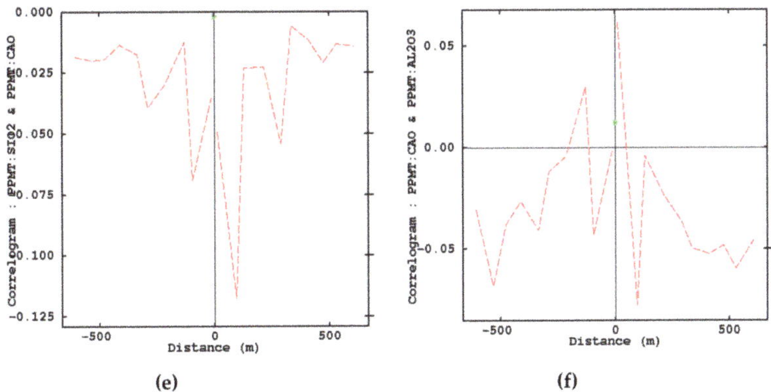

Figure 7. Correlograms of PPMT factors without MAF transformation ((**a**) PPMT Fe_2O_3 and PPMT Al_2O_3, (**b**) PPMT Fe_2O_3 and PPMT CaO, (**c**) PPMT SiO_2 and PPMT Fe_2O_3, (**d**) PPMT SiO_2 and PPMT Al_2O_3, (**e**) PPMT SiO_2 and PPMT CaO, (**f**) PPMT CaO and PPMT Al_2O_3).

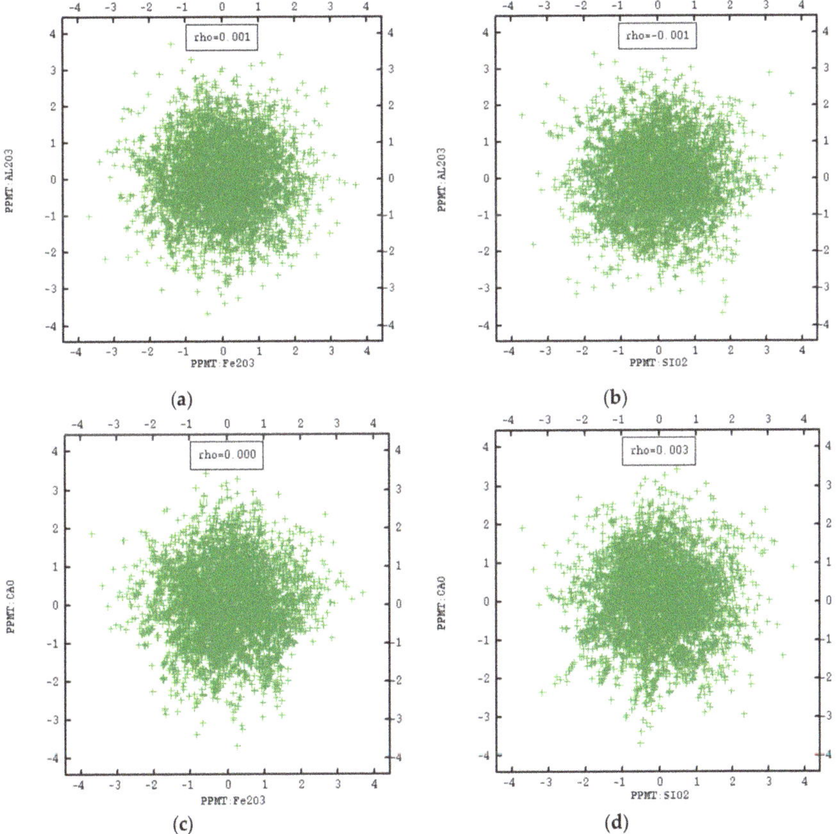

Figure 8. *Cont.*

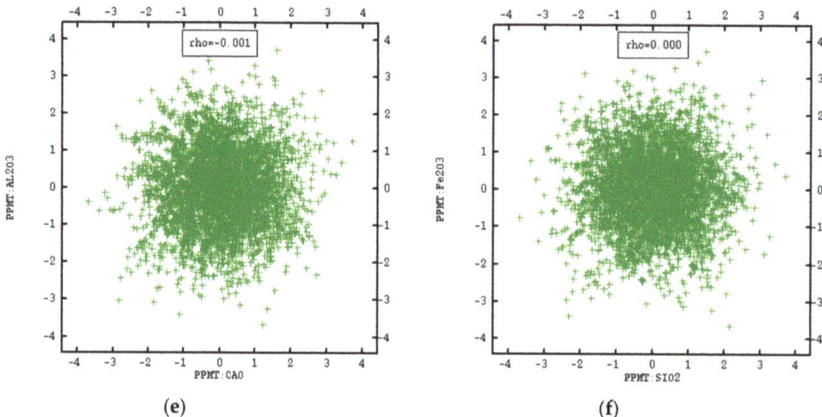

Figure 8. Scatter plots of transformed variables without integration of MAF: (**a**) PPMT_Al_2O_3 and PPMT_Fe_2O_3, (**b**) PPMT_Al_2O_3 and PPMT_SiO_2, (**c**) PPMT_CaO and PPMT_Fe_2O_3, (**d**) PPMT_CaO and PPMT_SiO_2, (**e**) PPMT_Al_2O_3 and PPMT_CaO, (**f**) PPMT_Fe_2O_3 and PPMT_SiO_2. Correlation at lag 0 is almost zero.

3.3. *Variogram Inference*

As mentioned in a previous section, the transformed factors retain the correlation at other lag distances except zero, and a further decorrelation transformation technique such as MAF cannot completely remove the inherent correlations. Following the proposed algorithm in this research, the next step is variogram analysis over the transformed variables. This step is needed to establish the co-kriging system in co-simulation algorithms, taking into account the linear model of co-regionalization. This latter introduces even small correlations in co-simulation algorithms. In this respect, inference of direct and cross-variograms for all four PPMT transformed variables is implemented. The sill of experimental cross-variogram as a measure of joint variability to some extent reflects the magnitude of the correlation between the variables [43]. In the case of standardized variables (variance equal to 1), the sill of cross-variograms must be the cross-correlation between collocated values of those variables [44].

It is also worth mentioning that anisotropy was examined through the original dataset (not transformed), and it was seen that the existence of anisotropy in the region was improbable. For fitting of theoretical variograms over experimental variograms, the linear model of co-regionalization [45] with semi-automatic technique is chosen, in which the semi-definiteness condition [36,46] is respected through the process of fitting. In this model, direct and cross-covariances are defined as the sum of basic covariances. Therefore, by this technique and after computation of experimental variograms, two nested structures (spherical, with the first scale factor as 54 m and the second scale factor as 216 m) are fitted accordingly, with no nugget effect, as can be seen in Figure 9 (Equation (7)).

$$
\begin{pmatrix}
\gamma_{Al_2O_3} & \gamma_{CaO/Al_2O_3} & \gamma_{Fe_2O_3/Al_2O_3} & \gamma_{SiO_2/Al_2O_3} \\
\gamma_{CaO/Al_2O_3} & \gamma_{CaO} & \gamma_{Fe_2O_3/CaO} & \gamma_{SiO_3/CaO} \\
\gamma_{Fe_2O_3/Al_2O_3} & \gamma_{Fe_2O_3/CaO} & \gamma_{Fe_2O_3} & \gamma_{SiO_2/Fe_2O_3} \\
\gamma_{SiO_2/Al_2O_3} & \gamma_{SiO_2/CaO} & \gamma_{SiO_2/Fe_2O_3} & \gamma_{SiO_2}
\end{pmatrix}
= \begin{pmatrix}
0.499 & 0.060 & -0.002 & 0.003 \\
0.006 & 0.499 & -0.001 & -0.001 \\
-0.002 & -0.001 & 0.499 & -0.001 \\
0.003 & -0.001 & -0.001 & 0.499
\end{pmatrix} \text{spherical } (54\text{ m}, 54\text{ m}, 54\text{ m})
+ \begin{pmatrix}
0.499 & 0.060 & -0.002 & 0.003 \\
0.006 & 0.499 & -0.001 & -0.001 \\
-0.002 & -0.001 & 0.499 & -0.001 \\
0.003 & -0.001 & -0.001 & 0.499
\end{pmatrix} \text{spherical } (216\text{ m}, 216\text{ m}, 216\text{ m}) \quad (7)
$$

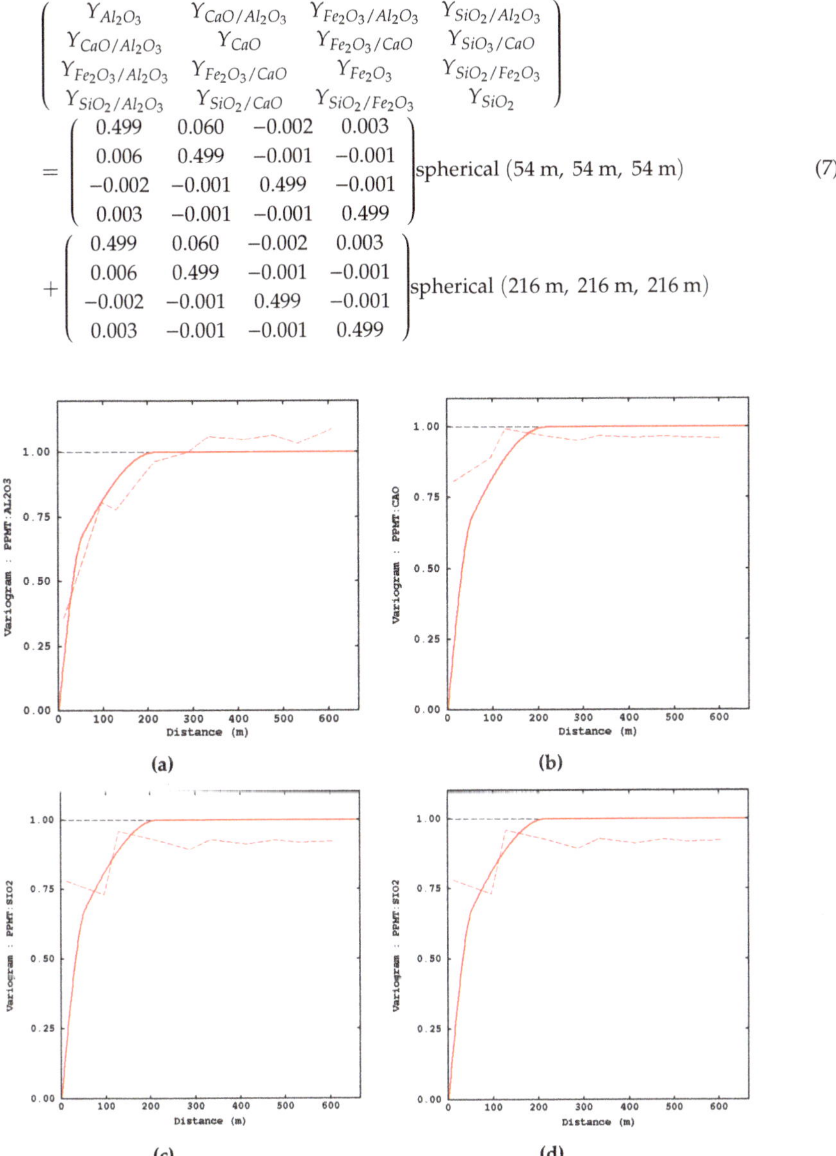

Figure 9. Fitted direct variograms of transformed variables: (**a**) PPMT Al_2O_3, (**b**) PPMT CaO, (**c**) PPMT SiO_2, (**d**) PPMT SiO_2. For brevity, only direct variograms are presented.

3.4. Stochastic Modeling in Limestone Deposit

After variogram inference over PPMT transformed covariates and following the proposed algorithm in this study, the step for implementing co-simulation takes into account grid nodes with dimensions of 48 m × 48 m × 11 m for each block to jointly model all transformed factors. Turning bands co-simulation (TBCOSIM) is selected because of its versatility and reliability in reproducing global and statistical parameters when compared to other Gaussian co-simulation algorithms [47]. In this method, each variable is first simulated nonconditionally and then through a post-processing step by co-kriging conditioned on the available information and borehole records [33]. In this regard, ordinary co-kriging is used with moving neighborhood ranges equal to the range of variograms (200 m) attending up to 50 nearest surrounding sample points in the process of conditioning. Multiple-search strategy is also selected for this purpose, since it shows better results compared to single-search strategy [48]. One of the main criticisms against TBCOSIM, however, is related to producing artifacts because of turning lines. To deal with this difficulty, the number of lines should be reasonably increased [33,49]. Therefore, 1000 lines was set for the turning bands simulation method to eliminate possible stripping effects. The number of realizations was considered to be 100. The realizations that interpret the spatial variability of each factor were then back-transferred to original scale though backward PPMT transformation. Next, E-type maps were generated by averaging through 100 realizations within each block in co-simulated element over the variability in original scale. E-type maps of CaO, Al_2O_3, Fe_2O_3, and SiO_2 were produced through 100 realizations, as shown in Figure 10. Before further analysis over the simulated results, it is necessary to check whether the outputs of this proposed algorithm are statistically valid.

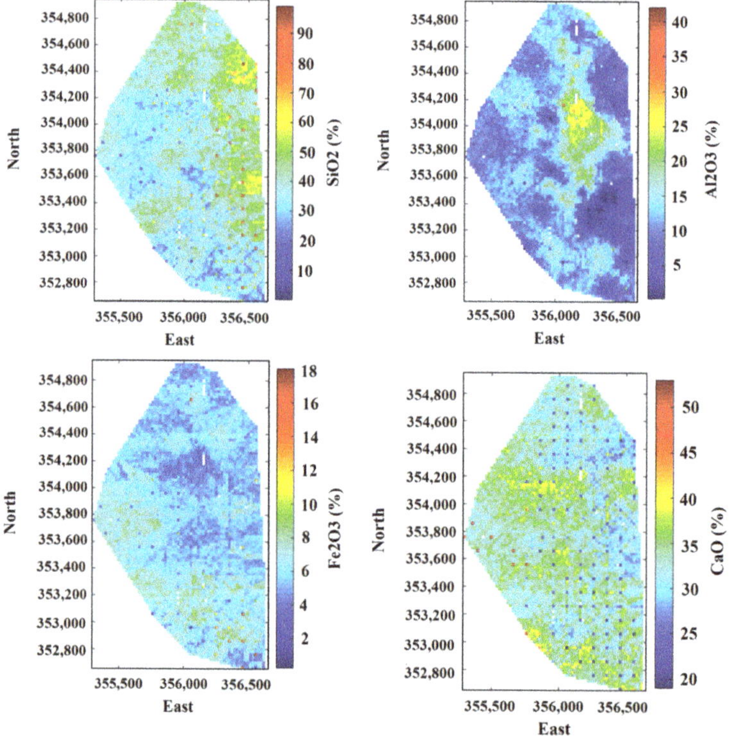

Figure 10. E-type maps of all underlying elements according to PPMT methodology.

3.5. Validation

Validation analysis in this section is concerned with the reproduction of original statistical characteristics such as mean, variance, correlation coefficient, and shape of bivariate relation. This type of validation process is required to give practitioners insight with regard to the level of reliability of the aforementioned approach in the section on mineral resource estimation and classification. A comparison between the means of original variables (Fe_2O_3, Al_2O_3, SiO_2, and CaO) and the means of simulated and back-transformed variables calculated from PPMT through 100 realizations is shown in Figure 11. As can be seen, PPMT is able to reproduce the original mean values of each variable.

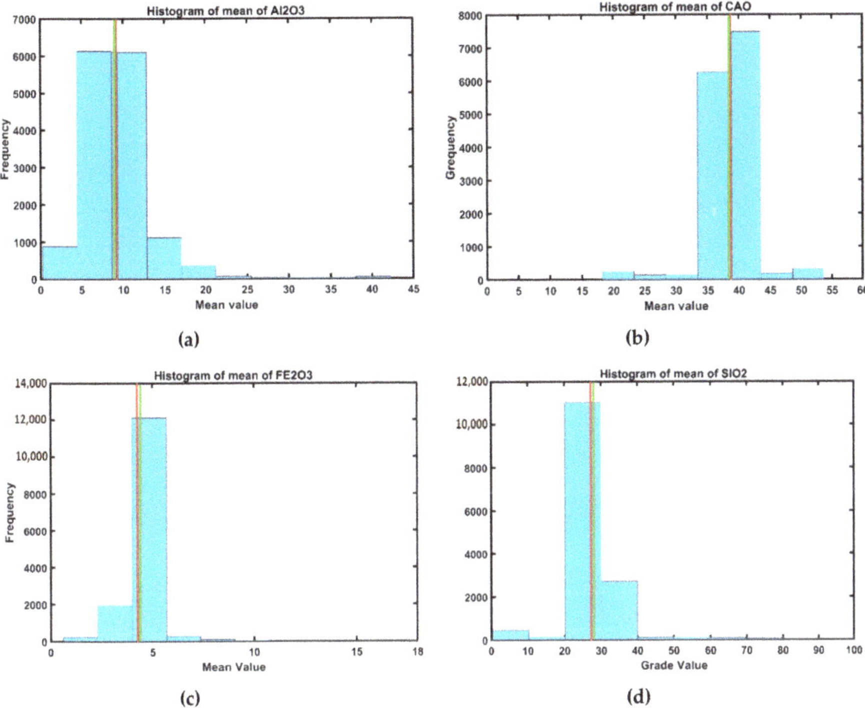

Figure 11. Histograms of mean values of: (**a**) aluminium oxide (Al_2O_3), (**b**) calcium oxide (CaO), (**c**) iron oxide (Fe_2O_3) and (**d**) silicon oxide (SiO_2) obtained from PPMT method. Green line is the average of mean values over 100 realizations; red line represents the original mean of variables.

The next comparison is the reproduction of global statistical analysis between variance of original variables (Fe_2O_3, Al_2O_3, SiO_2, and CaO) and simulated and back-transformed values obtained from the PPMT method through 100 realizations. For CaO and SiO_2 (Figure 12), it is intuitively determined that PPMT produces satisfactory outputs in terms of reproduction of variance. However, minor deviations, such as for Fe_2O_3, as can be seen from this figure, are referred to the influence of conditioning data [24,50,51]. However, this tiny departure of average of variances for 100 realizations from original variance is not remarkably significant.

Finally, yet importantly, comparisons of correlation coefficients provided by PPMT and original correlation coefficients are examined among all variables (Fe_2O_3, Al_2O_3, SiO_2, and CaO) through 100 realizations. As it can be seen from Figure 13, co-simulation methodology shows satisfactory results in the reproduction of original correlation coefficients among co-variables. This good performance among variables can be explained by the fact that co-simulation considers the intrinsic correlation between variables by incorporating the linear model of co-regionalization [51–53].

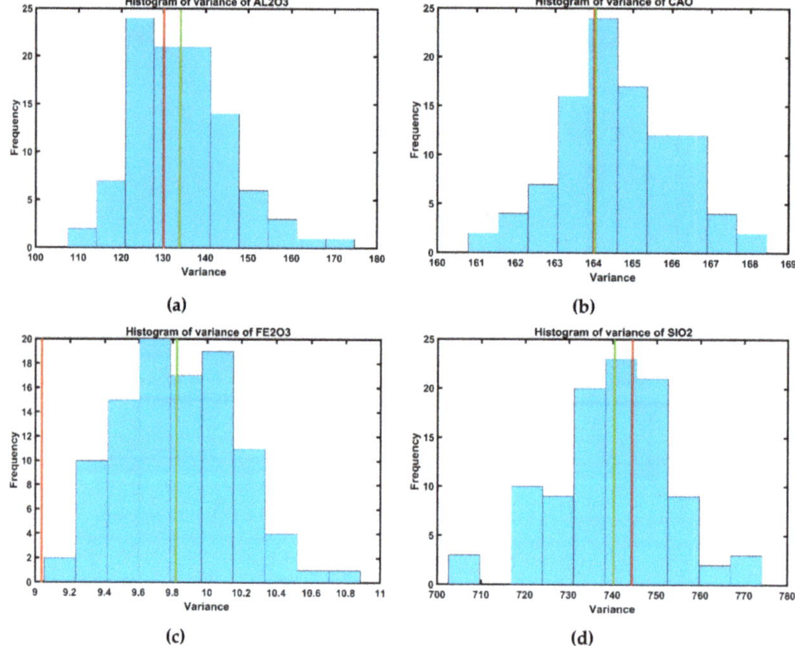

Figure 12. Histograms of mean variance of: (**a**) aluminium oxide (Al_2O_3), (**b**) calcium oxide (CaO), (**c**) iron oxide (Fe_2O_3), and (**d**) silicon oxide (SiO_2) obtained from PPMT method. Green line is the average mean of 100 realizations; red line represents the original mean of variables.

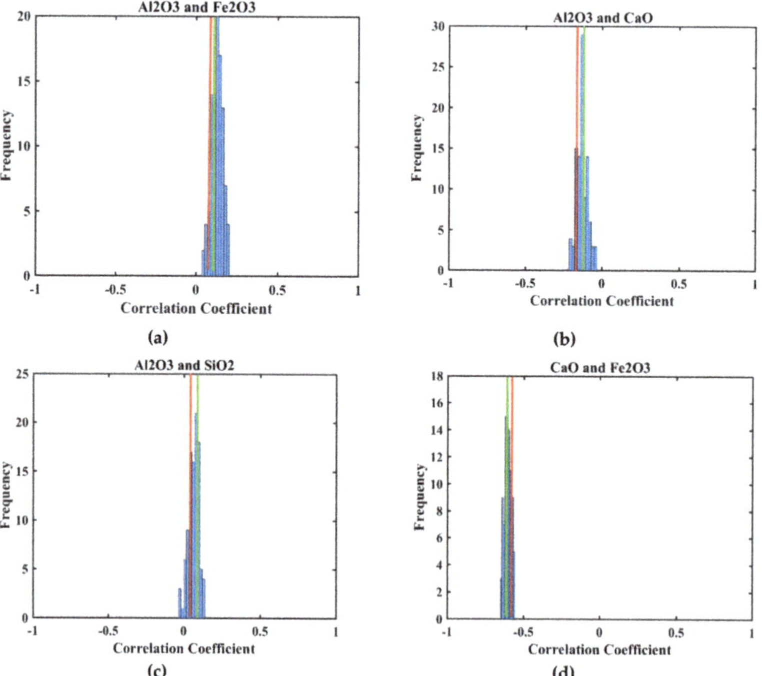

Figure 13. *Cont.*

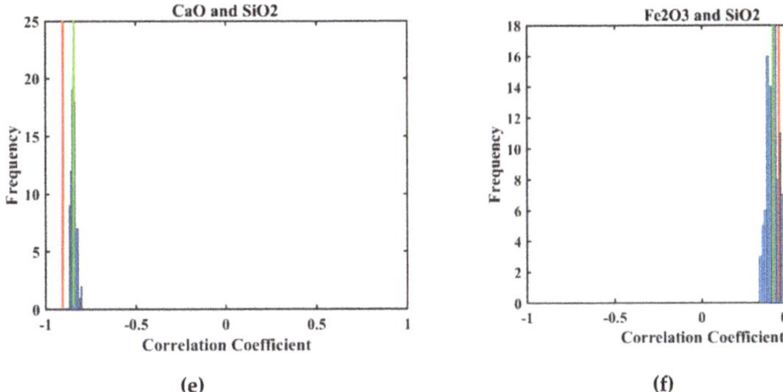

Figure 13. Graphs of correlation coefficients of: (**a**) Al_2O_3 and Fe_2O_3, (**b**) Al_2O_3 and CaO, (**c**) Al_2O_3 and SiO_2, (**d**) CaO and Fe_2O_3, (**e**) CaO and SiO_2, (**f**) and Fe_2O_3 and SiO_2 obtained by PPMT method. Green line is the average mean of 100 realizations; red line represents the original correlation coefficient between variables.

Another part of the comparison is related to examining the ability of the proposed method to reconstitute the shape of the original bivariate relations between pairs of chemical compounds. By visual inspection of reproductions of the shape of correlation for the underlying pairs of variables, one can see the difference between original and reproduced shapes in Figure 14. It should be mentioned that only one realization was taken to illustrate the reproduction of correlation shape, and it was randomly chosen as a first realization for all cases for the sake of fairness. As can be seen from the scatter plots of Fe_2O_3 and Al_2O_3, PPMT co-simulation successfully reproduces the shape of original correlation. The same features are evidently demonstrated in the reproduction of the shape of original correlation between Fe_2O_3 and CaO, Fe_2O_3 and SiO_2, and SiO_2 and CaO. However, inadequate results in the reproduction of the shape of correlation between SiO_2 and Al_2O_3, and CaO and Al_2O_3 can be seen. Overall, the proposed approach based on the combination of PPMT and co-simulation is capable of reconstructing the original shape of correlation.

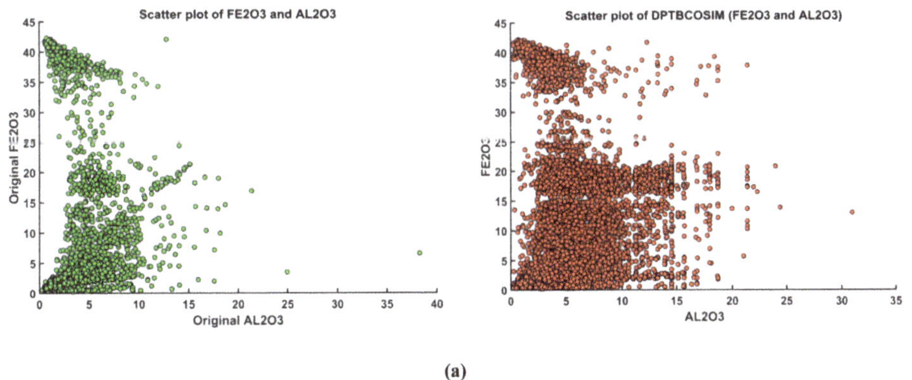

(a)

Figure 14. *Cont.*

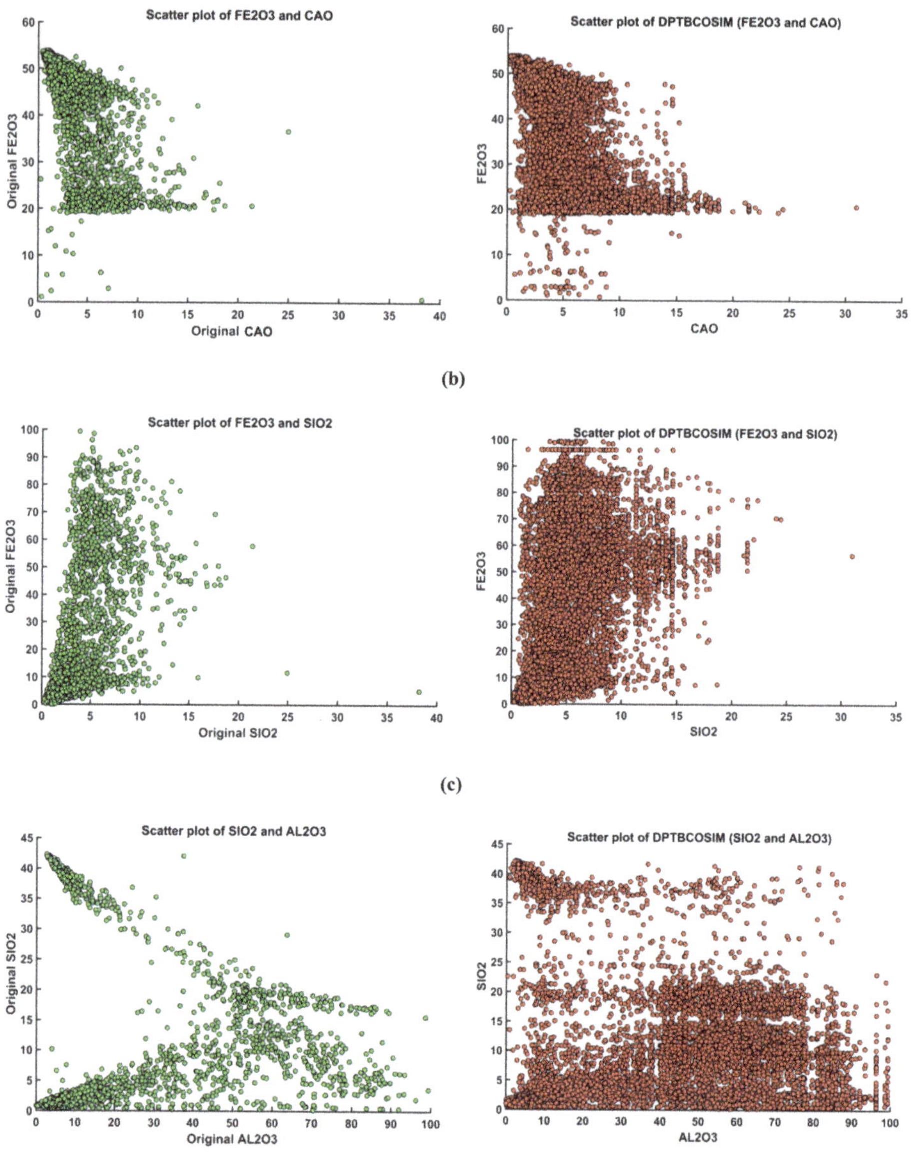

(d)

Figure 14. *Cont.*

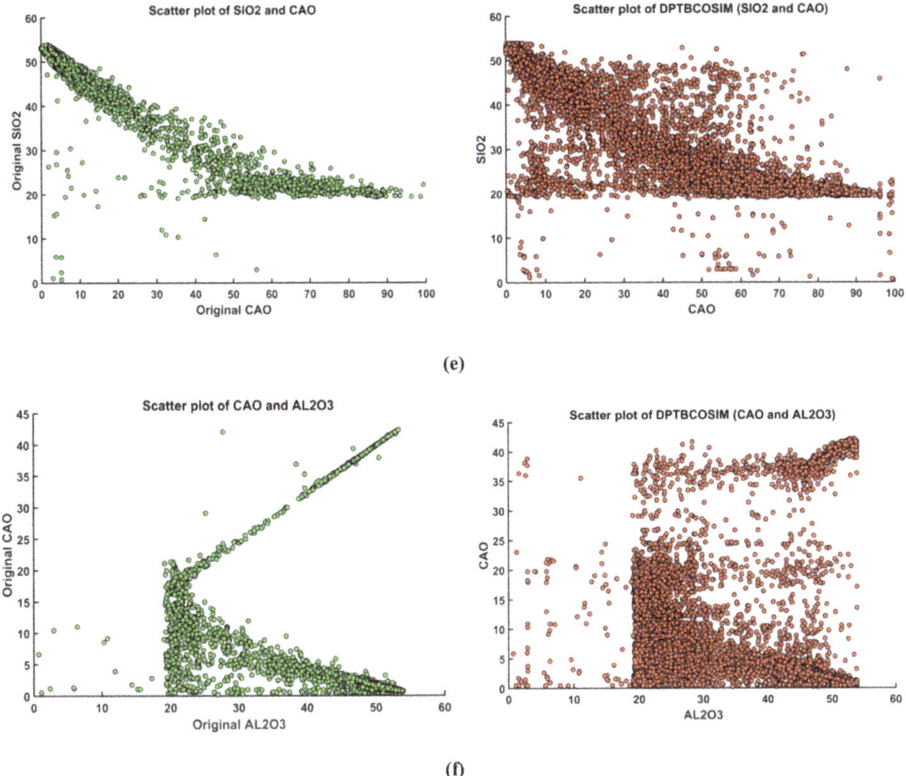

Figure 14. Reproduction of original correlation coefficient (green) between (**a**) Fe_2O_3 and Al_2O_3, (**b**) Fe_2O_3 and CaO, (**c**) Fe_2O_3 and SiO_2, (**d**) SiO_2 and Al_2O_3, (**e**) SiO_2 and CaO, (**f**) CaO and Al_2O_3 by PPMT method (red).

3.6. Mineral Resource Classification

Based on the JORC code definition (www.JORC.org), mineral resources can be classified into measured, indicated, or inferred based on the level of confidence. In fact, the guideline in this code was developed to assure transparency for investors in the declaration of mineral resources in order to prevent fraud [41]. A measured mineral resource is part of a deposit that presents a high level of confidence in the estimation of recovery functions such as tonnage, mean grade, and metal quantity above cutoff. Indicated mineral resources can be assigned to those areas that demonstrate a reasonable level of confidence in the estimation of recovery functions such as tonnage, mean grade, and metal quantity above the cutoff. Inferred mineral resources represent locations for which the recovery functions are evaluated at a low level of confidence. In this regard, there are mainly two algorithms for mineral resource classification, connected with either deterministic or stochastic paradigms. In the latter, one needs to employ, for instance, geostatistical simulation techniques to produce different scenarios of a mine deposit (i.e., realizations) with the aim of quantifying the uncertainty. This leads to probabilistic reporting of the mineral resource measures.

Mineral resource classification in this limestone deposit is done on the basis of two main targets, following the stochastic approach. One is related to classification based on ore zone definition (geological), in which mostly CaO and Fe_2O_3 are two critical components for the underlying zone. The second is mainly concerned with mining excavation units where CaO, SiO_2, and Al_2O_3 are vital variables for this zone definition. In the following, we present the results for both types of mineral resource classification that define the different categories based on ore zone definitions and mining units.

3.6.1. Ore Zone Definitions

One of the main objectives of this study is to employ the proposed algorithm based on a combination of PPMT transformation and co-simulation algorithm, and to show the results of mineral resource classification by demonstrating their distinctions. As shown previously in the section on validating simulation results with regard to reproducing original mean, variance, correlation coefficient, and shape of bivariate relations between pairs of variables, the PPMT and co-simulation methodology corroborates that the outputs of realizations are statistically sound and can be applied for further processing, such as mineral resource classification. Aktas-South deposit is grouped into three main ore zones based on the definitions of chemical cutoffs as indicated in Table 4. CaO in this limestone deposit is the main chemical compound. Resource estimation of this variable based on JORC code is shown in Table 5 for each ore zone. In order to show the distinctions between outputs of resource classification of CaO (measured, indicated, inferred, and total tonnage), the results are constructed in one unique graph, shown in Figure 15. As previously explained, the classification of resources is modelled according to uncertainty, in which measured resources are quantified 90% of the time within ±15%, indicated resources within ±15% and ±30%, and inferred resources within ±30% and ±100%, while other materials within more than ±100% are not distinguished as resources [41].

Table 4. Ore zone definitions based on chemical cutoffs.

Zone	Chemical Cutoffs
Marl-chert	≤40% and ≥20% of CaO
Pale yellow limestone	>40% of CaO and <3% of Fe_2O_3
Brown limestone	>40% of CaO and ≥3% and <4.5% of Fe_2O_3

Table 5. Resource estimation of CaO in Mt based on ore zone definitions.

Classification	Marl-Chert (Mt)	Pale Yellow Limestone (Mt)	Brown Limestone (Mt)
Measured	53	102.14288	52.56608
Indicated	30	10.79	5.4357
Inferred	125	67.99632	33.652
Total	208	180.9292	91.65378

As can be observed from Figure 15, marl-chert has the highest total tonnage of CaO, 208 Mt, according to the PPMT co-simulation method (proposed algorithm), followed by pale yellow limestone and brown limestone, with 181 Mt and 92 Mt, respectively. In the measured category, marl-chert ore and brown limestone ore tonnage are almost the same, 53 Mt, while pale yellow limestone has higher tonnage, 102 Mt. In the indicated and inferred categories, marl-chert has higher tonnages, 30 Mt and 125 Mt, respectively, followed by pale yellow limestone, 11 and 68 Mt, respectively, and brown limestone has the lowest tonnages, 5.5 Mt and 34 Mt, respectively. It should be mentioned that ore zone definition in this step has two restrictions (chemical cut-offs), and because of that results of resource classification differ.

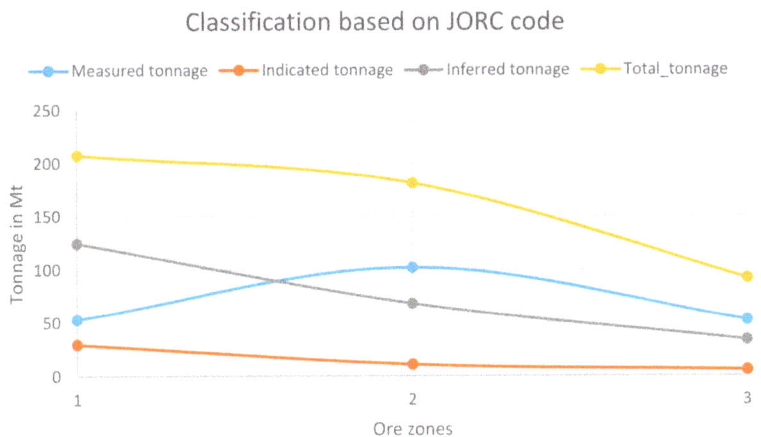

Figure 15. Resource classification of CaO in different ore zones: 1: marl-chert; 2: pale yellow limestone; 3: brown limestone.

3.6.2. Mining Units

In this deposit, there is another classification based on chemical cutoff definitions, but this time from the mining destination perspective, whereas the previous tonnage classification was based on the definitions of ore zones in favor of geological interpretations. The definitions of these mining units are presented in Table 6. In this paradigm, the units are mainly established in accordance with three main chemical compounds, CaO, SiO_2, and Fe_2O_3. As also shown in Table 3, there is strong correlation among these three variables. The results of the proposed approach, in fact, show that the correlation coefficients are reproduced at a satisfactory level of confidence. The importance of this is shown in this section; for instance, these three variables are the key factors in favor of grouping the Aktas-South limestone deposit for mining excavation, which consequently impacts the rigorous classification of mineral resources. The list of mining units based on chemical cutoffs is shown in Table 6. It should be mentioned that in this classification, the number of restrictions (chemical cutoffs) reaches four, which means that the results of classification will be according to seven restrictions. As can be seen, CaO and SiO_2 are two chemical compounds that define the ore/waste classification technique. The tonnage for each classification based on JORC code in this deposit is summarized in Table 7, and all results of the mentioned method with resource classification are summarized in single graphs for highly green limestone (HGLS), brown limestone (BROWNLS), ferrous limestone (FEROLS), cherty limestone (CHERTYLS), cherty limestone 2 (CHERTYLS2), MARL, and WASTE in Figure 16, following the same method as explained in the previous section [41].

Table 6. Mining units with restriction in chemical characteristics.

Mining Unit	Chemical Cutoffs
HGLS	CaO $\geq$ 40 and SiO_2 $\leq$15 and Fe_2O_3 < 3
BROWNLS	CaO $\geq$ 40 and SiO_2 $\leq$ 15 and Fe_2O_3 $\geq$ 3 and Fe_2O_3 < 4
FEROLS	CaO $\geq$ 40 and SiO_2 $\leq$ 15 and Fe_2O_3 $\geq$ 4
CHERTYLS	CaO < 50 and CaO > 20 and SiO_2 > 15 and SiO_2 $\leq$40
CHERTYLS2	CaO < 40 and CaO > 30 and SiO_2 $\leq$15
MARL	CaO < 45 and CaO > 10 and SiO_2 > 40
WASTE	CaO $\leq$ 10 and CaO > 0 and SiO_2 > 40

Table 7. Resource estimation of CaO in Mt based on mining unit.

Category	HGLS	BROWNLS	FEROLS	CHERTYLS	CHERTYLS2	MARL	WASTE
Measured tonnage	95.8	38.6	29.3	14.9	0.4	53.4	0.0
Indicated tonnage	9.3	2.8	3.4	13.5	1.5	20.4	0.0
Inferred tonnage	63.4	24.1	18.9	26.8	1.8	113.8	1.2
Total tonnage	168.5	65.6	51.5	55.1	3.7	187.5	1.2

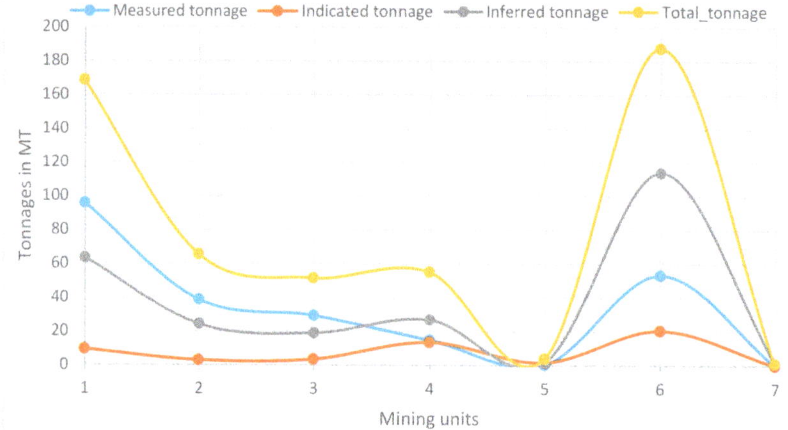

Figure 16. Resource classification of CaO in different mining units: 1: HGLS; 2: BROWNLS; 3: FEROLS; 4: CHERTYLS; 5: CHERTYLS2; 6: MARL; 7: WASTE.

Depending on the chemical restrictions of each lithology, resource tonnages have different trends of results by category. The highest total tonnages are shown in HGLS and MARL, followed by almost the same tonnages in BROWNS, FEROLS, and CHERTYLS (see Figure 16). The lowest total tonnages are computed in CHERTYLS2 and WASTE. Turning to the measured category, the highest tonnage is seen in HGLS, followed by MARL, BROWNS, FEROLS, and CHERTYLS, and the lowest tonnages are reported in CHERTYLS and WASTE. With an increased number of geological restrictions, the results of resource classification change more chaotically. However, it is clearly shown that CHERTYLS2 and WASTE, which have narrow restrictions in CaO, have the lowest tonnages in every category.

4. Conclusions

This contribution provides a technique for the multivariate geostatistical analysis of co-regionalized variables, particularly for datasets where complexity exists between bivariate relationships. The proposed algorithm in this study is based on the combination of a conventional stochastic simulation approach (Gaussian (co)-simulation) and a recently developed factorization technique (projection pursuit multivariate transformation (PPMT)). A main difference from other multivariate geostatistical approaches is that in PPMT, there is no need to further transform the factors to remove the correlation. Instead, a linear model of co-regionalization should be defined to introduce the small cross-spatial dependency among the transformed variables, in which it can be inferred from the factors right after PPMT forward transformation.

In addition, the real case study, limestone deposit, illustrates the effectiveness of the proposed algorithm for mineral resource classification based on the two-zone separation following the JORC code definition. To do so, the outputs of the realizations are first examined to determine whether they are statistically valid, then taken into account to classify resources as measured, inferred, or indicated. Through the validation step, reproduction of mean, variance, and global (cross)-correlation is examined, illustrating that all simulation results, on average, converge to the original statistical parameter, indicating the robustness of the proposed algorithm. Behind the seeming simplicity and great number of practical investigations, much effort is still needed to deal with more complex sampling configuration, particularly in the presence of partially or heterotopic sampling patterns. In such cases, using PPMT is restricted and an imputation technique may be an alternative to combine PPMT with (co)-simulation.

Author Contributions: N.B. implemented the experiments and wrote the paper, and N.M. wrote and reviewed the paper.

Funding: This research was funded by NAZARBAYEV UNIVERSITY via Faculty Development Competitive Research Grants for 2018–2020, grant number 090118FD5336.

Acknowledgments: The authors thank the Datamine Company for providing the dataset of limestone deposit for the underlying case study. We deeply thank Paul Alexandre, the two anonymous reviewers, and editorial board for their invaluable efforts towards improving the overall scientific quality of this manuscript.

Conflicts of Interest: The authors declare no conflict of interest. The founding sponsors had no role in the design of the study; in the collection, analyses, or interpretation of data; in the writing of the manuscript, and in the decision to publish the results.

References

1. Dimitrakopoulos, R. Stochastic Mine Planning—Methods, Examples and Value in an Uncertain World. In *Advances in Applied Strategic Mine Planning*; Springer: Cham, Switzerland, 2018; pp. 101–115.
2. Hustrulid, W.A.; Kuchta, M.; Martin, R.K. *Open Pit Mine Planning and Design, Two Volume Set & CD-ROM Pack: V1: Fundamentals, V2: CSMine Software Package, CD-ROM: CS Mine Software*; CRC Press: Boca Raton, FL, USA, 2013.
3. Kasmaee, S.; Raspa, G.; De Fouquet, C.; Tinti, F.; Bonduà, S.; Bruno, R. Geostatistical Estimation of Multi-Domain Deposits with Transitional Boundaries: A Sensitivity Study for the Sechahun Iron Mine. *Minerals* **2019**, *9*, 115. [CrossRef]
4. Joint Ore Reserves Committee (JORC) Code. The JORC Code and Guidelines. In *Australasian Code for Reporting of Exploration Results, Mineral Resources and Ore Reserves Prepared by The Australasian Institute of Mining and Metallurgy (AusIMM)*; Australian Institute of Geoscientists and Minerals Council of Australia: Crows Nest, Australia, 2012; Available online: www.jorc.org (accessed on 7 October 2015).
5. Monkhouse, P.H.L.; Yeates, G.A. Beyond Naive Optimisation. In *Advances in Applied Strategic Mine Planning*; Springer: Cham, Switzerland, 2018; pp. 3–18.
6. Mai, N.L.; Erten, O.; Topal, E. A new generic open pit mine planning process with risk assessment ability. *Int. J. Coal Sci. Technol.* **2016**, *3*, 407–417. [CrossRef]
7. Dimitrakopoulos, R. Stochastic optimization for strategic mine planning: A decade of developments. *J. Min. Sci.* **2011**, *47*, 138–150. [CrossRef]
8. Khosrowshahi, S.; Shaw, W.J.; Yeates, G.A. Quantification of Risk Using Simulation of the Chain of Mining—Case Study at Escondida Copper, Chile. In *Advances in Applied Strategic Mine Planning*; Springer: Cham, Switzerland, 2018; pp. 57–74.
9. Vallejo, M.N.; Dimitrakopoulos, R. Stochastic orebody modelling and stochastic long-term production scheduling at the KéMag iron ore deposit, Quebec, Canada. *Int. J. Mining Reclam. Environ.* **2018**, *33*, 462–479. [CrossRef]
10. Benndorf, J.; Dimitrakopoulos, R. Stochastic long-term production scheduling of iron ore deposits: Integrating joint multi-element geological uncertainty. *J. Min. Sci.* **2013**, *49*, 68–81. [CrossRef]
11. De-Vitry, C.; Vann, J.; Arvidson, H. Multivariate iron ore deposit resource estimation—A practitioner's guide to selecting methods. *Appl. Earth Sci.* **2010**, *119*, 154–165. [CrossRef]
12. Dimitrakopoulos, R.; Farrelly, C.T.; Godoy, M. Moving forward from traditional optimization: Grade uncertainty and risk effects in open-pit design. *Min. Technol.* **2002**, *111*, 82–88. [CrossRef]
13. Menabde, M.; Froyland, G.; Stone, P.; Yeates, G.A. Mining Schedule Optimisation for Conditionally Simulated Orebodies. In *Advances in Applied Strategic Mine Planning*; Springer: Cham, Switzerland, 2018; pp. 91–100.

14. Dimitrakopoulos, R. Conditional simulation algorithms for modelling orebody uncertainty in open pit optimisation. *Int. J. Surf. Min. Reclam. Environ.* **1998**, *12*, 173–179. [CrossRef]
15. Emery, X.; Lantue´joul, C. Tbsim: A computer program for conditional simulation of three-dimensional gaussian random fields via the turning bands method. *Comput. Geosci.* **2006**, *32*, 1615–1628. [CrossRef]
16. Isaaks, E.H. The Application of Monte Carlo Methods to the Analysis of Spatially Correlated Data: Unpublished. Ph.D. Thesis, Stanford University, Stanford, CA, USA, 1999; 213p.
17. Journel, A.G. Modeling Uncertainty: Some Conceptual Thoughts. In *Geostatistics Valencia 2016*; Springer: Cham, Switzerland, 1994; Volume 6, pp. 30–43.
18. Boisvert, J.B.; Rossi, M.E.; Ehrig, K.; Deutsch, C.V. Geometallurgical Modeling at Olympic Dam Mine, South Australia. *Math. Geosci.* **2013**, *45*, 901–925. [CrossRef]
19. Deutsch, C.V. Geostatistical Modelling of Geometallurgical Variables–Problems and Solutions. In Proceedings of the Second AUSIMM International Geometallurgy Conference/Brisbane, Brisbane, Australia, 30 September 2013; Volume 30, pp. 7–15.
20. Leuangthong, O.; Deutsch, C.V. Stepwise Conditional Transformation for Simulation of Multiple Variables. *Math. Geosci.* **2003**, *35*, 155–173.
21. Van den Boogaart, K.G.; Mueller, U.; Tolosana-Delgado, R. An affine equivariant multivariate normal score transform for compositional data. *Math. Geosci.* **2017**, *49*, 231–251. [CrossRef]
22. Barnett, R.; Manchuk, J.; Deutsch, C. Projection pursuit multivariate transform. *Math. Geosci.* **2014**, *46*, 337–359. [CrossRef]
23. Barnett, R.M.; Manchuk, J.G.; Deutsch, C.V. The Projection-Pursuit Multivariate Transform for Improved Continuous Variable Modeling. *SPE J.* **2016**, *21*, 2010–2026. [CrossRef]
24. Battalgazy, N.; Madani, N. Categorization of Mineral Resources Based on Different Geostatistical Simulation Algorithms: A Case Study from an Iron Ore Deposit. *Nat. Resour. Res.* **2019**, *28*, 1329–1351. [CrossRef]
25. Adeli, A.; Emery, X.; Dowd, P. Geological Modelling and Validation of Geological Interpretations via Simulation and Classification of Quantitative Covariates. *Minerals* **2017**, *8*, 7. [CrossRef]
26. Shao, H.; Sun, X.; Wang, H.; Zhang, X.; Xiang, Z.; Tan, R.; Chen, X.; Xian, W.; Qi, J. A method to the impact assessment of the returning grazing land to grassland project on regional eco-environmental vulnerability. *Environ. Impact Assess. Rev.* **2017**, *56*, 155–167. [CrossRef]
27. Vasylchuk, Y.V.; Deutsch, C.V. Improved grade control in open pit mines. *Min. Technol.* **2018**, *127*, 84–91. [CrossRef]
28. Rivoirard, J. *Introduction to Disjunctive Kriging and Non-Linear Geostatistics*; Clarendon Press: Oxford, UK, 1994; 181p.
29. Deutsch, C.V.; Journel, A.G. *GSLIB: Geostatistical Software Library and User's Guide*, 2nd ed.; Oxford University Press: New York, NY, USA, 1998.
30. Journel, A.G.; Huijbregts, C.J. *Mining Geostatistics*; Academic Press: London, UK, 1978.
31. Barnett, R.M. Projection Pursuit Multivariate Transform. In *Geostatistics Lessons*; Deutsch, J.L., Ed.; Centre for Computational Geostatistics, Department of Civil and Environmental Engineering, University of Alberta: Edmonton, AB, Canada, 2017; Available online: http://www.geostatisticslessons.com/lessons/lineardecorrelation.html (accessed on 24 October 2018).
32. Reed, M.; Simon, B. *Methods of Modern Mathematical Physics. I, II, III, IV*; Academic Press Inc.: New York, NY, USA, 1972; Volume 1975, p. 1979.
33. Emery, X. A turning bands program for conditional co-simulation of cross-correlated Gaussian random fields. *Comput. Geosci.* **2008**, *34*, 1850–1862. [CrossRef]
34. Verly, G.W. Sequential Gaussian Cosimulation: A Simulation Method Integrating Several Types of Information. In *Geostatistics Valencia 2016*; Springer: Cham, Switzerland, 1993; Volume 5, pp. 543–554.
35. Abildin, Y.; Madani, N.; Topal, E. A Hybrid Approach for Joint Simulation of Geometallurgical Variables with Inequality Constraint. *Minerals* **2019**, *9*, 24. [CrossRef]
36. Wackernagel, H. *Multivariate Geostatistics: An Introduction with Applications*; Springer: Berlin/Heidelberg, Germany, 2003.
37. Rivoirard, J.; Demange, C.; Freulon, X.; Lécureuil, A.; Bellot, N. A top-cut model for deposits with heavy-tailed grade distribution. *Math. Geosci.* **2013**, *45*, 967–982. [CrossRef]
38. Maleki, M.; Madani, N.; Emery, X. Capping and kriging grades with long-tailed distributions. *J. South. Afr. Inst. Min. Metall.* **2014**, *114*, 255–263.

39. Sinclair, A.J.; Blackwell, G.H. *Applied Mineral Inventory Estimation*; Cambridge University Press (CUP): Cambridge, UK, 2002.
40. Deutsch, J.L.; Deutsch, C.V. *Some Geostatistical Software Implementation Details*; Centre for Computational Geostatistics, University of Alberta: Edmonton, AB, Canada, 2010; Volume 412.
41. Rossi, M.E.; Deutsch, C.V. *Mineral Resource Estimation*; Springer: Berlin, Germany, 2014.
42. Madani, N.; Yagiz, S.; Adoko, A.C. Spatial Mapping of the Rock Quality Designation Using Multi-Gaussian Kriging Method. *Minerals* **2018**, *8*, 530. [CrossRef]
43. Goovaerts, P. *Geostatistics for Natural Resources Evaluation*; Oxford University Press: New York, NY, USA, 1997.
44. Leuangthong, O.; Khan, K.D.; Deutsch, C.V. *Solved Problems in Geostatistics*; John Wiley & Sons: Hoboken, NJ, USA, 2011.
45. Emery, X. Iterative algorithms for fitting a linear model of coregionalization. *Comput. Geosci.* **2010**, *36*, 1150–1160. [CrossRef]
46. Maleki, M.; Emery, X.; Mery, N. Indicator Variograms as an Aid for Geological Interpretation and Modeling of Ore Deposits. *Minerals* **2017**, *7*, 241. [CrossRef]
47. Paravarzar, S.; Emery, X.; Madani, N. Comparing sequential Gaussian simulation and turning bands algorithms for cosimulating grades in multi-element deposits. *Comptes Rendus Geosci.* **2015**, *347*, 84–93. [CrossRef]
48. Madani, N.; Emery, X. A comparison of search strategies to design the cokriging neighborhood for predicting coregionalized variables. *Stoch. Environ. Res. Risk Assess.* **2019**, *33*, 183–199. [CrossRef]
49. Lantue´joul, C. *Geostatistical Simulation, Models and Algorithms*; Springer: Berlin, Germany, 2002.
50. Madani, N.; Carranza, E.J.M. Co-simulated Size Number: An Elegant Novel Algorithm for Identification of Multivariate Geochemical Anomalies. *Nat. Resour. Res.* **2019**, in press. [CrossRef]
51. Eze, P.N.; Madani, N.; Adoko, A.C. Multivariate mapping of heavy metals spatial contamination in a Cu–Ni exploration field (Botswana) using turning bands co-simulation algorithm. *Nat. Resour. Res.* **2019**, *28*, 109–124. [CrossRef]
52. Madani, N. Multi-collocated cokriging: An application to grade estimation in the mining industry. In *Mining Goes Digital*; CRC Press: Wroclaw, Poland, 2019; pp. 158–167.
53. Maleki, M.; Madani, N. Multivariate geostatistical analysis: An application to ore body evaluation. *Iranian J. Earth Sci.* **2017**, *8*, 173–184.

Article

A Bat-Optimized One-Class Support Vector Machine for Mineral Prospectivity Mapping

Yongliang Chen [1,*], Wei Wu [2] and Qingying Zhao [3]

1 Institute of Mineral Resources Prognosis on Synthetic Information, Jilin University, Changchun 130026, China
2 Changchun Institute of Urban Planning and Design, Changchun 130033, China; wuwei2188@outlook.com
3 College of Earth Sciences, Jilin University, Changchun 130061, China; zhaoqy@jlu.edu.cn
* Correspondence: chenyl@jlu.edu.cn; Tel.: +86-431-8850-2410

Received: 5 April 2019; Accepted: 21 May 2019; Published: 23 May 2019

Abstract: One-class support vector machine (OCSVM) is an efficient data-driven mineral prospectivity mapping model. Since the parameters of OCSVM directly affect the performance of the model, it is necessary to optimize the parameters of OCSVM in mineral prospectivity mapping. Trial and error method is usually used to determine the "optimal" parameters of OCSVM. However, it is difficult to find the globally optimal parameters by the trial and error method. By combining OCSVM with the bat algorithm, the intialization parameters of the OCSVM can be automatically optimized. The combined model is called bat-optimized OCSVM. In this model, the area under the curve (AUC) of OCSVM is taken as the fitness value of the objective function optimized by the bat algorithm, the value ranges of the initialization parameters of OCSVM are used to specify the search space of bat population, and the optimal parameters of OCSVM are automatically determined through the iterative search process of the bat algorithm. The bat-optimized OCSVMs were used to map mineral prospectivity of the Helong district, Jilin Province, China, and compared with the OCSVM initialized by the default parameters (i.e., common OCSVM) and the OCSVM optimized by trial and error. The results show that (a) the receiver operating characteristic (ROC) curve of the trial and error-optimized OCSVM is intersected with those of the bat-optimized OCSVMs and (b) the ROC curves of the optimized OCSVMs slightly dominate that of the common OCSVM in the ROC space. The area under the curves (AUCs) of the common and trial and error-optimized OCSVMs (0.8268 and 0.8566) are smaller than those of the bat-optimized ones (0.8649 and 0.8644). The optimal threshold for extracting mineral targets was determined by using the Youden index. The mineral targets predicted by the common and trial and error-optimized OCSVMs account for 29.61% and 18.66% of the study area respectively, and contain 93% and 86% of the known mineral deposits. The mineral targets predicted by the bat-optimized OCSVMs account for 19.84% and 14.22% of the study area respectively, and also contain 93% and 86% of the known mineral deposits. Therefore, we have 0.93/0.2961 = 3.1408 < 0.86/0.1866 = 4.6088 < 0.93/0.1984 = 4.6875 < 0.86/0.1422 = 6.0478, indicating that the bat-optimized OCSVMs perform slightly better than the common and trial and error-optimized OCSVMs in mineral prospectivity mapping.

Keywords: one-class support vector machine; bat algorithm; mineral prospectivity mapping; receiver operating characteristic; area under the curve; Youden index

1. Introduction

One-class support vector machine (OCSVM) is an extended version of the support vector machine, which performs anomaly detection by modeling high-dimensional unlabeled data [1,2]. This method has high performance and efficiency in identifying anomalies from high-dimensional data of unknown population distribution and has been successfully applied in many research fields. Davy and Godsill established an OCSVM model to detect abrupt spectral changes from musical record data for audio signal segmentation [3]. Lengelle et al. established an OCSVM model to detect real-time abnormal events in audio surveillance [4]. Shin et al. established an OCSVM-based fast fault diagnosis model of manufacturing facilities [5]. Mahadevan and Shah established an OCSVM model to detect faults from process data in control systems [6]. Fergani et al. developed an OCSVM-based speaker diarization primary system [7]. Mourão-Miranda et al. established an OCSVM model to identify depressed patients from medical images [8]. Strobbe et al. conducted an automatic architectural style detection using the OCSVM model with graph kernels [9]. Roodposhti et al. established an OCSVM model to map drought sensitivity in atmospheric researches [10]. Saari et al. established an OCSVM model to detect windmill bearing faults [11]. Harrou et al. established an OCSVM model to detect anomalies in photovoltaic systems [12]. Chen and Wu applied OCSVM to mineral prospectivity mapping and geochemical anomaly detection [13,14].

The aforementioned applications reveal that the parameters of OCSVM directly affect the performance of anomaly detection. In these applications, trial and error method was used to select the optimal parameter values from a set of parameter values predefined by the user as the "optimal" parameter values of the OCSVM. It is difficult to find the globally optimal parameters by the trial and error method because the predefined parameter values most likely do not contain global optimal parameter values. Therefore, it is necessary to develop a more effective method to optimize the initialization parameters of OCSVM in anomaly detection.

The problem of improving the performance of OCSVM by adjusting the initialization parameters of the model can be reduced to the problem of objective function optimization. The swarm intelligence methods for solving large-scale optimization problems can be used to solve this problem. Particle swarm optimization (PSO) is one of the swarm intelligence methods widely used in machine learning to solve optimization problems [15]. The bat algorithm, another swarm intelligence method recently developed by Yang [16], has good convergence and performance in solving large-scale optimization problems [17–20]. In mineral prospectivity mapping, it is necessary to quickly determine the optimal values of OCSVM parameters, and the bat algorithm is especially suitable for this problem. Therefore, the bat algorithm was selected to automatically determine the optimal initialization parameters of OCSVM.

In order to use the bat algorithm to automatically optimize the initialization parameters of OCSVM in mineral prospectivity mapping, a model combining OCSVM with the bat algorithm is proposed in this study. The area under the curve (AUC) of OCSVM is calculated based on the OCSVM modeling result to measure the overall performance of the OCSVM model in mineral prospectivity mapping [13,14,21–23]. In the combined model, the AUC value of OCSVM is taken as the fitness value of the objective function optimized by the bat algorithm, the value ranges of OCSVM parameters are used to specify the search space of the bat population, and the iterative search process of the bat algorithm is used to automatically determine the optimal parameters of OCSVM. The combined model is hereafter called bat-optimized OCSVM.

The bat-optimized OCSVM was used to map mineral prospectivity of the Helong district, Jilin Province, China, and compared with both the OCSVM initialized by the default parameters (i.e., common OCSVM) and the OCSVM of which the optimal initialization parameters are determined by the trial and error method (i.e., trial and error-optimized OCSVM). The receiver operating characteristic (ROC) curves and AUC values were used to evaluate the performances of these OCSVMs [13,14,21–23]. Based on the data modeling results, the optimal threshold for extracting mineral targets is determined by using the Youden index. The main contribution of this paper is to propose a bat-optimized OCSVM, which can automatically optimize the initialization parameters of OCSVM, and improve the performance of OCSVM in mineral prospectivity mapping.

2. Materials and Methods

2.1. Geological and Geochemical Data

Geological and geochemical data for mineral prospectivity mapping came from the Digital Geological Survey recently conducted in the study area, which belongs to China's New-Round Land Resources Survey Project [24]. The research group from Jilin University carried out field work in the study area and collected the data of geological structures, metamorphic rocks, magmatic rocks, sedimentary rocks, and mineral deposits, and saved the data in the MAPGIS system developed by the China University of Geosciences (Wuhan, China). At the same time, the research group completed the stream sediment survey in the study area, which was conducted in accordance with the Geochemical Survey Criteria (No. DZ/T0011-91), covering 1320 km^2 with a sampling density of 1–2 samples per 0.25 km^2. The concentrations of 13 elements in each stream sediment sample were analyzed and tested by the Inner Mongolia Mineral Experiment Institute, China. The concentration of Au was analyzed by atomic absorption spectrometry (AAS), the concentrations of Hg and As were analyzed by atomic fluorescence spectrophotometry (AFS), and the concentrations of Ag, Sb, Mo, W, Cu, Pb, Zn, Bi, Ni, and Co were analyzed by inductively coupled plasma mass spectrometry (ICP-MS).

Geological data were preprocessed in MAPGIS and further processed in the MapInfo software platform. Firstly, the projection coordinates of the geological maps were converted into longitude and latitude coordinates, and then into MapInfo tables. Finally, known mineral deposits, regional faults, different geological formations, and different magmatic rocks as well as their boundaries, were extracted from the geological data in the MapInfo software platform as potential evidence layers. The regional faults and magmatic rock boundaries were transformed into areal entities by buffering them to the optimal width in the MapInfo software platform. Each potential evidence layer is saved as a MapInfo Interchange file, which is the input data for the Python program developed by Yongliang Chen (see Supplementary Materials). The Python program then generated a layer of 200 × 134 unit cells, called the unit cell layer. Each cell is a rectangular area of 0.2282 × 0.2296 km^2, which satisfies the condition that there is no more than one known mineral deposit in one cell.

Geochemical data were preprocessed using Surfer 12. For each geochemical element, its concentrations collected from the 6999 valid sampling points were used to generate a 200 × 134 grid data by using Inverse Distance to a Power. In geochemical data interpolation, an integer value of 2 was used as the power value of the Inverse Distance to a Power, and the number of samples used to estimate a grid point value was between 8 and 64. Figure 1 shows the concentrations of Au, Bi, Co, Cu, Mo, and Ni collected from the 6999 valid sampling locations in the study area. The concentrations of the remaining seven elements are not shown here because they were not selected for mineral prospecting mapping in Section 3.2. Figure 2 shows the interpolated data of Au, Bi, Co, Cu, Mo, and Ni produced by the Inverse Distance to a Power in Surfer 12. By comparing Figures 1 and 2, it can be found that the spatial distribution of element concentrations in the interpolated data is consistent with that in the vertical bar charts. Therefore, it is feasible to interpolate the geochemical data using Inverse Distance to a Power.

The grid map of each element generated above is consistent with the unit cell layer generated previously. Each grid point represents the corresponding unit cell in the unit cell layer. A grid point (unit cell) is defined as a true positive point if the cell represented by the grid point contains a known mineral deposit. Thus, the number of the true positive points defined in the study area is equal to the number of known mineral deposits. Except for these grid points which are defined as true positive points, all other grid points in the study area are defined as true negative points. The true positive and negative grid points defined in this section (Figure 3a) are hereafter used as the ground truth data to evaluate the performances of OCSVMs in subsequent sections.

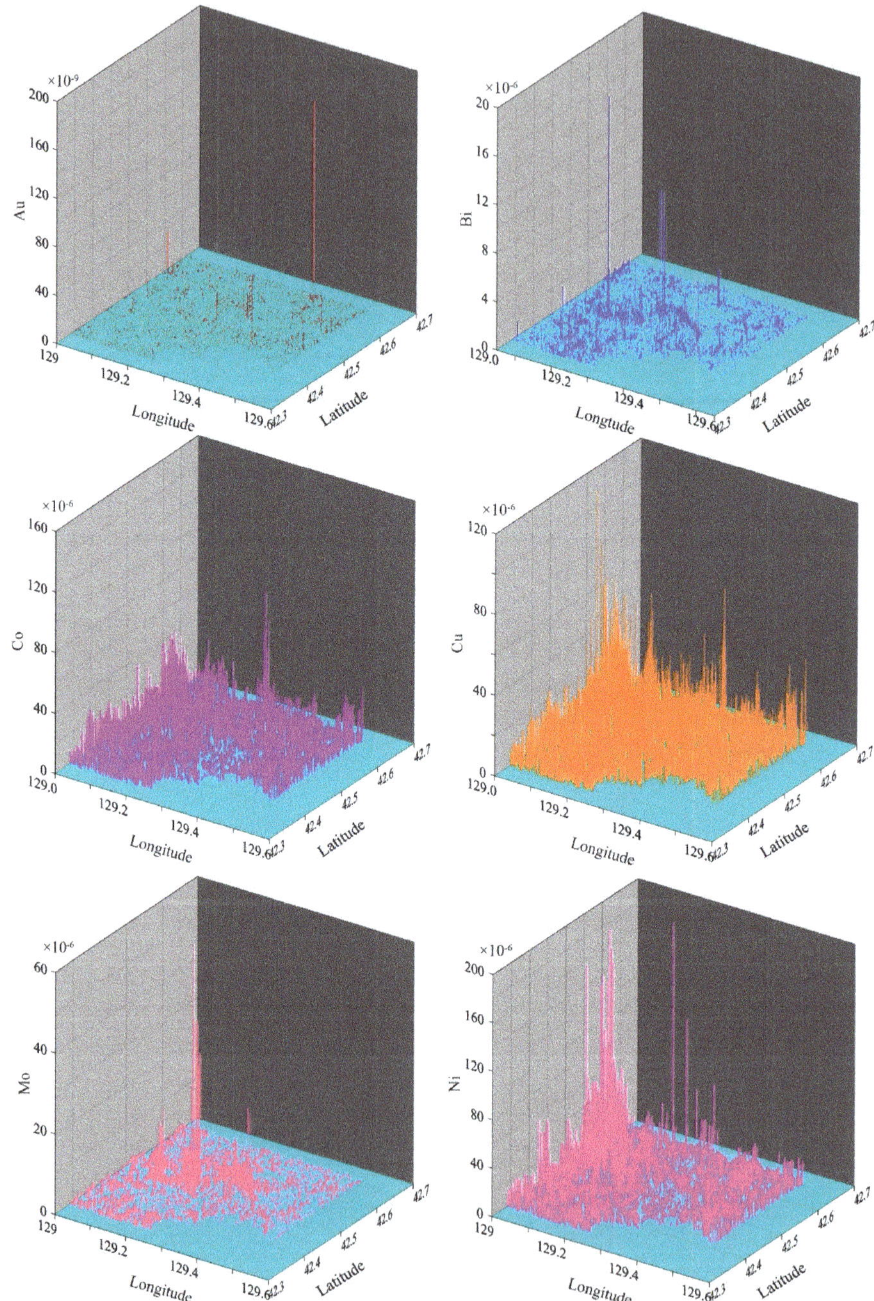

Figure 1. The concentrations of Au, Bi, Co, Cu, Mo, and Ni collected from the 6999 valid sampling locations in the study area.

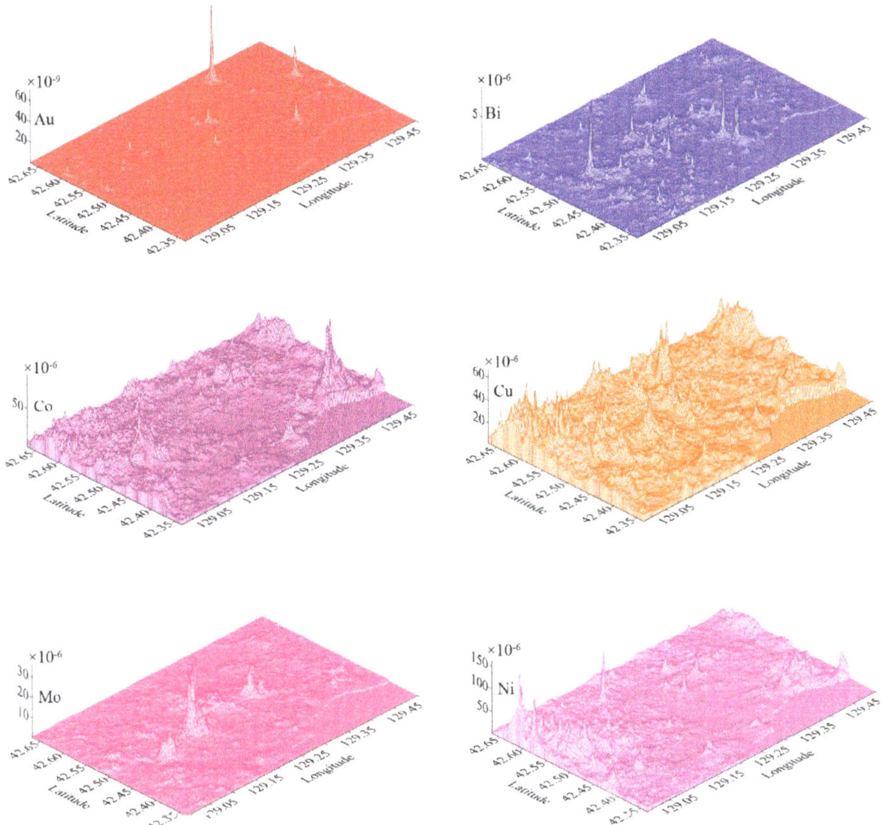

Figure 2. The grid data of Au, Bi, Co, Cu, Mo, and Ni produced by the interpolation method of Inverse Distance to a Power in Surfer 12.

The geological and geochemical evidences, spatially associated with known mineral deposits, were selected and converted into binary evidence map layers and used as the input data of OCSVM models. Binary geological evidences were selected by using the Youden index to evaluate spatial relationships between the geological evidences and known mineral deposits [13,14,21–23]. Continuous geochemical evidences were selected by statistically testing whether there exists significant spatial relationships between the geochemical evidences and the known mineral deposits [13,14,21–23]. The continuous geochemical evidences selected for mineral prospectivity mapping were then optimally converted into binary geochemical evidence layers by using the Youden index to evaluate the spatial relationship between the converted geochemical evidences and the known mineral deposits [13,14,21–23]. Figure 3b–r shows the 17 binary evidence maps selected for mineral prospectivity mapping in this study.

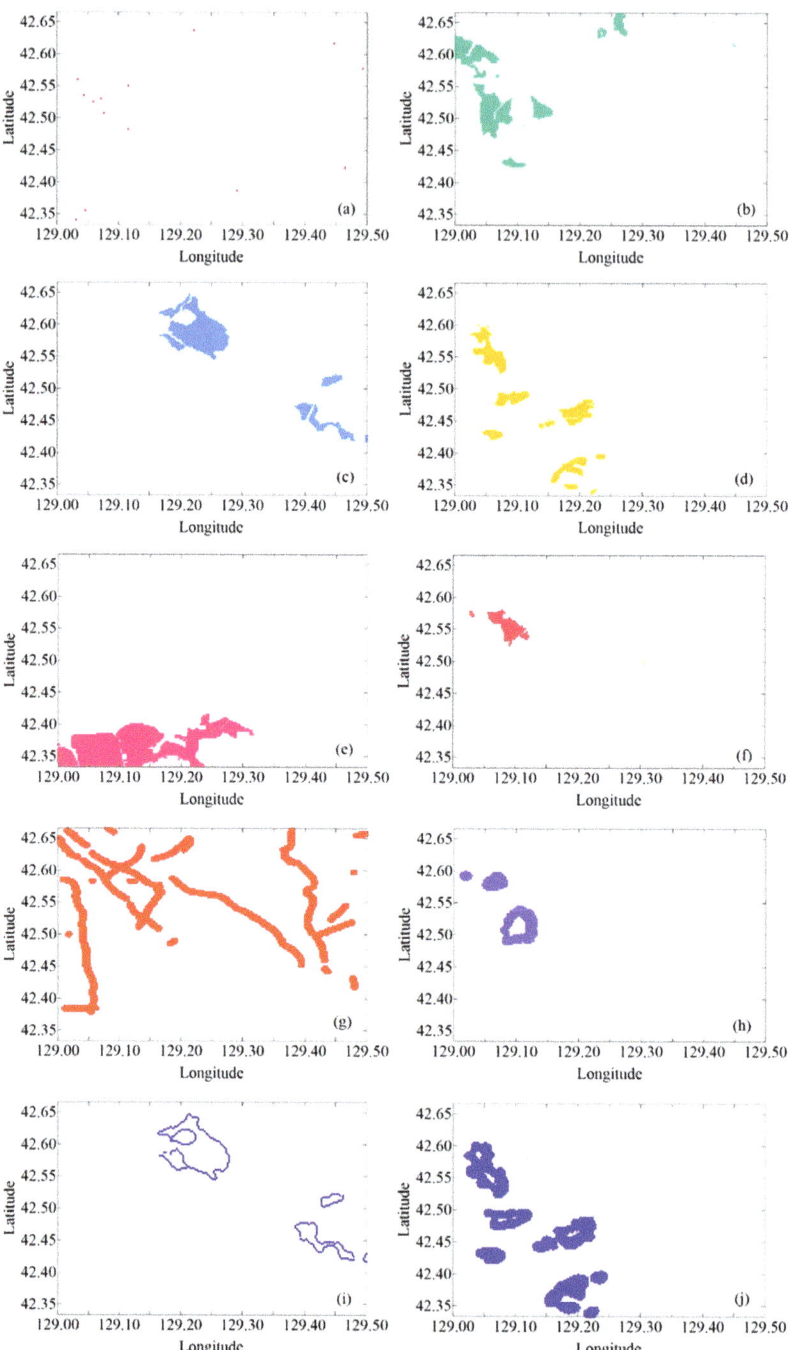

Figure 3. *Cont.*

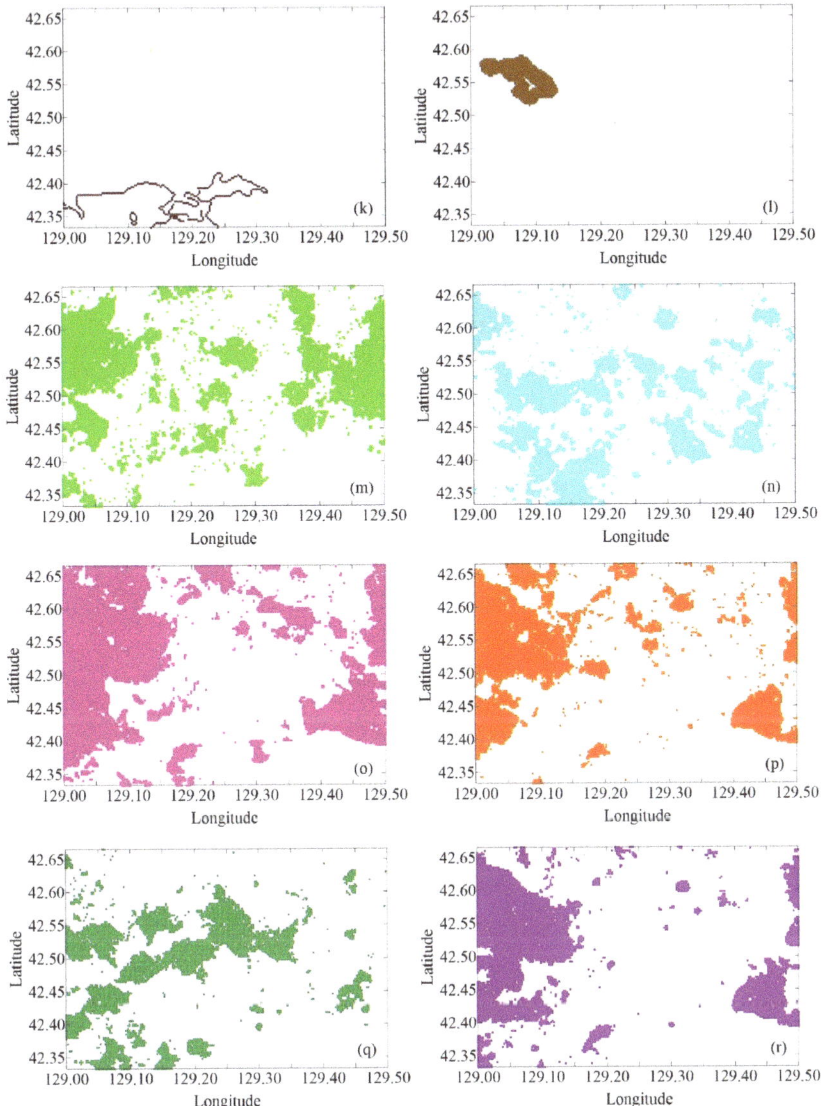

Figure 3. Mineral deposits and binary evidence map layers: (**a**) the unit cell layer containing known mineral deposits, (**b**) the Jinan Formation, (**c**) porphyritic biotite granodiorite, (**d**) porphyritic granodiorite, (**e**) fine-grained monzonite, (**f**) medium-fine-grained diorite, (**g**) fault with 0.5 km buffer, (**h**) troctolite boundary with 0.8 km buffer, (**i**) porphyritic biotite granodiorite boundary with 0.1 km buffer, (**j**) porphyritic granodiorite boundary with 0.6 km buffer, (**k**) fine-grained monzonite boundary with 0.1 km buffer, (**l**) medium-fine-grained diorite boundary with 1.0 km buffer, (**m**) gold concentration anomalies, (**n**) bismuth concentration anomalies, (**o**) cobalt concentration anomalies, (**p**) copper concentration anomalies, (**q**) molybdenum concentration anomalies, and (**r**) nickel concentration anomalies.

2.2. Receiver Operating Characteristic (ROC) Curve, Area under the Cuve (AUC), and Youden Index

The ROC curve of a continuous indicator is a graphical representation of the relationship between the continuous indicator and a binary target variable. Assuming that a study area has n grid points, the target variable divides the n grid points into true positive points and true negative points. A threshold is used to convert a continuous indicator into a binary indicator, which divides the n grid points into predicted positive points and predicted negative points. According to Chen and Wu [22], these classification results can be used to calculate benefit (that is, the percentage of the true positive points that are correctly predicted as positive points) and cost (that is, the percentage of the true negative points that are wrongly predicted as positive points). The computed benefits and costs vary with threshold. The ROC curve can represent the curve of benefit changing with cost under different threshold settings. A point on the ROC curve represents a threshold, and its vertical and horizontal coordinates represent the corresponding benefit and cost, respectively. The higher the relationship is between the continuous indicator and the binary target variable, the closer the ROC curve is to the upper left corner of the ROC space.

The AUC value of a continuous indicator is the area under the ROC curve of the continuous indicator, and it is a quantitative expression of the relationship between the continuous indicator and the binary target variable. Its value is in the range of 0.5 to 1, which corresponds respectively to the random and deterministic relationships between the continuous indicator and the binary target variable. Assume that there are t_p true positive and t_n true negative points in the study area. According to Chen [21], the AUC value of the continuous indicator can be expressed as

$$\mathrm{AUC} = \frac{1}{t_p t_n} \sum_{i=1}^{t_p} \sum_{j=1}^{t_n} \varphi\big(f(x_i), f(y_j)\big) \tag{1}$$

with

$$\varphi\big(f(x_i), (y_j)\big) = \begin{cases} 1, & f(x_i) > f(y_j) \\ 0.5, & f(x_i) = f(y_j) \\ 0, & f(x_i) < f(y_j) \end{cases}.$$

where $f(x_i)$ ($i = 1, 2, \ldots, p$) represents the observed value of the continuous indicator at the ith true positive point, and $f(y_j)$ ($j = 1, 2, \ldots, q$) represents the observed value of the continuous indicator at the jth true negative point.

AUC is a random variable, which can be used to construct the following random variable Z_{AUC} that conforms to the standard normal distribution [21]:

$$Z_{AUC} = \frac{AUC - 0.5}{S_{AUC}} \tag{2}$$

where S_{AUC} is the standard deviation of AUC, which can be calculated by

$$S_{\mathrm{AUC}} = \sqrt{\frac{AUC(1 - AUC) + (t_p - 1)\left(\frac{AUC}{2-AUC} - AUC^2\right) + (t_n - 1)\left(\frac{2AUC^2}{1+AUC} - AUC^2\right)}{t_p t_n}} \tag{3}$$

Z_{AUC} can be used to test whether there is a significant difference between the AUC value and 0.5 at the significance level of $\alpha = 0.05$ [13,14,21–23]. According to the unit normal loss function, at the significance level of $\alpha = 0.05$, the critical value of Z_{UAC} is 1.96. If the Z_{AUC} value calculated by Equation (2) is greater than the critical value of 1.96, the probability of a significant difference between the AUC value and the value of 0.5 is not less than 0.95.

The Youden index of a binary indicator is the quantitative expression of the relationship between the binary indicator and a binary target variable. It is defined as benefit minus cost (i.e., the difference between the vertical coordinate and the horizontal coordinate of a point on the ROC curve) [22]. The

Youden index is between −1 and +1, respectively, representing the deterministic negative and positive relationships. When the Youden index is close to zero, it means that there is little relationship between the binary indicator and the binary target variable.

In mineral prospectivity mapping, ROC curves and AUC values can be used to optimally select continuous evidence layers and to evaluate the performances of mineral prospectivity mapping methods [13,14,21,22]. The Youden index can be used to optimally select binary evidence map layers, as well as to determine the optimal threshold of a continuous evidence layer and the optimal buffering width of a linear evidence map layer [13,14,21,22].

2.3. OCSVM

Assume that m binary evidence layers are used for mineral prospectivity mapping in a study area of n unit cells. Data matrix $\{x_1, x_2, \cdots, x_n\}$ represents the observed evidence data of the n unit cells in the study area. Each column vector $x_i = (x_{i1}, x_{i2}, \cdots, x_{im})^T$ represents the observed values of the m evidence layers in the ith unit cell. Mapping mineral prospectivity using OCSVM is a binary classification process that classifies the n unit cells into single-class and outliers. An initialization parameter μ ($0 < \mu \leq 1$) is used to control the percentage of outliers among the n unit cells, that is, the n unit cells contain no more than μn outliers. Outliers usually account for only a small proportion of all the cells in the study area and are considered as mineral targets in mineral prospectivity mapping [13].

In OCSVM, the support vector machine (SVM) theory is used to estimate a hyperplane that maximumly separates single-class and outliers [2]. Due to the nonlinear separability between single-class and outliers, the following Gaussian kernel [25] is usually used in OCSVM:

$$K\left(x_i, x_j\right) = \exp\left(-\|x_i - x_j\|^2 / \sigma^{-2}\right) \tag{4}$$

where x_i and x_j, ($i, j = 1, 2, \ldots, n$), are respectively the ith and jth unit cells, and σ is the standard deviation of a Gaussian distribution, which is another initialization parameter of OCSVM.

The data set $\{x_1, x_2, \cdots, x_n\}$ is used to train the OCSVM model initialized with the parameters μ and σ. According to the trained OCSVM model, the anomaly score of each unit cell is calculated by

$$f(x) = \sum\nolimits_{i=1}^{n} \alpha_i \left[K\left(x_i, x_j\right) - K(x_i, x)\right], j \in (1, 2, \ldots, n) \tag{5}$$

where $f(x)$ is the anomaly score of cell x that denotes the degree of cell x being an outlier [13], α_i $(i = 1, 2, \ldots, n)$ is the Lagrange parameter, $K(\cdot,\cdot)$ is the Gaussian kernel.

The initialization parameters μ and σ directly affect the performance of the OCSVM model for mineral prospectivity mapping [13]. Determining the optimal values of the parameters μ and σ to improve the performance of OCSVM can be solved by combining the OCSVM model with the bat algorithm.

2.4. Bat-Optimized OCSVM

The bat algorithm is a heuristic search algorithm that simulates bats using sonar to detect prey and avoid obstacles. It maps L individuals in the bat population to L feasible solutions in a d-dimensional problem space, and uses the flying process of a bat in search for prey to simulate the optimization search process. The fitness value of solving the problem is used to evaluate the position of the bat, and the evolutionary process of survival of the fittest is used to simulate the iterative search process of the better feasible solution instead of the worse feasible solution. The bat algorithm dynamically controls the conversion between local search and global search to avoid the algorithm falling into the local optimum, and has good global convergence and superior performance in solving large-scale target optimization problems [17–20].

The bat algorithm is controlled by the following parameters: (a) L controlling the size of the bat population, (b) T controlling the number of iterations, (c) α and γ controlling convergence speed, (d)

A controlling sound loudness, (e) r controlling sound emission rate, and (f) f and λ controlling the detectable range. According to Yang and Gandomi [17], the values of A, r, and f can be set between 0 and 1, and the values of α and γ can be simply set to $\alpha = \gamma = 0.9$. $A_{\min} = 0$ means that a bat has just found the target and temporarily stops making any sound, and $r_{\min} = 0$ and $r_{\max} = 1$ respectively represent no pulse and the maximum emission rate. $f_{\min} = 0 \leq f \leq 1 = f_{\max}$ corresponds to $\lambda_{\max} \geq \lambda \geq \lambda_{\min}$. $\lambda_{\max}$ represents the detectable range, and adjusting it only needs to change f because $\lambda \times f$ is constant [16,17].

Each bat starts its heuristic search from a random location z_l in the d-dimensional search space after its loudness A_l, emission rate r_l, and frequency f_l are randomly initialized. Each bat l flies randomly with velocity v_l at location z_l, searching for prey with a fixed frequency f_l, varying wavelength λ_l and loudness A_l, and automatically adjusts wavelength λ_l according to the degree at which it is approaching the prey [16,17].

At each iteration t $(0 \leq t < T)$, a global search process is first conducted and the flying speed and spatial location of each bat are updated. The spatial coordinates of each bat l $(0 \leq l < L)$ is used to calculate the fitness value of the objective function, and then the spatial location corresponding to the largest fitness value is selected as the current optimal location z_*. According to Yang [16], the v_l^t and z_l^t are updated as follows:

$$f_l = f_{\min} + (f_{\max} - f_{\min})\beta \tag{6}$$

$$v_l^t = v_l^{t-1} + \left(z_l^t - z_*\right) f_l \tag{7}$$

$$z_l^t = z_l^{t-1} + v_l^t \tag{8}$$

where $\beta \in [0, 1]$ is a random number drawn from a uniform distribution, z_* is the current optimal location, and t is iteration number.

After the global search process described above, a local search is then performed around the current best location. According to Yang [16], during the local search, the new location is generated by the following local random walk and tested to see if it is the best among all the locations:

$$z_{\text{new}} = z_{\text{best}} + \varepsilon \left\langle A^t \right\rangle \tag{9}$$

where $\varepsilon \in [-1, 1]$ is a d-dimensional random vector, and $\left\langle A^t \right\rangle$ is the average loudness of the L bats at iteration t.

At the end of each iteration t, the loudness A_l and the emission rate r_l of each bat l are updated accordingly as follows:

$$A_l^{t+1} = \alpha A_l^t, \quad r_l^{t+1} = r_l^0[1 - \exp(-\gamma t)] \tag{10}$$

where $\alpha = \gamma = 0.9$ are constants [16,17].

The process of updating the velocities and the locations of bats is somewhat similar to that of PSO [20]. The pace and range of the movement are basically controlled by the frequency, just like the movement of the virtual birds in PSO. To some extent, the bat algorithm can be regarded as a balanced combination of PSO and the intensive local search governed by the frequency tuning ability and the variables of loudness and pulse rate. The loudness and pulse rate that influence the balance need to be updated in each iteration. However, PSO is slightly different from the global search process of the bat algorithm. In PSO, the velocity of each bird is updated by adding random perturbation to the optimal position of the bird and the optimal position of the population. While during the global search process of the bat algorithm, the velocity of each bat is updated according to the spatial difference between the current position of the bat and the current optimal position of the population. First the frequency of each bat is updated, and then the velocity of the bat is updated by adding the product of the spatial difference and the frequency. In both PSO and the global search process of the bat algorithm, the velocity of an individual is taken as the step length of updating the location of the individual.

In the bat-optimized OCSVM model, the search space of the bat algorithm is a two-dimensional space of which the coordinate axes are composed of μ and σ. The search range of the bat population is defined as $(0 < \mu \leq 1)$ and $(0 < \sigma < c)$. Here c is a positive constant given by the user. The fitness value maximized by the iterative search process of the bat algorithm is the AUC value of the OCSVM model. The iterative search process starts from L random locations within the search space. At each iteration, the two coordinates of the spatial location occupied by each bat are used as the values of μ and σ to initialize the OCSVM model, and then the model is trained on the data $\{x_1, x_2, \cdots, x_n\}$. The anomaly score of each unit cell is calculated using Equation (5) based on the trained OCSVM model. Finally, the AUC value of the OCSVM model is calculated using Equation (1) based on the anomaly scores and the ground truth data defined in Section 2.1. The location corresponding to the largest AUC value is selected as the current optimal location, and the spatial location of each individual bat is updated using Equations (6) to (8). After the locations of all the bats having been updated, the local search is implemented around the current optimal location using Equation (9). The loudness and the emission rate of each bat are updated accordingly using Equation (10). Table 1 outlines the pseudo code of the bat-optimized OCSVM model.

Table 1. The pseudo code of the bat-optimized one-class support vector machine (OCSVM) model.

The Algorithm for the Bat-Optimized OCSVM Model
Input:
Binary data $\{x_1, x_2, \ldots, x_n\}$;
Binary ground truth data $\{d_1, d_2, \ldots, d_n\}$.
Output:
Anomaly scores $\{f(x_1), f(x_2), \ldots, f(x_n)\}$.
Algorithm:
Initialization ():
Randomly initialize the location and velocity of each bat z_l and v_l, $(l = 1, 2, \ldots, L)$;
Define pulse frequency f_l at z_l, $(l = 1, 2, \ldots, L)$;
Initialize emission rate r_l and the loudness A_l, $(l = 1, 2, \ldots, L)$.
Evaluation ():
Initialize the OCSVM model using z_l, $(l = 1, 2, \ldots, L)$;
Train the OCSVM model on the binary data $\{x_1, x_2, \ldots, x_n\}$;
Compute the anomaly score of unit cell i using Equation (5), $(i = 1, 2, \ldots, n)$;
Compute the AUC of the OCSVM model initialized by z_l $(l = 1, 2, \ldots, L)$ using Equation (1).
While $(t < T)$:
Adjust the frequency of each bat f_l using Equation (6) $(l = 1, 2, \ldots, L)$;
Update the velocity and location of each bat z_l and v_l using Equations (7) to (8) $(l = 1, 2, \ldots, L)$;
Call Evaluation ().
If (random $< r_l$):
Select a location among the best locations;
Generate a local location around the selected best location;
Generate a new location according to Equation (9);
Call Evaluation ().
If (random $< A_l$ and the AUC for $\mathbf{z}_l$ $<$ the AUC for $\mathbf{z}_*$):
Accept the new locations;
Increase r_l and reduce A_l according to Equation (10);
Rank the bats and find the current best $\mathbf{z}_*$.
Output the results.

3. Mapping Mineral Prospectivity

3.1. *Geological Background and Mineralization*

The study area is a complex tectonic belt superimposed between the Paleo-Asian tectonic domain and the Circum-Pacific tectonic domain, which has undergone the ancient Asian ocean evolution and the subduction of the Mesozoic Pacific plate [26–28]. The northwest-trending Gudonghe tectono-magmatic complex belt runs through the whole study area and controls the spatial distribution of major geological formations since the Late Paleozoic. Widely exposed magmatic rocks account for 69.58% of the whole study area. Granite, granodiorite, diorite, and gabbro are mainly magmatic rocks, forming widely exposed batholiths and stocks (Figure 4). Zircon U-Pb ages of diorites are 173–175 Ma [29], indicating that the magmatic rocks were formed during the Yanshan tectonic period. The exposed strata account for 29.44% of the total study area. The main strata are the Jinan Formation of Late Archean, the Xindongchun and Changren Formations of Late Permian, the Changchai, Quanshuichun, and Dalazi Formations of Early Cretaceous, the Longjing Formation of Late Cretaceous, the Chuandishan basalt of Neocene, and the alluvium of Holocene.

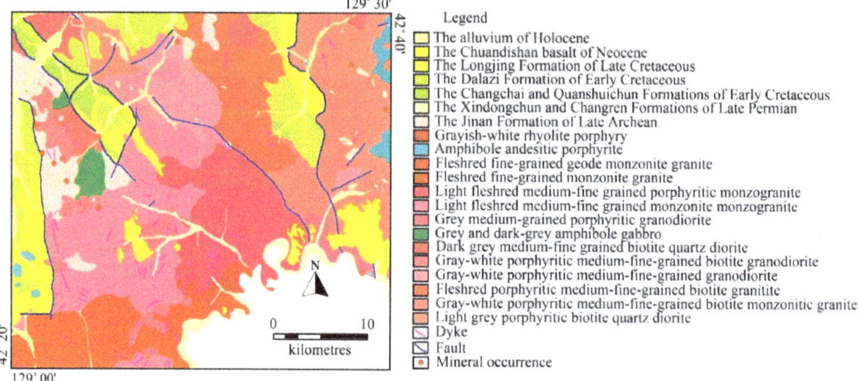

Figure 4. Simplified geologic map and known mineral deposits.

During the Yanshanian tectonic magmatism, a series of intermediate-acidic magmatic complexes were formed, which provided a continuous heat source and metallogenic materials for polymetallic mineralization [30]. There were 14 mineral deposits discovered in the study area. These mineral deposits are mainly hydrothermal and skarn type deposits which are closely related to multi-stage magmatic activities [30–32]. Most of the discovered mineral deposits are hosted in metamorphic rocks around or at the edges of magmatic intrusions (Figure 4). Regional structures, Archean formations, and the Yanshanian intermediate-acidic magmatic rocks are the three controlling factors for polymetallic mineralization.

3.2. Evidence Map Layers

In this section, geological and geochemical evidence layers are selected for mapping mineral prospectivity. Firstly, the optimal buffer width of each linear geological evidence is determined by using the Youden index, and then the linear evidence is converted into areal evidence through buffering to the optimal buffer width. Finally, the Youden indices of all the geological evidences are calculated, and the geological evidences with Youden indices larger than the predefined threshold are selected for mapping mineral prospectivity. The AUCs and Z_{AUC}s of all the geochemical elements are calculated, and those elements with Z_{AUC}s greater than the critical value of 1.96 are selected for mapping mineral prospectivity. The selected elements are finally optimally converted into binary evidences using the Youden index.

Faults and the boundaries of magmatic intrusions are linear evidences for mineral prospectivity mapping, which need to be converted into areal evidences by buffering in the MapInfo software platform. The optimal buffer width of one linear evidence can be determined by evaluating the spatial relationship between the buffered evidence and known mineral deposits using the Youden index [13,14,21–23]. For a linear evidence, the optimal buffer width maximizes its Youden index, meaning that the linear evidence buffered to the optimal buffer width has the highest spatial relationship with known mineral deposits. In this study, ten types of linear evidences were extracted in the study area. Figure 5 shows the curves of the Youden indices of various linear evidences varying with buffer width. The maximum Youden index and optimal buffer width of each linear evidence is listed in Table 2. The optimally buffered linear evidences were then used as areal evidences for mineral prospectivity mapping.

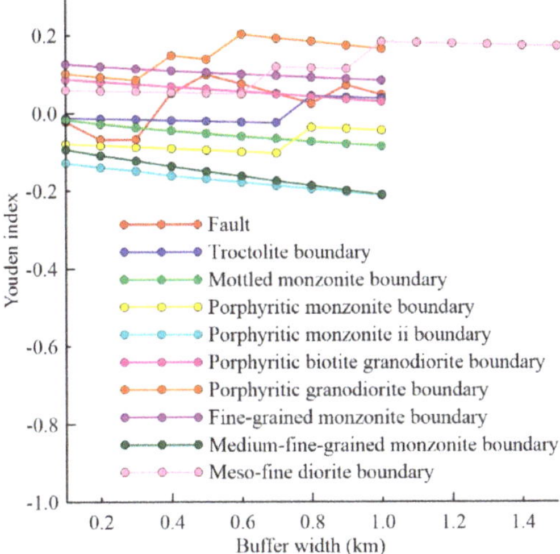

Figure 5. Curves of the Youden indices of the buffered linear evidences changing with buffer width.

Table 2. The maximum Youden indices and optimal buffer widths of the 10 linear evidences.

Linear Evidence	MYI	OBW (km)
Regional structure	0.09887	0.5
Troctolite boundary	0.04405	0.8
Mottled monzonite boundary	−0.01642	0.1
Porphyritic monzonite boundary	−0.03696	0.8
Stage II porphyritic monzonite boundary	−0.1287	0.1
Porphyritic biotite granodiorite boundary	0.08729	0.1
Porphyritic granodiorite boundary	0.2019	0.6
Fine-grained monzonite boundary	0.1264	0.1
Medium-fine-grained monzonite boundary	−0.09409	0.1
Medium-fine-grained diorite boundary	0.1831	1.0

Note: MYI denotes the maximum Youden index; OBW denotes the optimal buffer width.

After the above buffer analysis, a total of 26 areal geological evidence layers were derived as potential evidence layers for mineral prospectivity mapping in the study area. The Youden index of each layer was calculated to select the evidence layer with a higher spatial relationship with known mineral deposits. Theoretically, as long as the Youden index of an evidence layer is greater than zero, the evidence is considered to be spatially associated with known mineral deposits. However, there is no way to statistically test whether this spatial relationship is significant. Therefore, it is better to use a threshold slightly larger than zero when selecting the geological evidence layers in mineral prospectivity mapping. In this study, the following 11 geological evidence layers were selected using a threshold value = 0.01: (a) the Jinan Formation of Late Archean, (b) porphyritic biotite granodiorite, (c) porphyritic granodiorite, (d) fine-grained monzonite, (e) medium-fine-grained diorite, (f) fault with 0.5 km buffer, (g) troctolite boundary with 0.8 km buffer, (h) porphyritic biotite granodiorite boundary with 0.1 km buffer, (i) porphyritic granodiorite boundary with 0.6 km buffer, (j) fine-grained monzonite boundary with 0.1 km buffer, and (k) medium-fine-grained diorite boundary with 1.0 km buffer. These geological evidence layers are consistent with the metallogenic controlling factors discussed in Section 3.1. The evidence layer (a) is the Archean metamorphic formation, the evidence layers (b) through (e) are the Yanshanian magmatic intrusions, the evidence layer (f) is the regional structure, and the evidence layers (g) through (k) are the boundaries of the Yanshanian magmatic intrusions.

Geochemical evidence layers are selected by using the AUCs and Z_{AUC}s discussed in Section 2.2. Equation (1) was used to calculate the AUC value of each element according to the preprocessed data in Section 2.1. Then, Equation (2) was used to estimate the Z_{AUC} value according to the AUC value. If the estimated value of Z_{AUC} is greater than the critical value of 1.96 at the significance level of 0.05, the AUC value is considered to be significantly different from the value of 0.5. This means that the concentrations of the element are significantly spatially correlated to the known mineral deposits. In other words, the higher the concentration of the element in unit cells, the more likely the unit cells contain known mineral deposits. Table 3 lists the AUCs and Z_{AUC}s of 13 elements estimated in this study. As can be seen from Table 3, the Z_{AUC}s of Au, Co, Cu, Mo, Ni, and W are greater than the critical value of 1.96. Thus, the concentrations of these elements are significantly spatially correlated to the known mineral deposits.

Table 3. Area under the curves (AUCs) and Z_{AUC}s for 13 elements.

Element	AUC	Z_{AUC}	Element	AUC	Z_{AUC}	Element	AUC	Z_{AUC}
Ag	0.5268	0.3416	Cu	0.7222	2.8661	Sb	0.5802	1.0037
As	0.6195	1.4889	Hg	0.5949	1.1835	W	0.6561	1.9531
Au	0.6893	2.3958	Mo	0.7159	2.7727	Zn	0.6537	1.9217
Bi	0.6620	2.0295	Ni	0.7619	3.4986			
Co	0.7007	2.5540	Pb	0.4327	−0.9248			

According to the above statistical results, gold, Bi, Co, Cu, Mo, and Ni were selected as geochemical evidences for mineral prospectivity mapping. Among these elements, bismuth is an ore-forming associated element, and the other five elements are metallogenic elements. Thus, these statistical results are consistent with the mineralization characteristics of the study area. The optimal thresholds for extracting the concentration anomalies of the six elements were determined by evaluating the spatial relationship between the extracted anomalies and known mineral deposits using the Youden index. The higher the Youden index of the extracted anomalies (a binary map layer), the more likely the extracted anomalies spatially coincide with the known mineral deposits. The optimal threshold maximizes the Youden index of the extracted geochemical anomalies. Table 4 lists the maximum Youden indices and corresponding optimal thresholds of the six elements. Figure 6 shows the element concentration anomalies extracted from the grid data generated in Section 2.1. According to the extracted geochemical anomalies, six geochemical evidence layers were derived for mineral prospectivity mapping.

Table 4. The maximum Youden indices and optimal thresholds of the six indicator elements.

Element	MYI	OT	Element	MYI	OT	Element	MYI	OT
Au	0.3483	0.6421	Bi	0.3357	0.1996	Co	0.3989	7.3378
Cu	0.3889	11.0669	Mo	0.36406	1.1283	Ni	0.4715	10.5810

Note: MYI denotes maximum Youden index; OT denotes optimal threshold.

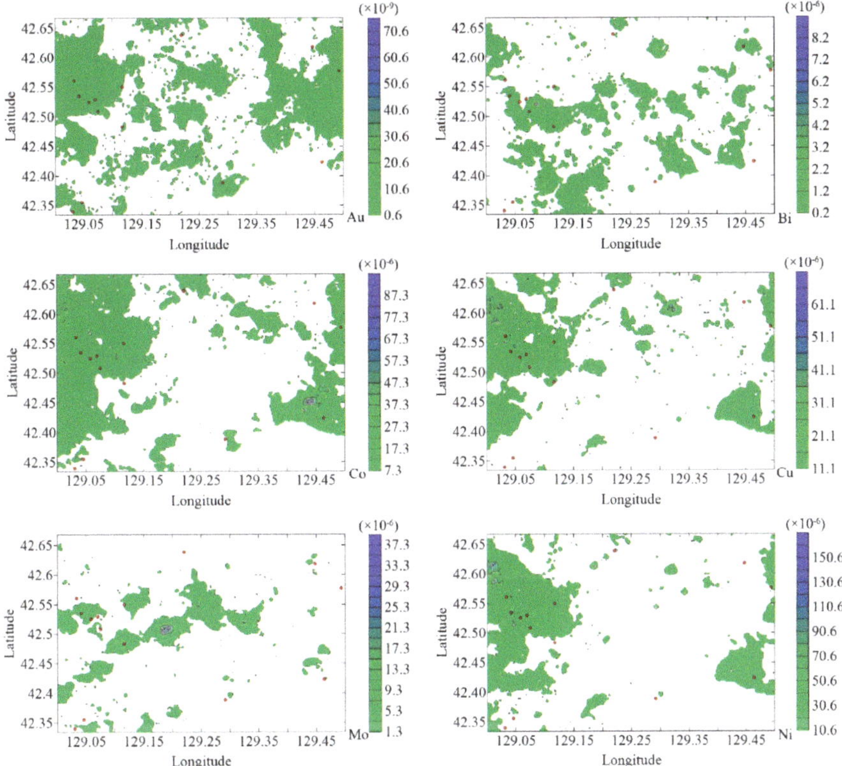

Figure 6. Contour maps of Au, Bi, Co, Cu, Mo, and Ni concentration anomalies.

3.3. Mineral Target Extraction

In mineral prospectivity mapping, an initialized OCSVM model is trained on binary evidence data and then used to calculate the anomaly score of each unit cell. According to the anomaly scores of all the unit cells, geological anomalies are extracted by the optimal threshold determined by using the Youden index [13,14,21–23]. The extracted geological anomalies are usually closely spatially related to known mineral deposits. Therefore, these geological anomalies can be used as mineral targets [13,14,21–23]. The optimal threshold for separating geological anomalies is usually determined by selecting the threshold corresponding to the maximum Youden index from all the potential thresholds. In this study, the OCSVMs either initialized with the default parameters or optimized by both the trial and error method and bat algorithm which were used to map mineral prospectivity in the study area. The optimal threshold values were determined by maximizing the corresponding Youden indices.

The OCSVM model was first initialized using the default parameter values μ = 0.5 and σ = 1.0/17 and then trained on the evidence data. Here, 17 is the number of evidence layers used for mineral prospectivity mapping. The AUC value of the OCSVM initialized with the default parameters is 0.8268. The trial and error method was then used to determine the "optimal" parameters of OCSVM. Firstly, we set μ = 0.5, and then set σ respectively to σ = 0.0588, 0.1, 0.5, 1.0, 5.0, and 10.0. Here, μ = 0.5 and σ = 0.0588 are the default values of μ and σ. The OCSVM initialized with each pair of μ and σ was used to map mineral prospectivity. Figure 7a shows the curve of the AUC value of the OCSVM model changing with σ. It can be seen from Figure 7a that the OCSVM model has the highest AUC value at σ = 1.0. Therefore, σ = 1.0 was selected as the "optimal" value of σ, and then we set μ = 0.1, 0.3, 0.5, 0.7, and 0.9. The OCSVM model initialized with each of the five pairs of μ and σ was used to map mineral prospectivity. The variation of the AUC value of the OCSVM model with μ is shown in Figure 7b. According to Figure 7a,b, the "optimal" values of σ and μ are $\sigma = 1.0$ and $\mu = 0.5$ respectively, and the corresponding maximum AUC value is 0.8567. The optimal threshold of the anomaly score was determined by using the Youden index and used to extract mineral targets. According to the value of the Youden index, the optimal threshold with respect to the maximum Youden index was selected from 1000 potential thresholds evenly distributed between the minimum and maximum values of anomaly scores. The optimal threshold for the common OCSVM model is optimal threshold OT0 = 89.8292 and the corresponding maximum Youden index is maximum Youden index MYI0 = 0.5092. The mineral targets extracted by the optimal threshold OT0 = 89.8292 are shown in Figure 8a. The optimal threshold for the "optimized" OCSVM is OT1 = 144.3031 and the corresponding maximum Youden index is MYI1 = 0.6214. The mineral targets extracted by the optimal threshold OT1 = 144.3031 are shown in Figure 8b.

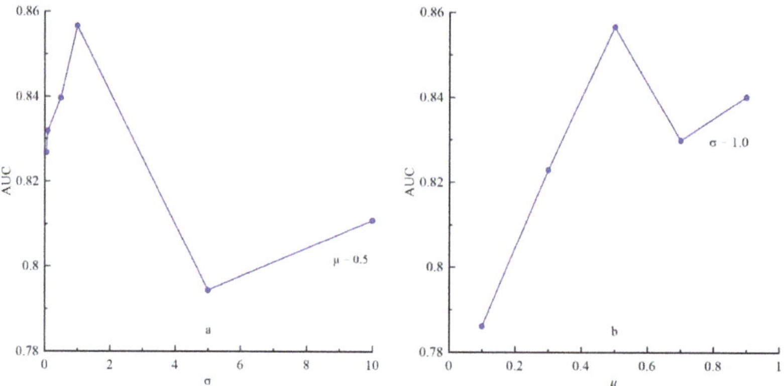

Figure 7. Curve of the AUC value of the OCSVM model changing with (**a**) σ and (**b**) μ.

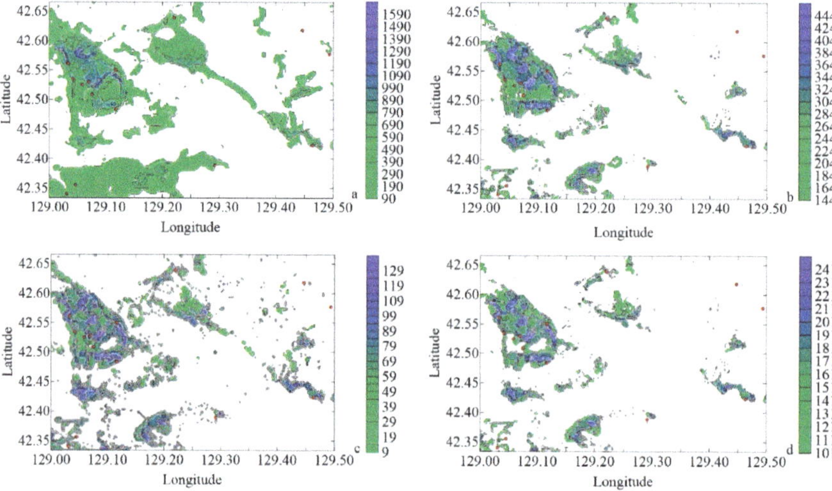

Figure 8. Mineral targets extracted by (**a**) the common OCSVM, (**b**) the trial and error-optimized OCSVM, (**c**) the bat-optimized OCSVM 1, and (**d**) the bat-optimized OCSVM 2.

In order to specify the search space of the bat population, the value ranges of μ and σ were empirically defined as (0, 1] and (0, 10], respectively. The eight initialization parameters of the bat algorithm were defined respectively as $L = 20$, $T = 30$, $f_{min} = 0$, $f_{max} = 1$, $A_{min} = 0$, $A_{max} = 1$, and $\alpha = \gamma = 0.9$. Figure 9a shows that in the optimization process of the bat algorithm, as the number of iterations increases, the AUC value of the OCSVM model becomes larger and larger. It can be seen from Figure 9a that after 23 iterations, the bat algorithm converges to AUC = 0.8649. The corresponding optimal values of μ and σ are respectively $\mu = 0.4276$ and $\sigma = 1.7559$. The optimal threshold determined by using the Youden index is OT2 = 9.2496, and the corresponding maximum Youden index is MYI2 = 0.5763. The mineral targets extracted by the optimal threshold OT2 = 9.2496 are shown in Figure 8c.

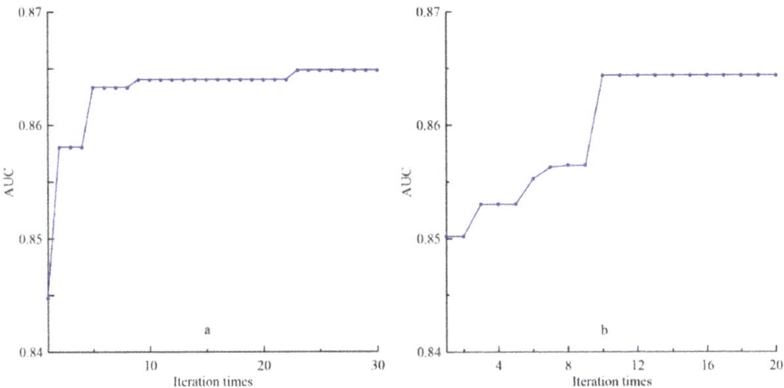

Figure 9. The AUC value of the OCSVM model varies with iterations: (**a**) the bat algorithm initialized with $L = 20$, $T = 30$, $f_{min} = 0$, $f_{max} = 1$, $A_{min} = 0$, $A_{max} = 1$, and $\alpha = \gamma = 0.9$; and (**b**) the bat algorithm initialized with $L = 30$, $T = 20$, $f_{min} = 0$, $f_{max} = 1$, $A_{min} = 0$, $A_{max} = 1$, and $\alpha = \gamma = 0.9$.

In order to compare the mineral prospectivity mapping results when different parameter values were used to initialize the bat algorithm, the initialization parameter values of the bat algorithm were changed to: $L = 30$, $T = 20$, $f_{min} = 0$, $f_{max} = 1$, $A_{min} = 0$, $A_{max} = 1$, and $\alpha = \gamma = 0.9$. Figure 9b shows that in the optimization process of the bat algorithm, the AUC value of the OCSVM model increases with the increase of the number of iterations. It can be seen from Figure 9b that after 10 iterations, the bat algorithm converges to AUC = 0.8644. The corresponding optimal solution is $\mu = 0.4764$ and $\sigma = 1.3602$. The optimal threshold determined by using the Youden index is OT3 = 101.4408, and the corresponding maximum Youden index is MYI3 = 0.5846. The mineral targets extracted by the optimal threshold OT3 = 101.4408 are shown in Figure 8d.

Figure 8a–d shows that the value ranges of anomaly scores generated by the four OCSVM models are quite different. However, these differences do not affect the validity of the OCSVM models in mineral prospectivity mapping, because we are only interested in the relative difference of anomaly scores between different unit cells and do not care about the absolute value of the anomaly score of a unit cell. The value range of the anomaly score is mainly affected by σ, and the smaller σ is, the larger it is. To illustrate the relationship between the value range of the anomaly score and σ, we set $\mu = 0.5$, and then set $\sigma = 0.05$, 0.1, 0.5, 1.0, 5.0, 10.0, 50.0, 100.0, and 500.0, respectively. The OCSVM model initialized with $\mu = 0.5$ and different values of σ was used to map mineral prospectivity. The corresponding minimum and maximum values of the anomaly score are listed in Table 5. The relationship between the value range of the anomaly score and σ can also be explained theoretically. Parameter σ is the responding width of the Gaussian kernel function in Equation (4). Reducing the value of σ is equivalent to amplifying the difference between samples in the kernel space. As a result, the value range of the anomaly score is enlarged. In order to make the anomaly scores generated by different OCSVM models have the same data change interval, the anomaly scores generated by OCSVM can be further transformed as follows:

$$\tilde{f}(x) = \frac{f(x) - \min\{f(x)\}}{\max\{f(x)\} - \min\{f(x)\}} \quad (11)$$

where $\tilde{f}(x)$ is the transformed anomaly score, $\min\{f(x)\}$ and $\max\{f(x)\}$ are the minimum and maximum values of $f(x)$. The transformed anomaly score $\tilde{f}(x)$ is between 0 and 1. This transformation does not affect the performance of OCSVM in mineral prospectivity mapping.

Table 5. The minimum and maximum values of the anomaly score generated by the OCSVM initialized with different values of σ.

Score \ σ	0.05	0.1	0.5	1.0	5.0	10.0	50.0	100.0	500.0
Minimum	−300	−300	−200	−60	−5	−5	−5	−5	−5
Maximum	2000	1600	1200	460	95	95	95	95	95

Note: parameter $\mu = 0.5$.

4. Results

In this section, the mineral prospectivity mapping results in Section 3.3 was statistically evaluated. The ROC curve and AUC value were used to evaluate whether the anomaly scores generated by the OCSVM model in Section 3.3 are effective for predicting known mineral deposit locations in the study area. The ROC curve and the AUC value are respectively the graphical and overall representations of the spatial relationship between the anomaly scores and known mineral deposit locations [13,14,21–23]. The higher the spatial relationship is between the anomaly scores and known mineral deposit locations, the more effective the anomaly scores are in predicting the known mineral deposit locations. Accordingly, the closer the ROC curve of the anomaly scores is to the upper left corner of the ROC space, the closer the AUC value of the anomaly scores is to 1.0, indicating that the corresponding OCSVM model performs better in mineral prospectivity mapping. The AUC value can be used to further estimate the Z_{AUC} value, and check whether there is a significant spatial relationship between the anomaly scores and the known mineral deposit locations [13,14,21–23].

In this study, each of the 1000 potential thresholds evenly distributed between the minimum and maximum anomaly scores was used to extract geological anomalies from the unit cell population. The benefit and cost for the threshold were calculated according to the predicted positive (anomaly) and negative (normal) points (unit cells) as well as the ground truth data defined in Section 2.1. The ROC curve of the anomaly scores was finally drawn based on the 1000 pairs of costs and benefits. Figure 10 shows the ROC curves of the anomaly scores generated by the common and optimized OCSVMs. As can be seen from Figure 10, although the four ROC curves are intersected, the ROC curves of the optimized OCSVMs slightly dominate that of the common OCSVM in the ROC space. Therefore, according to the ROC curve analysis results, the optimized OCSVMs perform better than the common OCSVM in mineral prospectivity mapping.

The AUC values of the common and optimized OCSVMs were calculated using Equation (1) according to the anomaly scores and the ground truth data defined in Section 2.1. The Z_{AUC}s for the common and optimized OCSVMs were calculated using Equation (2). Table 6 lists the performance evaluation statistics of the common and optimized OCSVMs in mineral prospectivity mapping.

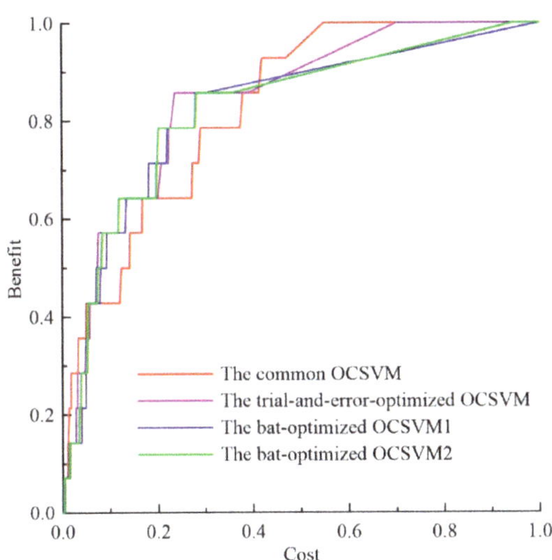

Figure 10. The receiver operating characteristic (ROC) curves of the common and optimized OCSVMs.

Table 6. Performance evaluation statistics of the common and optimized OCSVM models.

Statistics	AUC	Z_{AUC}	MYI	OT	PGA (%)	Benefit (%)	PMT (s)
OCSVM0	0.8268	4.8032	0.5092	89.8292	29.61	93	47.73
OCSVM1	0.8567	5.6029	0.6214	144.3031	18.66	86	n/a
OCSVM2	0.8649	5.8639	0.5763	9.2496	19.84	93	24,856.56
OCSVM3	0.8644	5.8483	0.5846	101.4408	14.22	86	39,314.25

Note: MYI denotes maximum Youden index; OT denotes optimal threshold; PGA denotes the percentage of geological anomalies; PMT denotes program modeling time; OCSVM0 denotes the OCSVM initialized with the default parameters; OCSVM1 denotes the OCSVM optimized by trial and error; and OCSVM2 and OCSVM3 denote the OCSVMs optimized respectively by the bat algorithm with $L = 20$, $T = 30$ and the bat algorithm with $L = 30$, $T = 20$.

As can be seen from Table 6, the AUC value of the common and trial and error-optimized OCVSMs are 0.8268 and 0.8566, while the bat-optimized OCSVMs are respectively 0.8649 and 0.8644. Therefore, according to the estimated AUC values, the bat-optimized OCSVMs perform slightly better than the common and trial and error-optimized OCSVMs in mineral prospectivity mapping. The Z_{AUC}s of the common and trial and error-optimized OCSVMs are 4.8032 and 5.6029, and the Z_{AUC}s of the bat-optimized OCSVMs are 5.8639 and 5.8483, respectively. These four Z_{AUC}s are far higher than the critical value of 1.96. Therefore, both the common and optimized OCSVMs are significantly effective in predicting the known mineral deposit locations in the study area. In other words, the mineral targets predicted by the common and optimized OCSVMs are significantly spatially associated with known mineral deposits in the study area.

According to the anomaly scores generated by the common and trial and error-optimized OCVSMs, the optimal thresholds OT0 = 89.83 and OT1 = 144.3031 extract 29.61% and 18.66% of the study area as mineral targets respectively, and 93% and 86% of known mineral deposits are located in these mineral targets. According to the anomaly scores generated by the bat-optimized OCVSMs, the optimal threshold OT2 = 9.2496 and OT3 = 101.4408 extract 19.84% and 14.22% of the study area as mineral targets respectively, and 93% and 86% of known mineral deposits are located in the corresponding mineral targets. Therefore, we have $0.93/0.2961 = 3.1408 < 0.86/0.1866 = 4.6088 < 0.93/0.1984 = 4.6875 < 0.86/0.1422 = 6.0478$. We call these ratios unit benefit values. Therefore, the bat-optimized

OCSVMs perform slightly better than the common and trial and error-optimized OCSVMs in mineral prospectivity mapping.

By comparing Figures 4 and 8, it can be concluded that the mineral targets predicted by the common and optimized OCSVMs spatially coincide with the Late Archean Jinan Formation and the Yanshanian magmatic rocks (including fleshy red fine-grained monzonite, gray-white porphyritic biotite granodiorite, and porphyritic monzonite). The mineral targets predicted are spatially controlled by the regional northwest-trending structures. These results are consistent with the regional geological characteristics of the study area discussed in Section 3.1.

5. Discussion

When the bat-optimized OCSVMs are used to map mineral prospectivity, the parameters L, T, $f_{\min}$, $f_{\max}$, $A_{\min}$, $A_{\max}$, α, and γ need to be defined for the bat algorithm. Among these parameters, only L and T maybe significantly affect the performance of the bat-optimized OCSVM. The other six parameters are usually defined as the default values suggested by Yang and Gandomi [17].

In order to test the influences of L and T on the performance of bat-optimized OCSVM, the performance evaluation statistics listed in Table 6 and the ROC curves shown in Figure 10 were used to compare the performances of the two bat-optimized OCSVMs. As can be seen from Table 6 and Figure 10, although the performance evaluation statistics of the two models are different, the ROC curves of the two models shown in Figure 10 are almost coincident in the ROC space, and the AUC values of the two models (0.8649 and 0.8644) are approximately equal. The Pearson's correlation coefficient between the anomaly scores generated by the two models is $R = 0.9792$ (the number of samples is 18,905), indicating a high correlation between the anomaly scores generated by the two models. As can be seen from Figure 9, the two iterative processes of the bat algorithm converge to AUC = 0.8644 and AUC = 0.8649, respectively. Therefore, the values of L and T have no significant influence on the performance of the bat-optimized OCSVM in mineral prospectivity mapping. As long as L and T values are within an appropriate range ($20 \leq L, T \leq 30$), the bat-optimized OCSVM can achieve similar good results in mineral prospecting mapping.

The bat algorithm initialized by $L = 20$, $T = 30$, $f_{\min} = 0$, $f_{\max} = 1$, $A_{\min} = 0$, $A_{\max} = 1$, and $\alpha = \gamma = 0.9$ was used to repeatedly optimize OCSVM parameters three times, to verify whether the bat algorithm can always converge to the global optimum. The results show that in each data modeling process, the bat algorithm converges to the same maximum AUC = 0.8649. Therefore, the bat algorithm can generally converge to the global maximum in OCSVM parameter optimization in mineral prospectivity mapping.

In this study, the bat-optimized OCSVM is established to map mineral prospectivity. The model can also be used to solve other similar anomaly detection problems. For example, the bat-optimized OCSVM can be established to detect multivariate geochemical anomalies. The application condition of the model is that there are a certain number of known mineral deposits in the study area to define the ground truth data, so that the ROC curve analysis can be implemented.

It should be pointed out that compared with the common OCSVM, the bat-optimized OCSVM needs more time to search for the optimal parameter values of OCSVM. This leads to the low efficiency of the bat-optimized OCSVM in data modeling. It can be seen from this study that the data modeling time of the bat-optimized OCSVM 1 (24856.56 s) and the bat-optimized OCSVM 2 (39314.25 s) is much longer than that of the common OCSVM (47.73 s).

6. Conclusions

I. A bat-optimized one-class support vector machine is developed by combining one-class support vector machine with the bat algorithm. The combined model can automatically optimize the parameter values of a one-class support vector machine, so as to improve the performance of the model in mineral prospectivity mapping. The method proposed in this paper only needs to set the search space, and then the algorithm automatically searches the optimal parameters. Compared to the trial and error method, the proposed method has more opportunities to find the global optimal parameters.

II. The bat-optimized one-class support vector machine requires a certain number of known mineral deposits in the study area, which can be used to define the true positive and negative points (cells), and be used as the ground truth data for the receiver operating characteristic curve analysis. As long as a study area meets this application condition, the bat-optimized model can also be established to solve other similar anomaly detection problems in geosciences, such as multivariate geochemical anomaly detection. The bat-optimized one-class support vector machine can be used as a semi-supervised machine learning model to handle anomaly detection problems in other application fields.

III. The case study shows that the AUC values of the bat-optimized one-class support vector machine models are greater than those of the common and trial and error-optimized one-class support vector machine models. The mineral targets predicted by the bat-optimized one-class support vector machine models have larger unit benefit values compared to those predicted by the common and trial and error-optimized one-class support vector machine models. Therefore, the bat-optimized one-class support vector machine models are a mineral prospectivity mapping method with high performance.

IV. The Z_{AUC} values of both the common and optimized one-class support vector machine models calculated in the case study are much higher than the critical value 1.96 at the significant level of 0.05. Therefore, the mineral targets predicted by both the common and optimized one-class support vector machine models are significantly spatially associated with known mineral deposits in the study area.

V. The mineral targets predicted by both the common and optimized one-class support vector machine models are spatially consistent with geological and metallogenic characteristics of the study area. The predicted mineral targets spatially coincide with the Late Archean Jinan Formation and the Yanshanian magmatic rocks, and are obviously controlled by northwest-trending structures. The two formations and the structures are the three regional mineralization controlling factors in the study area.

Supplementary Materials: The following are available online at http://www.mdpi.com/2075-163X/9/5/317/s1. Supplementary Materials: Python code.

Author Contributions: Y.C. developed the Python code, completed the main research, and wrote the text; W.W. processed geological and geochemical data and plotted all figures; Q.Z. established the geological and geochemical data bases in MapGIS and MapInfo.

Funding: This study was supported by the National Natural Science Foundation of China (Grant No. 41672322 and 41872244).

Acknowledgments: The authors thank Sheli Chai of Jilin University for his assistance in the collection of geological and geochemical data. The authors are also grateful to the two anonymous reviewers for their constructive comments, which greatly improved the manuscript.

Conflicts of Interest: The authors declare no conflict of interest.

References

1. Hayton, P.; Schölkopf, B.; Tarassenko, L.; Anuzis, P. Support vector novelty detection applied to jet engine vibration spectra. In Proceedings of the Advances in Neural Information Processing Systems 13 (NIPS' 2000), Denver, CO, USA, 27 November–2 December 2000; pp. 946–952.
2. Schölkopf, B.; Platt, J.; Shawe-Taylor, J.; Smola, A.; Williamson, R. Estimating the support of a high-dimensional distribution. *Neural Comput.* **2001**, *13*, 1443–1471. [CrossRef]
3. Davy, M.; Godsill, S.J. Detection of abrupt spectral changes using support vector machines—An application to audio signal segmentation. In Proceedings of the International Conference on Acoustics, Speech, and Signal Processing (IEEE ICASSP-02), Orlando, FL, USA, 13–17 May 2002; pp. 1313–1316.
4. Lengelle, R.; Capman, F.; Ravera, B. Abnormal events detection using unsupervised one-class SVM–Application to audio surveillance and evaluation. In Proceedings of the 8th IEEE International Conference on Advanced Video and Signal-Based Surveillance (AVSS 2011), Klagenfurt, Austria, 30 August–2 September 2011; pp. 124–129.
5. Shin, H.J.; Eom, D.H.; Kim, S.S. One-class support vector machines—An application in machine fault detection and classification. *Comput. Ind. Eng.* **2005**, *48*, 395–408.

6. Mahadevan, S.; Shah, S.L. Fault detection and diagnosis in process data using one-class support vector machines. *J. Process Control* **2009**, *19*, 1627–1639. [CrossRef]
7. Fergani, B.; Davy, M.; Houacine, A. Speaker diarization using one-class support vector machines. *Speech Commun.* **2008**, *50*, 355–365. [CrossRef]
8. Mourão-Miranda, J.; Hardoon, D.R.; Hahn, T.; Marquand, A.F.; Williams, S.C.R.; Shawe-Taylor, J.; Brammer, M. Patient classification as an outlier detection problem: An application of the One-Class Support Vector Machine. *NeuroImage* **2011**, *58*, 793–804. [CrossRef]
9. Strobbe, T.; Wyffels, F.; Verstraeten, R.; De Meyer, R.; Van Campenhout, J. Automatic architectural style detection using one-class support vector machines and graph kernels. *Autom. Constr.* **2016**, *69*, 1–10. [CrossRef]
10. Roodposhti, M.S.; Safarrad, T.; Shahabi, H. Drought sensitivity mapping using two one-class support vector machine algorithms. *Atmos. Res.* **2017**, *193*, 73–82. [CrossRef]
11. Saari, J.; Strömbergsson, D.; Lundberg, J.; Thomson, A. Detection and identification of windmill bearing faults using a one-class support vector machine (SVM). *Measurement* **2019**, *137*, 287–301. [CrossRef]
12. Harrou, F.; Dairi, A.; Taghezouit, B.; Sun, Y. An unsupervised monitoring procedure for detecting anomalies in photovoltaic systems using a one-class support vector machine. *Sol. Energy* **2019**, *179*, 48–58. [CrossRef]
13. Chen, Y.L.; Wu, W. Mapping mineral prospectivity by using one-class support vector machine to identify multivariate geological anomalies from digital geological survey data. *Aust. J. Earth Sci.* **2017**, *44*, 639–651. [CrossRef]
14. Chen, Y.L.; Wu, W. Application of one-class support vector machine to quickly identify multivariate anomalies from geochemical exploration data. *Geochem. Explor. Environ. Anal.* **2017**, *17*, 231–238. [CrossRef]
15. Poli, R.; Kennedy, J.; Blackwell, T. Particle swarm optimization-An overview. *Swarm Intell.* **2007**, *1*, 33–57. [CrossRef]
16. Yang, X.S. A new metaheuristic bat-inspired algorithm. In Proceedings of the Nature Inspired Cooperative Strategies for Optimization (NICSO 2010), Granada, Spain, 12–14 May 2010; González, J.R., Pelta, D.A., Cruz, C., Terrazas, G., Krasnogor, N., Eds.; Springer: Berlin, Germany, 2010; pp. 65–74.
17. Sharawi, M.; Emary, E.; Saroit, I.A.; El-Mahdy, H. Bat swarm algorithm for wireless sensor networks lifetime optimization. *Int. J. Sci. Res.* **2012**, *3*, 655–664.
18. Yang, X.S.; Gandomi, A.H. Bat algorithm: A novel approach for global engineering optimization. *Eng. Comput.* **2012**, *29*, 464–483. [CrossRef]
19. Goyal, S.; Patterh, M.S. Wireless sensor network localization based on bat algorithm. *Int. J. Emerg. Technol. Comput. Appl. Sci. IJETCAS* **2013**, *4*, 507–512.
20. Yang, X.S.; Karamanoglu, M.; Fong, S. Bat algorithm for topology optimization in microelectronic applications. In Proceedings of the First International Conference on Future Generation Communication Technologies, London, UK, 12–14 December 2012; pp. 150–155.
21. Chen, Y.L. Mineral potential mapping with a restricted Boltzmann machine. *Ore Geol. Rev.* **2015**, *71*, 749–760. [CrossRef]
22. Chen, Y.L.; Wu, W. A prospecting cost-benefit strategy for mineral potential mapping based on ROC curve analysis. *Ore Geol. Rev.* **2016**, *74*, 26–38. [CrossRef]
23. Chen, Y.L.; Wu, W. Mapping mineral prospectivity using an extreme learning machine regression. *Ore Geol. Rev.* **2017**, *80*, 200–213. [CrossRef]
24. Liu, F.S.; Zhang, M.L. Complete quality management of the new-round land resources survey. *Chin. Geol.* **1999**, *267*, 20–21. (In Chinese)
25. Zhang, J.; Marszalek, M.; Lazebnik, S.; Schmid, C. Local features and kernels for classification of texture and object categories: A comprehensive study. *Int. J. Comput. Vis.* **2007**, *73*, 213–238. [CrossRef]
26. Zhang, Y.B.; Wu, F.Y.; Wilde, S.A.; Zhai, M.G.; Lu, X.P.; Sun, D.Y. Zircon U-Pb ages and tectonic implications of Early Paleozoic granitoids at Yanbian, Jilin Province, northeast China. *Island Arc* **2004**, *13*, 484–505. [CrossRef]
27. Wu, F.; Lin, J.; Wilde, S.A.; Zhang, Q.; Yang, J. Nature and significance of early Cretaceous giant igneous event in eastern China. *Earth Planet. Sci. Lett.* **2005**, *233*, 103–119. [CrossRef]
28. Yu, J.J.; Wang, F.; Xu, W.L.; Gao, F.H.; Pei, G.P. Early Jurassic mafic magmatism in the Lesser Xing'an-Zhangguangcai Range, NE China, and its tectonic implications: Constraints from zircon U-Pb chronology and geochemistry. *Lithos* **2012**, *142–143*, 256–266. [CrossRef]

29. Wu, P.F.; Sun, D.Y.; Wang, T.H.; Gou, J.; Li, R.; Liu, W.; Liu, X.M. Chronology, geochemical characteristic and petrogenesis analysis of diorite in Helong of Yanbian area, northeastern China. *Geol. J. China Univ.* **2013**, *19*, 600–610. (In Chinese)
30. Yan, D.; Li, N.; Xu, M.; Miao, M.M. Mineralization characteristics and genesis of the Bailiping silver deposit in Helong City, Jilin Province. *Jilin Geol.* **2015**, *34*, 36–41. (In Chinese)
31. Wan, W.Z.; Wang, J.B.; Feng, X.Y.; Zhang, H.; Jia, N.; Zhang, Y.L. Geological features and prospecting directions of the Heanhe gold deposit in the Helong area, Jilin Province, China. *Jilin Geol.* **2010**, *29*, 71–75. (In Chinese)
32. Pan, Y.D.; Xu, B.J.; Sun, Y.; Hou, L. Geological features of the Jinchengdong gold deposit in Helong City, Jilin Province, China. *Jilin Geol.* **2016**, *35*, 30–35. (In Chinese)

Review

Tools and Workflows for Grassroots Li–Cs–Ta (LCT) Pegmatite Exploration

Benedikt M. Steiner

Camborne School of Mines, University of Exeter, Penryn TR10 9FE, UK; b.steiner@exeter.ac.uk

Received: 22 July 2019; Accepted: 17 August 2019; Published: 20 August 2019

Abstract: The increasing demand for green technology and battery metals necessitates a review of geological exploration techniques for Li–Cs–Ta (LCT) pegmatites, which is applicable to the work of mining companies. This paper reviews the main controls of LCT pegmatite genesis relevant to mineral exploration programs and presents a workflow of grassroots exploration techniques, supported by examples from central Europe and Africa. Geological exploration commonly begins with information gathering, desktop studies and Geographic Information System (GIS) data reviews. Following the identification of prospective regional areas, initial targets are verified in the field by geological mapping and geochemical sampling. Detailed mineralogical analysis and geochemical sampling of rock, soil and stream sediments represent the most important tools for providing vectors to LCT pegmatites, since the interpretation of mineralogical phases, deportment and liberation characteristics along with geochemical K/Rb, Nb/Ta and Zr/Hf metallogenic markers can detect highly evolved rocks enriched in incompatible elements of economic interest. The importance of JORC (Joint Ore Reserves Committee) 2012 guidelines with regards to obtaining geological, mineralogical and drilling data is discussed and contextualised, with the requirement of treating LCT pegmatites as industrial mineral deposits.

Keywords: LCT; pegmatite; lithium; exploration; targeting

---

1. Introduction

Over the last four decades research into rare metal Li–Cs–Ta (LCT) pegmatite mineralisation has predominantly focused on the understanding of late-stage magmatic fractionation and hydrothermal alteration processes enriching incompatible elements of potential economic interest in felsic peraluminous melts. In particular, the mineralogy and geochemistry of LCT pegmatites have been studied in great detail and are presented in case studies from Canada and the US [1–4], Ireland [5,6], the European Variscides [7–12], the Sveconorwegian orogeny [13,14], Central-East Africa [15–18] and Australasia [19–22]. These case studies have resulted in the definition of tectonic, mineralogical and geochemical factors controlling the location of mineralisation and zonation within the roof of peraluminous S-type granite plutons, pegmatitic offshoots into metasedimentary country rock or anatectic melts with no established link to S-type granites.

On the other hand, from an economic and industrial perspective, only a limited number of case studies are available, describing relevant mineral exploration techniques which are directly applicable to the daily business of mining and exploration companies. In the recent past, aspects of mineral exploration techniques applied to LCT pegmatite mineralisation have been published, covering (i) hyperspectral remote sensing [23], (ii) whole-rock geochemistry [1,24–26], (iii) stream sediment geochemistry [27–29], (iv) soil geochemistry [30], and (v) best practice reporting of LCT pegmatite exploration results according to the JORC (Australian Joint Ore Reserves Committee) standard [31]. In the current literature, a unified approach for industrial grassroots LCT pegmatite exploration is largely lacking. Therefore, this paper attempts to fill this gap by providing a workflow for early stage target generation, identification and testing of LCT pegmatites.

2. An Overview of LCT Pegmatites and Implications for Mineral Exploration

In light of the recent green technologies and battery metals "boom", rare element LCT pegmatites form an important aspect of sustainable Li, Cs and "coltan" (Nb–Ta) supplies (approx. 40% of the global Li production), and are therefore explored for and mined globally [10,32,33]. Major pegmatite occurrences, such as Tanco (Canada) and Greenbushes (Australia) contain important resources of these metals and new deposits are currently explored for around the world. Available literature on the nature and genesis of pegmatite mineralisation is vast and often disputed, and will be reviewed to highlight the formation, occurrence, composition, zoning, and rare metal enrichment of pegmatites, with a view to identifying physical and chemical properties relevant to mineral exploration and mining.

2.1. Definition and Classification

London (2008) [34] defines a pegmatite as an igneous rock of predominantly granitic composition, which is distinguished from other igneous rocks by (1) the extremely coarse and systematically variable size of its crystals, typically increasing by a ~10^2 from margins to centre, or (2) by an abundance of crystals with skeletal, graphic or other strongly directional growth habits, or (3) by a prominent spatial zonation of mineral assemblages, including monomineralic zones. Pegmatites have traditionally been linked to a parental granite [1,3] and can either occur in close proximity to the latter or as off-shoot dyke swarms up to 20 m wide and several kilometres long in metasedimentary country rocks (Figure 1). However, pegmatites with a missing apparent parental granite are common [3], and it is suspected that the source granite occurs at depth [30]. On the other hand, studies published in the 2010s on the anatectic origin of granitic pegmatites in Europe and North America demonstrate that there are "thousands of pegmatites without parental granites" [13], i.e., pegmatite fields can be unrelated to a source granite and instead form by partial melting and anatexis of crustal material [4,13,14,35–39], or energy and melt circulation along deep lithospheric fault zones [40]. In the current literature, there are numerous descriptions of pegmatite sub-classifications, but to date there is no universally accepted model explaining the diverse features and genesis of granitic pegmatites [39]. An exhaustive discussion and critical evaluation of historic pegmatite classifications was published by Müller et al. (2018) [41] and currently represents a significant summary of pegmatite classification studies. From an economic point of view rare-metal pegmatites, as opposed to barren pegmatites, contain beryl, lithium aluminosilicates, phosphates and rare metal oxides [42]. Černý and Ercit (2005) [43] and Černý et al. (2012) [44] use mineralogical, geochemical and structural (zonation) criteria along with necessary knowledge of P-T conditions to broadly divide rare-element pegmatites into two geochemical groups or families, taking into account the composition and volume of source material from which the granitic melt is originated: Nb-Y-F (NYF) and LCT enriched pegmatites, which can be further divided into an array of sub-classes and sub-types. NYF pegmatites are sourced by alkaline A-type granites originating from the melting of lower continental crust and mantle, whilst LCT pegmatites are commonly related to the partial melting of subducted continental crust consisting of metamorphosed sediments, leading to the emplacement of peraluminous S-type granites [44]. Although this classification is widely accepted in the literature, a number of aspects, such as the necessity for consideration of the tectonic histories and settings of the pegmatites, are disputed. For example, Dill (2015) [10] describes the spatial association of Variscan NYF and LCT pegmatites at Pleystein and Hagendorf (Germany), respectively, which would imply two drastically different geodynamic settings across a distance of several kilometres. Studies of pegmatite occurrences in the Sveconorwegian orogen have shown that NYF pegmatites do not only form in anorogenic (A-type magmatism) settings, but also in compressional or extensional orogenic settings unrelated to plutonic magmatism [13]. Additional shortcomings of the Černý and Ercit classification [43] are the dependence on the crystallisation depth and pressure, which is commonly difficult to obtain, and a possible lack of characteristic REE and F-minerals in pegmatites, which do not allow to fit pegmatites into the NYF classification [41]. Rare-element pegmatites that were formed in a pluton-unrelated setting can therefore lack the characteristic F ("NY" instead of "NYF") or Li ("CT" instead of "LCT") signature [13]. These aspects question the validity of the currently accepted classification. Müller et al. (2018) [41] advocate for a revised classification of pegmatites relying on measurable criteria, such as (1) indicative minerals and mineral assemblages of primary magmatic origin, (2) quantitative mineralogy and geochemistry of quartz, garnet, topaz, beryl and columbite-group minerals, (3) pegmatite structure (zonation and graphic quartz-feldspar intergrowths) and (4) crystallisation age. Despite the ongoing discussion on pegmatite classification, this paper will focus exclusively on economically relevant and sought-after pegmatites with LCT signature and their associated mineralisation and exploration potential.

Figure 1. Typical exposure of Li–Cs–Ta (LCT) pegmatite in the Ruhanga area, Rwanda. This pegmatite is currently (in 2018) exploited for columbite-tantalite mineralisation. The pegmatite body is 15–20 m wide and has a lateral extension of at least 2.5 km. Complete kaolinisation of feldspars, resulting in an unzoned appearance, characterise most Rwandan pegmatites. This creates issues with slope stability [45], but allows simple artisanal open-cast excavation of the orebody to a depth of 30 m. Note the rafts of biotite schist in the weathered pegmatite illustrating the assimilation of country rock material, which can have a detrimental influence on ore grade and dilution, and therefore needs to be appropriately modelled.

2.2. Formation, Rare Metal Enrichment, and Geodynamic Setting

Pegmatite-forming melts are produced by (1) fractional crystallisation of granitic magma plutons (Figure 2), and (2) partial melting of crustal or mantle rocks involving the presence of fluids [46]. In the first scenario, siliceous melt is separated from parental granites by filter pressing and intrudes into metasedimentary country rocks, where the melt crystallises and forms unzoned or zoned pegmatites [10]. The concentration of fluxing elements, such as F, B, P, Li and H_2O, reduces the viscosity and solidus temperature of the melt, but increases the solubility of large-ion-lithophile elements (LILE) and high-field-strength elements (HSFE) [3]. Depending on temperature, LILE precipitate and form minerals when fluxing elements are removed from the melt by chemical quenching [2]. However, fractional crystallisation is not regarded as the sole factor of pegmatite formation and element mobility in peraluminous melts. Firstly, several studies suggest that felsic pegmatites can be derived by partial melting and anatexis rather than fractional crystallisation of a parent granite (Figure 3) [13,39,41]. This is particularly evident in the pegmatite fields of Mt. Mica [4] and Southern Norway [14,47] where geophysical gravity surveys, radiometric dating, isotopes, whole-rock and mineral chemistry indicate that pegmatites are unrelated to nearby S-type granite plutons. Secondly, there is evidence that sub-solidus hydrothermal alteration, specifically the transition between pure magmatic fractionation and hydrothermal alteration, plays a significant role in localising economically significant mineralisation of Sn, W, Li, Cs, Ta and Nb [7]. Whole-rock geochemical data presented in Ballouard et al. (2016) [7] demonstrated that K/Rb, Nb/Ta and Zr/Hf ratios represent valuable geochemical and metallogenic markers for this magmatic–hydrothermal transition zone. K/Rb < 150, Nb/Ta < 5 and Zr/Hf < 18 define pegmatitic–hydrothermal evolution characterised by increased substitution of K with Rb in micas and feldspars, fractionation of Nb over Ta due to hydrothermal sub-solidus reactions enriching Ta in F-rich residual melts and leading to secondary muscovitisation [48], and increasing Kd values of Hf in zircon [8,49]. The removal of Li (petalite or spodumene) and Li-F (lepidolite) minerals from the silicate melt, along with the diffusion of these fluxing elements into adjacent host rocks, will lead to a sudden drop in Ta solubility and the subsequent precipitation of columbite-tantalite group minerals. Therefore, the Nb/Ta ratio decreases with increasing fractionation, and Nb-Ta mineral phases evolve from manganocolumbite to manganotantalite [26]. In this context, binary plots of K/Rb vs. Li, Cs, or Ta and K/Rb vs. Nb/Ta or Zr/Hf present a useful tool to visualise magmatic fractionation and hydrothermal alteration processes and to determine the most fractionated/prospective samples. Whilst these petrogenetic ratios were initially developed on the mineral chemistry of muscovites and other host silicates, the concept has meanwhile been successfully implemented in mineral exploration campaigns [26,28,29,45], underlining the significance of lithogeochemistry in applied aspects of peraluminous rare metal granite and pegmatite exploration.

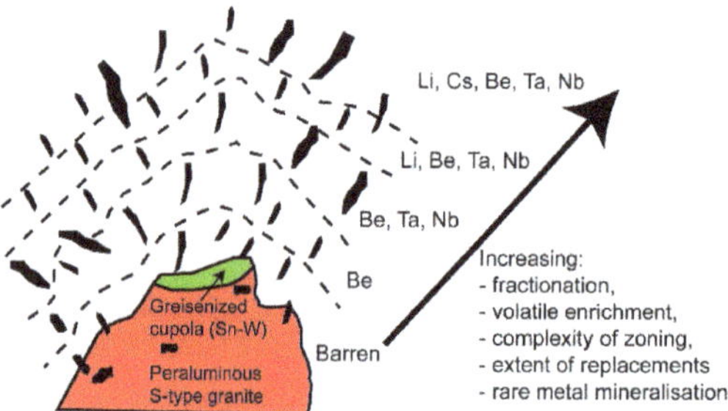

Figure 2. Traditional model of regional zonation and rare metal enrichment of pegmatites and pegmatite fields related to S-type granite plutons implying that zoned pegmatite fields are related to and have a similar age as a nearby granite intrusion (modified from [26]). This model is currently challenged by a number of authors who outline evidence for an anatectic origin of pegmatite melts, e.g., [4,13,36,37,47]. Whilst pegmatite fields can be zoned, the zonation is in some cases not spatially related to a pluton, but to individual pods of pegmatitic melt.

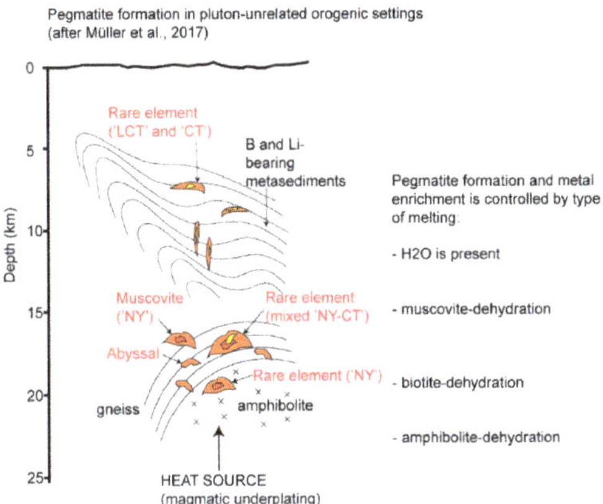

Figure 3. Simplified model of anatectic pegmatites formed in orogenic settings characterised by a distinct lack of nearby granite plutons (modified from [13]).

LCT pegmatites are related to orogenic belts, typically involving high-grade metamorphic terrains and reworked or exhumed deep continental crust. Examples are found around the world, such as Palaeoproterozoic granulite belts of northern Scotland [50], Archean–Proterozoic greenstone belts of Canada, Africa and Australia [10,20,22], and Phanerozoic mountain belts, particularly the European Variscides (also termed the Hercynian Fold Belt), and its extension along the Alleghanian Orogen into the New England province of North America [10]. Previous regional studies have demonstrated that LCT pegmatites predominantly occur in continent-continent collision zones [51], whilst Sn and W deposits can be decoupled and occur in post-orogenic extension and continental arc settings [12]. Based on European pegmatite occurrences, Dill (2015) [10] divided the geodynamic setting of rare metal pegmatites into primary "Variscan" and reworked "Alpine" types, whilst "Andean" and "Island Arc" type pegmatites are considered barren. The fertility of LCT pegmatites is determined by the presence of (i) intensely chemically weathered, and therefore Li-, Cs-, Sn- and W-enriched sedimentary protolith, accumulating along continental margins as a result of sedimentary and tectonic processes, and (ii) an orogenic and mantle-derived heat source during subduction and melting processes resulting in the preferential formation of Sn, W and LCT mineralisation across major suture zones [12]. In essence this means that, from a geodynamic perspective, prospective LCT pegmatites are found and should be explored for in areas that underwent crustal growth and reworking of voluminous metasedimentary source material which, coupled with a heat source, mobilised the ore elements from the source rock to form peraluminous melts. The awareness and supporting data for applicable geodynamic settings, mineralisation system architecture, fluid and metal sources, pathways, reservoirs and traps have been available for several decades, however the integration of these factors into a mineral systems model originally defined by Wyborn et al. (1994) [52] has not been taken into account until recently. Using examples from Archean rare metal pegmatites of the Archean Yilgarn and Pilbara cratons of Western Australia, Sweetapple (2017) [21] demonstrated that integrating (a) sources (precursor granitoids generated by partial melting of Archean trondjemite–tonalite–granodiorite rocks), (b) host rocks (mafic or ultramafic rocks in greenstone belts), (c) fluid flow sources, reservoirs, and heat drivers (younger granites) and (d) melt and fluid pathways (foliations associated with regional shear and fault zones, fractures generated in host rock due to magmatic overpressure) can be implemented at craton to deposit scale exploration targeting.

2.3. Composition, Grade/Tonnage and Geometry

The bulk composition of pegmatites contains subequal proportions of quartz, alkali feldspar, and plagioclase, comprising greater than 80% of the rock, and therefore plots in the granite field of the

Quartz, Alkali feldspar, Plagioclase, Feldspathoid (QAPF) diagram [3]. However, the composition is inherently poorly understood due to issues with sampling heterogeneous, coarse-grained geological bodies. The Tanco pegmatite at Bernic Lake, Manitoba, is probably one of the best studied economic pegmatites and is classified as granitic considering a combined quartz, plagioclase and alkali feldspar abundance of 82% [53]. Dill (2015) [10] provides bulk geochemical data for pegmatitic rocks of the Hagendorf–Pleystein Pegmatite Province (Germany), supporting the overall granitic composition of pegmatites. LCT pegmatites are generally devoid of magnetic or conductive minerals and consequently lack a density, gamma-ray and magnetic contrast to metasedimentary host rocks [54]. Of interest to the mineral exploration geologist are pegmatites that contain abundant and economic quantities of lithium aluminosilicates and phosphates, pollucite, beryl and Sn–Nb–Ta–REE oxides. In the case of the Leinster rare-element pegmatite belt in Ireland, spodumene pegmatites contain spodumene (10–40%), albite (25–35%), quartz (15–20%), Li-muscovite (10–15%), spessartine garnet (5%), and <5% K-feldspar, apatite, cassiterite and sphalerite [5]. Similar compositions are observed in other economic deposits, such as King Mountain (US), Wolfsberg (Austria) and Kaustinen (Finland). However, only a minor fraction of pegmatites in any given district are enriched in these elements and minerals, whilst the majority of pegmatites contains these elements at minor (<5%) or accessory (<1%) levels [34].

Ore grades in rare-element pegmatites are commonly irregularly distributed [31]. Bradley et al. (2017) [55] present comparative data of major, economic LCT pegmatites and demonstrate that economic pegmatites should generally contain minimum grades of 1 wt. % Li_2O and 0.1 wt. % Ta_2O_5 at tonnages of 7 Mt and 0.01 Mt, respectively. Importantly, this study implies that pure world-class Li pegmatites are less abundant (10 used for the study) than Ta deposits (35 used in the study). Pegmatites can take the shape of flat-lying or variably dipping dikes, sills, pods, tabular and lenticular-shaped bodies. In a study of the Evje-Iveland pegmatite field (southern Norway), Snook (2013) [47] outlined the dimensions of zoned, anatectic pegmatites in amphibolites and other metasedimentary country rocks, commonly taking the shape of pods and lenses with a thickness of 4 to 15 m and a strike length of 10 to 50 m. In contrast, the Mt. Mica pegmatite (US) is thought to consist of batches of anatectic melt which accumulated and coalesced into a larger volume [4]. On the other hand, the Kaustinen pegmatites (Finland) consist of a flat-lying dike swarm with a length of 450 m and a maximum individual dike width of 10 m [56]. The pegmatite dikes of the giant Greenbushes deposit (Australia) are up to 3 km long, 40 to 250 m wide, 400 m deep and variably dip 40–50° [20]. This means that grade and tonnage of economic metals in rare-element pegmatites can be variable. Consequently, distinguishing barren from fertile or rare element pegmatites, and importantly outlining economic grades and tonnages using mineralogical and geochemical techniques is a key aspect in commercial pegmatite exploration.

2.4. *Regional and Internal Zoning*

Pegmatites are typically characterised by regional and internal zonation patterns, manifested by distinctive mineralogical and geochemical signatures, which are routinely used in geological and geochemical exploration programs [1,26]. In pegmatite swarms and groups, zonation as a function of cooling, chemical fractionation and hydrothermal alteration can be observed with increasing distance from the granitic source, forming a sequential appearance of exotic minerals [3,34]. Figure 2 shows an idealised sequence of barren proximal to Be–, Be–Nb–Ta, Li–Be–Ta–Nb and Li–Cs–Be–Ta–Nb–enriched pegmatites related to S-type granite plutons. This outward sequence of mineral and element zonation patterns has been challenged by a number of authors who outline evidence for an anatectic origin of pegmatite melts [4,13,14]. Whilst pegmatite fields can be zoned, the zonation is in some cases not spatially related to a pluton, but to individual pods of pegmatitic melt in metasedimentary country rock which were likely to be affected by secondary hydrothermal alteration processes [47]. Internally, the structure of pegmatites is variable and can take the form of either zoned or unzoned patterns. London (2018) [3] provides a detailed account of internal zonation patterns which will be summarised here. Unzoned pegmatites are located within or adjacent to granite cupolas and are characterised by their isotropic rock fabric and coarse grain size (several cm or larger) in relation to plutonic rocks.

Plagioclase and quartz are anhedral and of uniform size, whereas K-feldspar is porphyritic and exhibits skeletal or graphic granite intergrowths. Unzoned pegmatites can form economic sources of quartz, feldspar, spodumene and mica. On the other hand, the internal fabric of zoned pegmatites depends on the habit and orientation of crystals, as well as the spatial segregation of minerals by zonal mineral assemblages. Zoned pegmatites comprise border, wall, intermediate, and core zones which have unique mineralogical and geochemical properties. Border zones comprise 1–3 cm-wide cooling margins of fine-grained and granophyric granitic rock at the contact with the host rock. The typical mineralogy comprises plagioclase-quartz-muscovite, with accessory tourmaline, beryl, garnet, cassiterite and columbite. The wall zone consists of inward-orientated K-feldspar, micas, beryl, and tourmaline, with considerably larger crystal sizes. The intermediate zone follows inward from the wall zones and is principally formed of more abundant monophase K-feldspar, quartz, muscovite, sodic plagioclase and lithium aluminosilicates. The centre or core of zoned pegmatites is pure quartz. Historically, it was assumed that the quartz core formed during the last stage of the pegmatite crystallisation process. However, experiments by London and Morgan (2017) [57] demonstrated that in Li-rich melts, albite, and lepidolite can follow the crystallisation of pure quartz. From a practical perspective, the recognition of regional and internal pegmatite zonation is a fundamental task during geological and geochemical sampling and fieldwork.

3. Suggested Grassroots Exploration Strategy

3.1. Literature and Desktop Study

Grassroots exploration programs commonly commence with a detailed literature review of regional geology, mineralisation occurrences, and styles (Figure 4). Exploration geologists unfamiliar with LCT and Sn(–W) mineralisation are advised to become familiar with general pegmatite geology and mineralogy, in order to develop an understanding of element associations, fractionation patterns, orebody geometry and grade–tonnage relationships [2,3,10,34]. Furthermore, a desktop study should aim to identify evidence of suitable regional geo-tectonic settings involving deep crustal reworking and exhumation processes, leading to partial melting of crustal, metasedimentary source rocks, such as along the European Variscan belt [12]. In particular, consideration should be put on different orogenic and anorogenic settings and the presence of anatectic melts [13]. Supporting evidence in the form of exhumed granulites, migmatites and two-mica peraluminous granites with accessory minerals, for example tourmaline and fluorite indicating the presence of B and F fluxing elements, can be obtained from regional geological maps and databases. For example, Dill (2015) [10] provides an exhaustive list of pegmatite sites and mineralogy, which is useful and applicable to generating conceptual exploration targets. National geological survey departments provide summary reports of historic company exploration work and represent a valuable resource. For example, the Finnish Geological Survey carried out investigative work into the occurrence of LCT pegmatites in the Kaustinen Province and other permissive tracts, and published geological, geophysical and mineralogical data in support of a growing local mining industry [56,58]. Similarly, the Ontario Geological Survey regularly produces multi-commodity exploration reports covering summaries of recommended areas prospective for LCT mineralisation [59].

Figure 4. Workflow for early-stage LCT pegmatite target generation, definition and testing.

3.2. *Review of GIS Public Domain Datasets and Regional Target Selection*

A literature and desktop study is followed by a review of public domain GIS data and leads to regional target selection. Mineral occurrences and abandoned mine GIS databases support the spatial delineation of ore and pathfinder minerals, along with formerly productive mines, and so represent a tool to delineate prospective mineral belts. Hyperspectral remote sensing data obtained from Landsat, ASTER and Sentinel-2 satellite missions work particularly well in arid environments characterised by a limited abundance of vegetation to delineate granitic intrusions, hydrothermal alunite and clay minerals, and Li-bearing silicate minerals (spodumene and petalite), using RGB band combinations, ratios and selective principal component analysis [23,60]. In a case study from the Fregeneda-Almendra region in Spain and Portugal Cardoso-Fernandes et al. (2019) [23] utilized and developed a variety of spectral signatures to identify currently excavated LCT pegmatites through hydrothermal alteration halo mapping and direct identification of Li-bearing minerals. In this context, the importance of thermal bands in the discrimination of Li-bearing pegmatites using remote sensing data was demonstrated, as silicate minerals (spodumene and petalite) have distinctive emission bands in the thermal spectrum. The most applicable band combinations and ratios for the determination of areas enriched in Li minerals were determined as RGB 2113 and 7/6 (ASTER); RGB 3211 and 3/5 (Landsat-8); RGB 216 and 2/4 (Landsat-5); RGB 3212 and 3/8 (Sentinel-2). In the same study, selective PCA of two-band subsets (e.g., bands 1 and 3 for ASTER) highlighted Li-bearing areas as the subset used two target bands and two principal components only, and therefore reduced the overall noise of the dataset.

Regional airborne radiometric surveys are another applicable tool for identifying granite intrusions and stocks, i.e., the metal and heat source of the mineralisation system, through analysis of K–Th–U spectrometer data [61,62]. Airborne radiometric and magnetic surveys are usually conducted simultaneously and, whilst LCT pegmatites often do not contain magnetic or conductive sulphide minerals [54,56], magnetic surveys allow geologists to establish structural control on granite emplacement and link regional shear, fault, and dilational zones to known mineral and geochemical anomalies, therefore establishing a fluid migration pathway and potential mineralisation traps.

Regional geochemistry datasets outline the enrichment of incompatible and pathfinder elements (Sn, W, Nb, Ta, Li, K, Cs, Rb, Zr, Hf, Cu, Be and B) in rock, soil and stream sediment data. However, historically, most governmental geological surveys have employed aqua regia digests as part of their analytical procedures [63]. This choice significantly decreases the ability to dissolve silicate and oxide minerals, such as cassiterite, petalite, spodumene, columbite, tantalite and zircon. Therefore, concentrations of incompatible and pathfinder elements might be significantly lower in aqua regia digested samples, thereby restricting the usability of these datasets in rare element pegmatite exploration [64]. Conversely, other governmental surveys in Ireland employed non-digested pressed pellet X-ray fluorescence (XRF) analyses [65], which have clear advantages as the analytical technique is able to measure most major rock-forming elements, as well as Ta, Nb, Zr and Hf, but excluding Li. Nevertheless, XRF techniques still prove to be valid exploration tools, particularly when pathfinder elements and K/Rb, Nb/Ta and Zr/Hf ratios are analysed [28]. However, depending on the quality and age of datasets, not all desirable pathfinder elements will be available. Regional geochemical data can be employed to conduct univariate anomaly mapping [66] and dimension reduction techniques (principal component analysis), in order to identify relevant incompatible and pathfinder element associations and subsequently link the results with geological maps and published literature. For example, the assessment of the 1980s regional rock geochemical analyses provided by the French Bureau de Recherches Géologiques et Minières (BRGM) led to the initial delineation of a 200 km^2 extensive W and LCT exploration target area in the central and northern Vosges Mountains [29].

3.3. Targeted Fieldwork, Sampling, Mineralogy and Follow-Up

Once targets and elements of a mineralisation system have been identified by desktop and data studies, investigations should focus on field visits and a detailed description of the geological setting. Geological mapping, structural analysis of pegmatite emplacement structures, lithogeochemical analysis and radiometric age dating aim to (1) reveal the regional distribution of pegmatite bodies and (2) locate and confirm evidence for highly fractionated and evolved melts, located in the cupolas of granite stocks, pegmatitic off-shoots into country rock [10] or anatectic melts unrelated to S-type granites [4,36,67]. In other words, at this early stage, the provenance of the pegmatitic melts should be determined in order to link the overall geotectonic (crustal depth) setting with pegmatite occurrences and metal sources leading to a better understanding of the spatial dimensions and chemical composition ("LCT" vs. "CT" in orogenic/ pluton-unrelated settings) of expected pegmatite fields [13]. Supportive mineralogical evidence for fractionated melts are the visual presence of abundant muscovite and cassiterite, along with tourmaline, topaz, and fluorite, which act as fluxing indicators [9,68]. Geochemical analysis of rock samples collected during mapping campaigns aims to provide concentrations of fractionation pathfinder elements and evidence for a genetic relationship of the pegmatite with either a metasedimentary-sourced S-type granite pluton or anataxis-derived migmatite. The analysis of bulk pegmatite composition, in particular, is often challenging due to the very large grain size and zonation encountered. As a result, representative bulk grab or hand drill samples of either exploration trenches or active development benches or stopes need to be collected across the pegmatite from the hanging-wall contact to the foot-wall contact with pulverised sub-samples mixed and analysed as one representative sample [4]. Despite potential accuracy issues, varying detection limits for W and Ta, and the absence of Li and Cs analyses, in-situ portable XRF (pXRF) analysis of Sn, W, Ta and K/Rb proves to be a valid primary geochemical investigation technique for determining the enrichment of ore elements [69], whilst REE, B, Li, Sn and Ta determined by fusion ICP (Inductively Coupled Plasma) and DCP (Direct Current Plasma) spectroscopy is required for the reliable determination of overall grade, source material and provenance [4].

Numerous case studies underline the significance of automated mineralogical techniques in the rare metal granite and pegmatite exploration process [27,70–73]. Rapid identification of ore mineral associations, intergrowths, and liberation factors is achieved by the use of UV light for identifying Li silicates, scanning electron microscopy (SEM), electron microprobe analysis (EPMA), automated SEM-EDS mineralogy (QEMSCAN® or Mineralogic®), laser induced breakdown spectroscopy (LIBS), Raman spectroscopy and X-ray diffraction (XRD) analysis. Li-bearing minerals, such as spodumene, eucryptite, lepidolite, zinnwaldite or petalite, are particularly challenging to identify visually and to distinguish from common rock-forming silicate minerals, therefore emphasising the importance of mineralogical techniques. XRD techniques provide information on ore grade Nb–Ta oxides, and Li-bearing silicate minerals, whilst QEMSCAN®, SEM and EPMA techniques can quantify the purity/ contamination of the ore by deleterious elements (F, Fe and P) and determine potential recovery and processing methods, such as conventional flotation, gravity, or magnetic separation technologies [31]. As part of the EU Horizon 2020 Flexible and Mobile Economic processing technologies (FAME) project, Simons et al. (2018) [74] developed a new method for characterisation of Li minerals using SEM/ QEMSCAN®-based automated mineralogy. This approach is based on distinct Al:Si ratios, and the absence of other major elements, but also used a sample that has been optically validated for the minerals of interests to act as a reference, in granite-relate pegmatites and greisens, such as spodumene, petalite, beryl and zinnwaldite. Whilst spodumene and petalite were successfully identified on samples from Li-prospects at Cínovec (Czech Republic), Gonçalo (Portugal) and the Kaustinen pegmatite field (Finland), using characteristic Al:Si ratios of 1:2 and 1:4, respectively, Li micas could not be uniquely identified from Li-absent micas as the micas did not show any distinct chemical difference other than the Li which is not detectable by automated SEM-EDS mineralogy. For this reason, bespoke characterisation and QEMSCAN® calibration/ database development for each individual prospect is required.

Geochemical sampling and analysis of soil, till and stream sediments has proven to be a very effective method for localising metal anomalies in geological materials affected by secondary dispersion processes [75,76]. Considering the exploration for LCT pegmatite deposits, a preliminary orientation study will determine the most suitable sample location, weight, size fraction, grid and analytical requirements. Previous pXRF orientation studies of stream sediments have demonstrated that optimum field sample weights are 0.5–1.5 kg and that the silt to very fine sand grain size fraction (<75 μm) yields the best signal to noise ratio [29]. In contrast, soil surveys may necessitate the collection of the coarser sand fraction (>600 μm) to achieve optimum chemical analysis results [64]. The obvious difference in the choice of grain size fractions is a result of secondary dispersion processes, where comminution of heavy mineral grains, such as cassiterite or columbite-tantalite, is less pronounced in residual soils than in active streams. The presence of target metals in silicate minerals requires the utilisation of either pressed pellet XRF or strong chemical digests before subsequent ICP-MS analysis. A total sodium-peroxide fusion has proven effective in targeted granite and pegmatite exploration [45], whilst near-total four acid digest (HCl, HNO_3, HF and $HClO_4$) is useful in regional multi-commodity and granite-pegmatite geochemical surveys [29]. Partial aqua regia (HNO_3 and HCl at a 1:3 molar ratio) digests lack the ability to attack silicate and heavy mineral grains and are consequently of limited use [64]. One to three stream sediment samples per 4 km^2, collected in first and second order streams in mountainous terrain, along with panning and visual inspection of a heavy mineral concentrate, proved sufficient to recognise and narrow down mineralised zones within a prospective granite [29]. Black heavy mineral aggregates, comprising ilmenite, rutile, and containing inclusions of columbite-(Fe) and pyrochlore group minerals, were successfully used to outline pegmatite mineralisation in southeast Germany [27]. Detailed targeted geochemistry of residual B horizon soils (40 cm depth), with a sample grid spacing of 15 m × 100 m and perpendicular to geological strike led to the delineation of Sn, Cs and W anomalies related to vein mineralisation in slates and meta-siltstones along the southern flank of the Carnmenellis granite of Cornwall, United Kingdom [64]. Concentrations of Li (20–170 ppm), Cs (0.64–15.7 ppm), Ta (0.2–4.4 ppm), and Sn (2–6 ppm) represent a positive soil anomaly over the LCT pegmatite field of the Little Nahani Pegmatite Group, eastern Yukon Territory, Canada [77]. In the Ruhanga area of Rwanda, large muscovite flakes, present in the B-horizon of residual, deeply weathered tropical soils overlying pegmatite swarms, were analysed by pXRF, and returned very low values of $4 < K/Rb < 4.8$ and Ta values ranging from 150–450 ppm, implying highly fractionated peraluminous bedrock (Figure 5) [45]. Meanwhile, soils overlying adjacent unmineralised biotite schists are characterised by $270 < K/Rb < 320$. Exploration in glacial drift terrain, however, will require the collection of bottom-of-till C horizon samples, in order to minimise the effects of glacial dispersion on lithogeochemical signatures [76]. Conversely, weak-leach enzyme leaches (selective leach extraction which employs an enzyme reaction to preferentially dissolve amorphous manganese dioxides) on transported B horizon soils have been successfully tested in Ontario and Manitoba, and supported the generation of exploration targets [24,30].

Figure 5. Portable XRF (pXRF) equipment is a valuable screening tool in grassroots LCT pegmatite exploration. At Ruhanga (Rwanda), B horizon soils overlying kaolinised LCT pegmatite contained large flakes of muscovite. Analysis by pXRF provided K/Rb values of 4.8, indicating highly fractionated bedrock and therefore a possible target [45].

Whilst the Canadian case studies utilised a univariate anomaly detection approach, lithogeochemical fingerprinting of regional and highly fractionated target lithologies using immobile major and trace elements, K/Rb, Nb/Ta, and Zr/Hf ratios, along with Li, W and Sn analyses, demonstrated that mineralisation processes and systems can be delineated in areas with soil cover and other geological media affected by secondary dispersion processes [29,45]. Applying geochemical and metallogenic markers to stream sediment analyses, for example, facilitated the mapping of Sn, W, Li and Ta anomalies in hydrothermally-altered granite source material characterised by 70 < K/Rb < 225 and Nb/Ta < 5. Furthermore, depending on the size of the sample grid, the lithogeochemical classification process can lead to the production of regional lithogeochemical maps [28], which can potentially improve existing governmental datasets and streamline exploration in greenfields exploration programs. Local targets outlined by geological mapping and geochemical surveys are subsequently explored by channel sampling or trenching the bedrock [24], followed by detailed mapping and composite bulk sampling, in order to account for the potential irregular distribution of ore elements and minerals. Importantly, selective sampling of the pegmatite ore body should not be carried out, as isolated and thin, high grade veins of limited width or length are unlikely to be of economic interest [31]. The "nugget effect" conundrum implies that mapping and sampling should aim to identify multiple pegmatites or pegmatite swarms, providing sufficient tonnage over the course of mine life.

Similar to outcrop observations, visual mineralogy, pegmatite zonation and structural orientation data are important attributes exploration geologists need to recognise, describe and record in databases. New geological, mineralogical and geochemical data and evidence obtained from field exploration are used to continuously update the mineralisation system model previously established for the area of investigation.

3.4. Drilling

After exploration targets have been identified through desktop studies, geochemical and geological fieldwork, and sampling, prospective targets will have to be drill tested in order to obtain information about key geological factors rendering a LCT pegmatite economic, i.e., orebody geometry (lateral and vertical extension, dilution through country rock rafts, Figure 1), grade, and tonnage in order to produce a geological and grade 3D model. Reverse circulation (RC) drilling is the preferred method as, unlike traditional diamond core (DD) drilling, this technique facilitates the recovery of large diameter bulk samples and therefore minimises the "nugget effect", often experienced with narrow-vein tin mineralisation and large pegmatitic minerals [31]. However, DD holes allow geotechnical parameters to be collected early in an exploration campaign. Operating in steep, mountainous terrain such as Rwanda, particularly requires information about geomechanical properties (RQD, RMR and Q-factors, hydrogeological parameters), to ensure safe mine and excavation designs during the mine operation phase [45]. The geometry of pegmatites is of particular importance as the shape and orientation of the pegmatite body will determine not only the best exploratory drillhole orientation and resulting intersection with the pegmatite body, but also inform on potential mining methods. Whilst the sub-horizontal pegmatite body of the Tanco deposit (Canada) requires room and pillar underground mining operation below Bernic Lake [53], the 40–50° inclined Greenbushes pegmatite is currently excavated using a large open pit operation following the pegmatite along the strike and dip of the Donnybrook-Bridgetown shear zone [20]. Furthermore, clause 49 in the JORC guidelines [78] essentially requires LCT pegmatites to be treated as industrial mineral deposits. For this reason, drillhole samples are required to undergo chemical and mineralogical analysis for determination of lithium product grades (Li_2O or Li_2CO_3), along with a specification of the nature of lithium minerals present in the deposit [31].

4. Case Studies of LCT Pegmatite Exploration

Two previously published examples highlighting key aspects of LCT pegmatite exploration (Figure 4) will be presented in the following two sections. The first case study describes conceptual target generation and a geochemical exploration survey in the Vosges Mountains (France) [29], whilst the second case study principally outlines the outcomes of an exploration programme involving geochemistry, geophysics and petrology leading to the discovery of economic LCT pegmatites at Kaustinen (Finland) [56,58].

4.1. Conceptual Target Generation, Field Mapping and Petrogenetic Exploration Indicators

The increasing interest and upswing of the battery metals industry during the mid-2010s led to a screening study of underexplored, and potentially prospective, Variscan terrains in Central Europe [29]. An initial review of French Geological Survey (BRGM) data [79] identified W anomalies of >50 ppm in I-type granites of the central, highly metamorphosed, domain of the Variscan Vosges Mountains as a potential exploration area for Sn–W and Li–Cs–Ta mineralisation. Even though the area around Sainte-Marie-aux-Mines (Figure 6) is known to host post-Variscan base metal deposits in highly metamorphosed gneisses, no granite-related mineralisation systems were previously described.

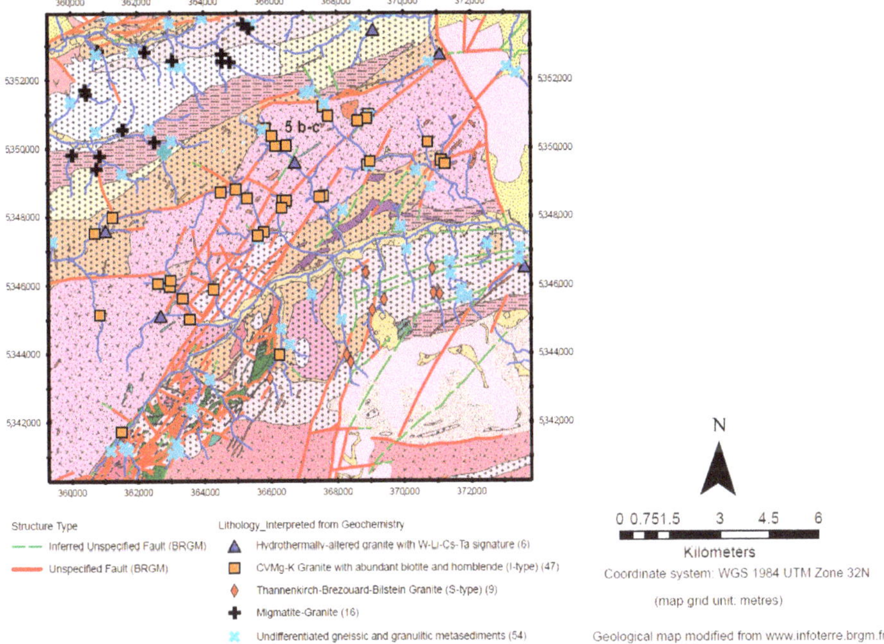

Figure 6. Lithologies interpreted from stream sediment geochemistry on modified Bureau de Recherches Géologiques et Minières (BRGM) geological map and structures, modified from Steiner (2019) [29] and BRGM (2019) [79]. The illustrated area covers the SW–NE trending Central Vosges Mg–K (CVMg–K granite), surrounded by highly metamorphic rocks of Paleozoic age. Observed leucogranitic and pegmatitic rocks with visible muscovite and tourmaline (location indicated as 5 b–c) along with interpreted "hydrothermally-altered granites with W–Li–Cs–Ta signature" are principally located along SW–NE orientated fault systems in the CVMg–K granite.

Following this desktop review, a decision was made to investigate the area in closer detail in order to prove whether the historic anomalies are (1) isolated occurrences, (2) related to the seemingly ubiquitous I-type (Central Vosges Mg–K) granite and (3) part of a potentially larger late-stage magmatic–hydrothermal mineralisation system. Initial reconnaissance work consisted of scout mapping of an approx. 200 km^2 area, supported by regional stream sediment sampling of first and second order streams. A pXRF screening and orientation study conducted on a selected number of sieved and homogenised stream sediment samples determined the −75µm fraction to be the optimum sample medium as the fraction returned the highest values for W, Cu and Ta (Figure 7). Additional multi-commodity targeted four acid digest followed by ICP-MS analysis resulted in the confirmation of pXRF results and the delineation of distinct Li (830 ppm), Ta (28.9 ppm) and W (578 ppm) anomalies in the host granite. In particular, the analysis of petrogenetic ratios, i.e, Nb/Ta < 5, 36 < Zr/Hf < 39 and

70 < K/Rb < 225 (Figure 8), revealed the presence of apparently strongly fractionated granite material in streams forming the topographic expression of large-scale shear zones and faults. Detailed mapping of these zones led to the discovery of distinct muscovite-tourmaline bearing leucogranites near previous W and Li anomalies in otherwise mantle-derived I-type granite host rock, implying that the stream sediment anomalies are presumably related to a secondary melting and fractionation event producing highly evolved felsic rocks. Ongoing unpublished research using SEM and QEMSCAN techniques investigate the mineralogy of both stream sediment samples and leucogranites with the aim to determine economic deportment of target elements in minerals. This case study shows that conceptual target generation using regional geological knowledge and historic data from national geological surveys can assist the identification of preliminary target areas. Furthermore, geological mapping and observations along with regional multi-element geochemistry and the utilisation of petrogenetic ratios can successfully narrow down target areas and reveal zones of increased magmatic fractionation and hydrothermal alteration, that may potentially be prospective for Li, Ta and W mineralisation.

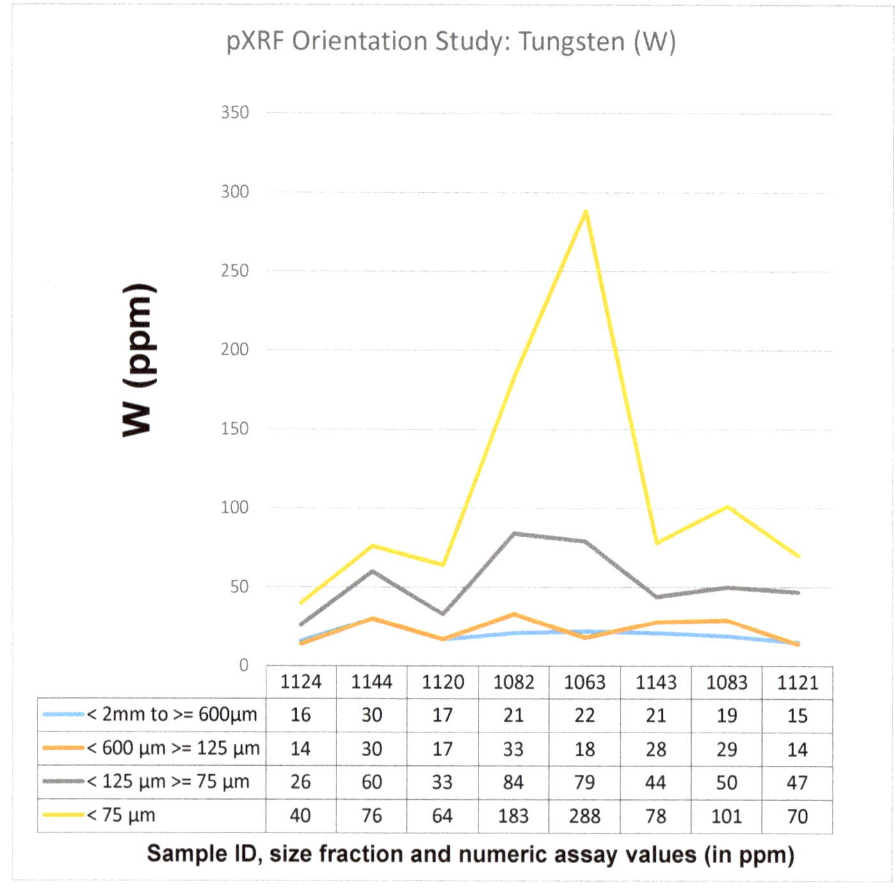

	1124	1144	1120	1082	1063	1143	1083	1121
< 2mm to >= 600µm	16	30	17	21	22	21	19	15
< 600 µm >= 125 µm	14	30	17	33	18	28	29	14
< 125 µm >= 75 µm	26	60	33	84	79	44	50	47
< 75 µm	40	76	64	183	288	78	101	70

Figure 7. Results of an orientation study [29] carried out using portable pXRF equipment in order to determine the most suitable grain size fraction for the present stream sediment sampling campaign.

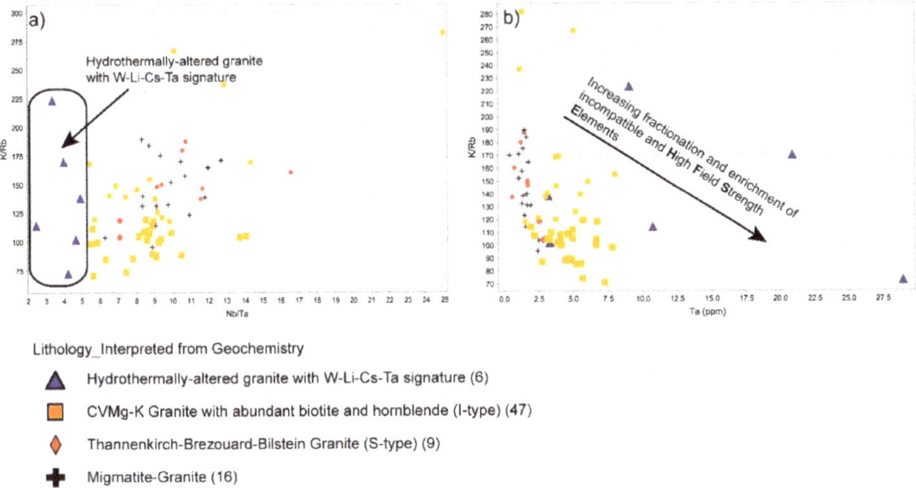

Figure 8. Examples from a stream sediment study [29] illustrating the use of (**a**) K/Rb vs. Nb/Ta and (**b**) K/Rb vs. Ta bivariate plots to determine highly fractionated granitic lithologies enriched in incompatible elements. In the central Vosges Mountains (France), peraluminous melts affected by hydrothermal alteration are thought to occur within a larger I-type granite province. In stream sediment samples, hydrothermally-altered granites are characterised by 70 < K/Rb < 225 and Nb/Ta < 5.

4.2. Till Geochemistry, Geophysical Surveys and Petrology

The Kaustinen LCT pegmatite field, consisting of a number of exploration areas, is one of the most prospective rare element pegmatite terrains in Western Finland [58], located in the Pohjanmaa (Ostrobothnia) Schist belt (Figure 9). The area was explored by the Finnish Geological Survey (GTK) for more than a decade since the early 2000s [29]. After the governmental exploration phase had finished in 2013 the deposit contained 1.3 Mt with 1.08 wt. % Li_2O. The deposit is currently owned and developed by Keliber Oy, a Finnish industrial minerals company. The Kaustinen pegmatite swarms cross-cut intercalated Paleoproterozoic mica schists and metavolcanic host rocks and form relatively flat-lying, tabular bodies with dimensions of up to 500 m × 20 m [80]. The pegmatite swarms are thought to be related to the nearby Kaustinen "pegmatite" granite. Pegmatites are mildly zoned and generally contain albite, quartz, K-Feldspar, spodumene and muscovite along with accessory amounts of Nb–Ta oxides, cassiterite and tourmaline (Figure 10). Initial exploration by the GTK in the 2000s involved a follow-up of a 1950s spodumene pegmatite boulder discovery, surface mapping, glacial boulder tracing, and bottom-of-till geochemical sampling along 50 m × 50 m grids and selected lines with a 15 m sample spacing. The distinct occurrences of spodumene pegmatite rocks in glacial boulder fans along with Li anomalies of up to 5770 ppm in till led to the definition of linear trends and exploration targets in metasedimentary and meta volcanic country rocks (Figure 11). Subsequently, 15.5 line km of gravity and 4.4 km^2 of gravity and magnetic ground geophysical surveys were conducted in order to determine a potential geophysical signature of the pegmatites. The results, however, did not confirm a geophysical signature of the pegmatites as the petrophysical properties of pegmatites and metasedimentary and metavolcanic host rocks are similar (Figure 12). Drill targets were therefore principally generated by geological observations and geochemical surveys. A later GTK drilling programme comprised a combined 155 RC and DD drillholes (17 km), geochemical multi-element assaying using XRF, acid and fusion techniques coupled with ICP-MS. Detailed logging of drill core allowed to outline the dimensions of the Kaustinen pegmatites. In addition, the application of optical petrography (Figure 10) and SEM on visible mineralisation intersections in core resulted in a detailed understanding of pegmatite mineralogy and zonation.

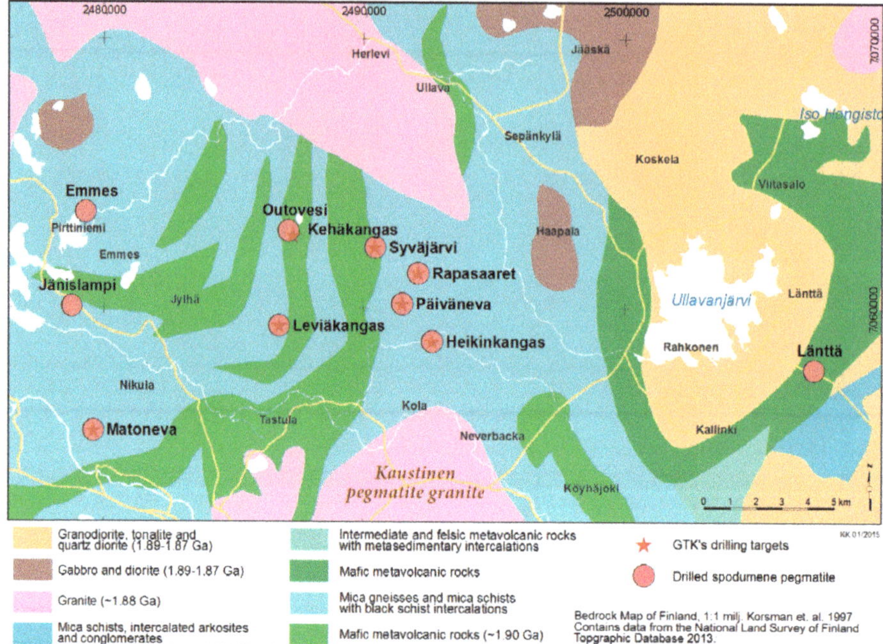

Figure 9. The Kaustinen Li pegmatite province, western Finland, showing the locations of the drilled spodumene pegmatites between 2004 and 2012. From Ahtola et al. (2015) [56].

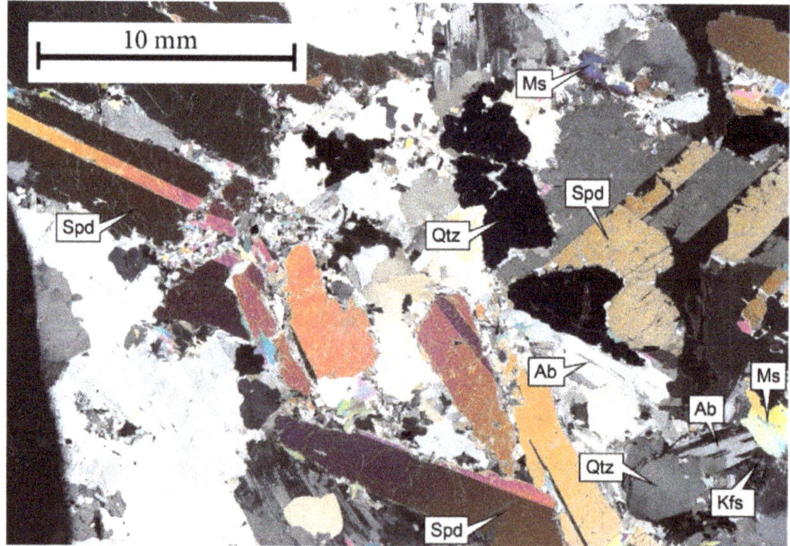

Figure 10. Thin section of spodumene pegmatite from drill core in the Leviäkangas exploration area, Kaustinen Li pegmatite province. Mineral abbreviations: Spd (Spodumene), Qtz (Quartz), Kfs (Potassium feldspar, Ab (Albite), Ms (Muscovite). From Ahtola et al. (2015) [56].

Figure 11. Map of the till sampling grid (from Ahtola et al. (2015) [56]) with plotted Li anomalies (in ppm) in the Rapasaaret target area, Kaustinen Li pegmatite province. The highest Li values (>1000 ppm) indicate the presence of Li pegmatite. Note: this study has not applied petrogenetic fractionation ratios as exploration vectors.

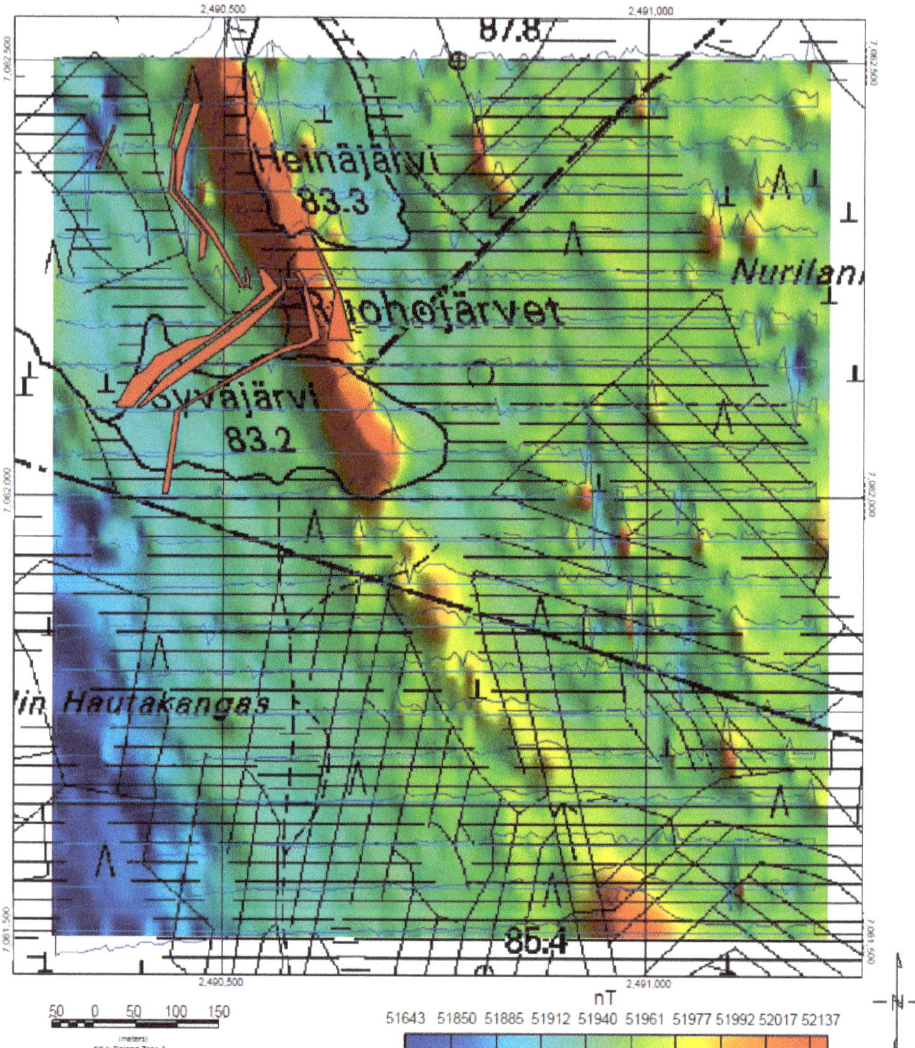

Figure 12. Ground magnetic map of the Syväjärvi exploration area, Kaustinen Li pegmatite Province. Modified from Ahtola et al. (2015) [56]. The blue lines show the magnetic profiles, whereas the red polygons mark the spodumene pegmatite. The positive magnetic anomaly (red colour) is caused by intermediate volcanic rocks. Basemap © National Land Survey of Finland, license number MML/VIR/TIPA/217/10.

In summary, the Kaustinen case study clearly demonstrates that the most reliable techniques for LCT pegmatite exploration are detailed geological field observations, along with geochemical and drilling surveys and mineralogical studies leading to the delineation of pegmatite geometry, zonation and chemical composition.

5. Conclusions

Petrological and mineralogical aspects of LCT pegmatite-related mineralisation have been extensively studied over the last four decades, leading to a detailed understanding of tectonic and geochemical controls on pegmatite emplacement and ore genesis. This paper highlights a field trialled early-stage, grassroots exploration workflow applicable to the scope and aims of a mineral exploration company. Desktop literature and GIS studies form the starting point of an exploration campaign and provide a synthesis of historical information, data, regional prospectivity along with the identification of LCT mineralisation system elements. In a follow-up field campaign, geological outcrop and trench mapping, as well as detailed mineralogy and geochemical rock, soil and stream sediment sampling campaigns, are designed to continuously delineate and test areas of increased prospectivity, which are eventually drill tested in order to obtain additional information on orebody dimensions, grade, tonnage, and geomechanical properties. In light of clause 49 of the JORC reporting guidelines, the importance of determining ore mineral species requires the implementation of mineralogical methods at all stages of a LCT pegmatite exploration campaign.

Funding: This research received no external funding.

Acknowledgments: The author would like to thank Paul Alexandre and three anonymous reviewers for providing constructive comments on the manuscript.

Conflicts of Interest: The authors declare no conflict of interest.

References

1. Černý, P. Exploration strategy and methods for pegmatite deposits of tantalum. In *Lanthanides, Tantalum, and Niobium*; Moller, P., Černý, P., Saupe, F., Eds.; Springer: New York, NY, USA, 1989; pp. 274–302.
2. Linnen, R.L.; Van Lichterfelde, M.; Černý, P. Granitic Pegmatites as Sources of Strategic Metals. *Elements* **2012**, *8*, 275–280. [CrossRef]
3. London, D. Ore-forming processes within granitic pegmatites. *Ore Geol. Rev.* **2018**, *101*, 349–383. [CrossRef]
4. Simmons, W.; Falster, A.; Webber, K.; Roda-Robles, E.; Boudreaux, A.P.; Grassi, L.R.; Freeman, G. Bulk composition of Mt. Mica Pegmatite, Maine, USA: Implications for the origin of an LCT type pegmatite by anatexis. *Can. Mineral.* **2016**, *54*, 1053–1070. [CrossRef]
5. Barros, R.; Menuge, J.F. The origin of spodumene pegmatites associated with the Leinster Granite in southeast Ireland. *Can. Mineral.* **2016**, *54*, 847–862. [CrossRef]
6. Kaeter, D.; Barros, R.; Menuge, J.F.; Chew, D.M. The magmatic–hydrothermal transition in rare-element pegmatites from southeast Ireland: LA–ICP–MS chemical mapping of muscovite and columbite–tantalite. *Geochim. Cosmochim. Acta* **2018**, *240*, 98–130. [CrossRef]
7. Ballouard, C.; Poujol, M.; Boulvais, P.; Branquet, Y.; Tartèse, R.; Vigneresse, J.L. Nb-Ta fractionation in peraluminous granites: A marker of the magmatic–hydrothermal transition. *Geology* **2016**, *44*, 231–234. [CrossRef]
8. Breiter, K.; Škoda, R. Zircon and whole-rock Zr/Hf ratios as markers of the evolution of granitic magmas: Examples from the Teplice caldera (Czech Republic/Germany). *Mineral. Petrol.* **2017**, *111*, 435–457. [CrossRef]
9. Breiter, K.; Ďurišová, J.; Hrstka, T.; Korbelová, Z.; Vašinová Galiová, M.; Müller, A.; Simons, B.; Shail, R.K.; Williamson, B.J.; Davies, J.A. The transition from granite to banded aplite–pegmatite sheet complexes: An example from Megiliggar Rocks, Tregonning topaz granite, Cornwall. *Lithos* **2018**, *302*, 370–388. [CrossRef]
10. Dill, H.G. Pegmatites and aplites: Their genetic and applied ore geology. *Ore Geol. Rev.* **2015**, *69*, 417–561. [CrossRef]
11. Roda, E.; Pesquera, A.; Velasco, F.; Fontan, F. The granitic pegmatites of the Fregeneda area (Salamanca, Spain): Characteristics and petrogenesis. *Mineral. Mag.* **1999**, *63*, 535–558. [CrossRef]
12. Romer, R.L.; Kroner, U. Phanerozoic tin and tungsten mineralization—Tectonic controls on the distribution of enriched protoliths and heat sources for crustal melting. *Gondwana Res.* **2016**, *31*, 60–95. [CrossRef]
13. Müller, A.; Romer, R.L.; Pedersen, R.-B. The Sveconorwegian Pegmatite Province—Thousands of Pegmatites Without Parental Granites. *Can. Mineral.* **2017**, *55*, 283–315. [CrossRef]
14. Müller, A.; Ihlen, P.M.; Snook, B.; Larsen, R.B.; Flem, B.; Bingen, B.; Williamson, B.J. The Chemistry of Quartz in Granitic Pegmatites of Southern Norway: Petrogenetic and Economic Implications. *Econ. Geol.* **2015**, *110*, 1737–1757. [CrossRef]

15. Dittrich, T.; Seifert, T.; Schulz, B.; Hagemann, S.; Gerdes, A.; Pfänder, J. *Archean Rare-Metal Pegmatites in Zimbabwe and Western Australia: Geology and Metallogeny of Pollucite Mineralisations*, 1st ed.; Springer: Heidelberg, Germany, 2019; 125p.
16. Hulsbosch, N.; Van Daele, J.; Reinders, N.; Dewaele, S.; Jacques, D.; Muchez, P. Structural control on the emplacement of contemporaneous Sn–Ta–Nb mineralized LCT39] pegmatites and Sn bearing quartz veins: Insights from the Musha and Ntunga deposits of the Karagwe–Ankole Belt, Rwanda. *J. Afr. Earth Sci.* **2017**, *134*, 24–32. [CrossRef]
17. Lehmann, B.; Halder, S.; Munana, J.R.; Ngizimana, J.d.l.P.; Biryabarema, M. The geochemical signature of rare–metal pegmatites in Central Africa: Magmatic rocks in the Gatumba tin–tantalum mining district, Rwanda. *J. Geochem. Expl.* **2014**, *144*, 528–538. [CrossRef]
18. Melcher, F.; Graupner, T.; Oberthür, T.; Schütte, P. Tantalum-(niobium-tin) mineralisation in pegmatites and rare-metal granites of Africa. *S. Afr. J. Geol.* **2017**, *120*, 77–100. [CrossRef]
19. Heng, C.L.; Chand, F.; Singh, S.D. Primary Tin Mineralization in Malaysia: Aspects of Geological Setting and Exploration Strategy. In *Geology of Tin Deposits in Asia and the Pacific*; Hutchison, C.S., Ed.; Springer: Berlin/Heidelberg, Germany, 1988; pp. 593–613.
20. Partington, G.A.; McNaughton, N.J.; Williams, I.S. A review of the geology, mineralization, and geochronology of the Greenbushes Pegmatite, Western Australia. *Econ. Geol.* **1995**, *90*, 616–635. [CrossRef]
21. Sweetapple, M.T. Granitic pegmatites as mineral systems: Examples from the Archaean. *NGF Abstr. Proc.* **2017**, *2*, 139–142.
22. Sweetapple, M.; Collins, P.L.F. Genetic Framework for the Classification and Distribution of Archean Rare Metal Pegmatites in the North Pilbara Craton, Western Australia. *Econ. Geol.* **2002**, *97*, 873–895. [CrossRef]
23. Cardoso-Fernandes, J.; Teodoro, A.C.; Lima, A. Remote sensing data in lithium (Li) exploration: A new approach for the detection of Li-bearing pegmatites. *Int. J. Appl. Earth Obs. Geoinf.* **2019**, *76*, 10–25. [CrossRef]
24. Dimmell, P.M.; Morgan, J.A. The Aubry Pegmatites: Exploration for Highly Evolved Lithium-Cesium-Tantalum Pegmatites in Northern Ontario. *Explor. Min. Geol.* **2005**, *14*, 45–59. [CrossRef]
25. Möller, P.; Morteani, G. Geochemical exploration guide for tantalum pegmatites. *Econ. Geol.* **1987**, *82*, 1888–1897. [CrossRef]
26. Selway, J.B.; Breaks, F.W.; Tindle, A.G. A Review of Rare–Element (Li–Cs–Ta) Pegmatite Exploration Techniques for the Superior Province, Canada, and Large Worldwide Tantalum Deposits. *Explor. Min. Geol.* **2005**, *14*, 1–30. [CrossRef]
27. Dill, H.G.; Weber, B.; Melcher, F.; Wiesner, W.; Müller, A. Titaniferous heavy mineral aggregates as a tool in exploration for pegmatitic and aplitic rare-metal deposits (SE Germany). *Ore Geol. Rev.* **2014**, *57*, 29–52. [CrossRef]
28. Steiner, B. Using Tellus stream sediment geochemistry to fingerprint regional geology and mineralisation systems in southeast Ireland. *Ir. J. Earth Sci.* **2018**, *36*, 45–61.
29. Steiner, B.M. W and Li–Cs–Ta signatures in I–type granites—A case study from the Vosges Mountains, NE France. *J. Geochem. Expl.* **2019**, *197*, 238–250. [CrossRef]
30. Galeschuk, C.R.; Vanstone, P.J. Exploration Techniques for Rare-Element Pegmatite in the Bird River Greenstone Belt, Southeastern Manitoba. In *Proceedings of Exploration 07: Fifth Decennial International Conference on Mineral Exploration*; Milkereit, B., Ed.; Decennial Mineral Exploration Conferences: Toronto, ON, Canada, 2007; pp. 823–839.
31. Scogings, A.; Porter, R.; Jeffress, G. Reporting Exploration Results and Mineral Resources for lithium mineralised pegmatites. *AIG News Issue* **2016**, *125*, 32–36.
32. Dessemond, C.; Lajoie-Leroux, F.; Soucy, G.; Laroche, N.; Magnan, J.-F. Spodumene: The Lithium Market, Resources and Processes. *Minerals* **2019**, *9*, 334. [CrossRef]
33. Kesler, S.E.; Gruber, P.W.; Medina, P.A.; Keoleian, G.A.; Everson, M.P.; Wallington, T.J. Global lithium resources: Relative importance of pegmatite, brine and other deposits. *Ore Geol. Rev.* **2012**, *48*, 55–69. [CrossRef]
34. London, D. Pegmatites. *Can. Mineral.* **2008**, *10*, 368.
35. Fuchsloch., W.C.; Nex, P.A.M.; Kinnaird, J.A. Classification, mineralogical and geochemical variations in pegmatites of the Cape Cross-Uis pegmatite belt, Namibia. *Lithos* **2018**, *296*, 79–95. [CrossRef]

36. Konzett, J.; Schneider, T.; Nedyalkova, L.; Hauzenberger, C.; Melcher, F.; Gerdes, A.; Whitehouse, M. Anatectic Granitic Pegmatites from the Eastern Alps: A Case of Variable Rare-Metal Enrichment During High-Grade Regional Metamorphism–I: Mineral Assemblages, Geochemical Characteristics, and Emplacement Ages. *Can. Mineral.* **2018**, *56*, 555–602. [CrossRef]
37. Konzett, J.; Hauzenberger, C.; Ludwig, T.; Stalder, R. Anatectic Granitic Pegmatites from the Eastern Alps: A Case of Variable Rare Metal Enrichment During High-Grade Regional Metamorphism. II: Pegmatite Staurolite As an Indicator of Anatectic Pegmatite Parent Melt Formation—A Field and Experimental Study. *Can. Mineral.* **2018**, *56*, 603–624. [CrossRef]
38. Schuster, R.; Ilickovic, T.; Mali, H.; Huet, B.; Schedl, A. Permian pegmatites and spodumene pegmatites in the Alps: Formation during regional scale high temperature/low pressure metamorphism. *NGF Abstr. Proc.* **2017**, *2*, 122–125.
39. Simmons, W.B.; Webber, K.L. Pegmatite genesis: State of the art. *Eur. J. Mineral.* **2008**, *20*, 421–438. [CrossRef]
40. Zagorsky, V.Y.; Vladimirov, A.G.; Makagon, V.M.; Kuznetsova, L.G.; Smirnov, S.Z.; D'yachkov, B.A.; Annikova, I.Y.; Shokalsky, S.P.; Uvarov, A.N. Large fields of spodumene pegmatites in the settings of rifting and postcollisional shear–pull-apart dislocations of continental lithosphere. *Russ. Geol. Geophys.* **2014**, *55*, 237–251. [CrossRef]
41. Müller, A.; Simmons, W.; Beurlen, H.; Thomas, R.; Ihlen, P.M.; Wise, M.; Roda-Robles, E.; Neiva, A.M.R.; Zagorsky, V. A proposed new mineralogical classification system for granitic pegmatites; Part I, History and the need for a new classification. *Can. Mineral.* **2018**, *56*, 1–25. [CrossRef]
42. London, D. A petrologic assessment of internal zonation in granitic pegmatites. *Lithos* **2014**, *184*, 74–104. [CrossRef]
43. Černý, P.; Ercit, T.S. The classification of granitic pegmatites revisited. *Can. Mineral.* **2005**, *43*, 2005–2026. [CrossRef]
44. Černý, P.; London, D.; Novak, M. Granitic pegmatites as reflections of their sources. *Elements* **2012**, *8*, 289–294. [CrossRef]
45. Steiner, B. *Rwanda Pegmatites and Exploration Strategy*; Unpublished Report Mila Resources; 2018; 23p.
46. London, D. Granitic pegmatites: An assessment of current concepts and directions for the future. *Lithos* **2005**, *80*, 281–303. [CrossRef]
47. Snook, B.R. Towards Exploration Tools for High Purity Quartz: An Example from the South Norwegian Evje-Iveland Pegmatite Belt. Ph.D. Thesis, Camborne School of Mines, University of Exeter, Exeter, UK, 2013. Unpublished. 284p.
48. Linnen, R.L.; Keppler, H. Columbite solubility in granitic melts: Consequences for the enrichment and fractionation of Nb and Ta in the Earth's crust. *Contrib. Mineral. Petrol.* **1997**, *128*, 213–227. [CrossRef]
49. Fujimaki, H. Partition-coefficients of Hf, Zr, and REE between zircon, apatite and liquid. *Contrib. Mineral. Petrol.* **1986**, *94*, 42–45. [CrossRef]
50. Shaw, R.A.; Goodenough, K.M.; Roberts, N.M.W.; Horstwood, M.S.A.; Chenery, S.R.; Gunn, A.G. Petrogenesis of rare-metal pegmatites in high-grade metamorphic terranes: A case study from the Lewisian Gneiss Complex of north-west Scotland. *Precambrian Res.* **2016**, *281*, 338–362. [CrossRef]
51. Černý, P. Fertile granites of Precambrian rare-element pegmatite fields: Is geochemistry controlled by tectonic setting or source lithologies? *Precambrian Res.* **1991**, *51*, 429–468. [CrossRef]
52. Wyborn, L.A.I.; Heinrich, C.A.; Jaques, A.L. Australian Proterozoic Mineral Systems: Essential Ingredients and Mappable Criteria. In *The AusIMM Annual Conference*; The Australasian Institute of Mining and Metallurgy: Carlton, Australia, 1994; pp. 109–115.
53. Stilling, A.; Černý, P.; Vanstone, P.J. The Tanco pegmatite at Bernic Lake, Manitoba. XVI. Zonal and bulk compositions and their petrogenetic significance. *Can. Mineral.* **2006**, *44*, 599–623. [CrossRef]
54. Trueman, D.L.; Černý, P. Exploration for rare-metal granitic pegmatites. In *Granitic Pegmatites in Science and Industry*; Cerný, P., Ed.; Short Course Handbook; Mineralogical Association of Canada: Quebec, QC, Canada, 1982; pp. 463–493.
55. Bradley, D.C.; McCauley, A.D.; Stillings, L.M. Mineral-Deposit Model for Lithium-CesiumTantalum Pegmatites. In *Scientific Investigations Report 2010–5070–O*; US Geological Survey: Reston, WV, USA, 2017; 32p.

56. Ahtola, T.; Kuusela, J.; Käpyaho, A.; Kontoniemi, O. Overview of lithium pegmatite exploration in the Kaustinen area in 2003–2012. In *Report of Investigation 220*; Geological Survey of Finland: Espoo, Finland, 2015; 28p.
57. London, D.; Morgan, G.B., VI. Experimental Crystallization of the Macusani Obsidian, with Applications to Lithium-rich Granitic Pegmatites. *J. Petrol.* **2017**, *58*, 1005–1030. [CrossRef]
58. Rasilainen, K.; Eilu, P.; Ahtola, T.; Halkoaho, T.; Kärkkäinen, N.; Kuusela, J.; Lintinen, P.; Törmänen, T. *Quantitative Assessment of Undiscovered Resources in Lithium–Caesium–Tantalum Pegmatite-Hosted Deposits in Finland*; Geological Survey of Finland: Espoo, Finland, 2018; 31p.
59. Ontario Geological Survey. *Recommendations for Exploration 2017–2018*; OGS Resident Geologist Program; Ontario Geological Survey: Thunder Bay, ON, Canada, 2018; 100p.
60. Kalinowski, A.; Oliver, S. ASTER Mineral Index Processing Manual. *Remote Sens. Appl. Geosci. Aust.* **2004**, *37*, 36.
61. Cook, S.E.; Corner, R.J.; Groves, P.R.; Grealish, G.J. Use of airborne gamma radiometric data for soil mapping. *Aust. J. Soil Res.* **1996**, *34*, 183–194. [CrossRef]
62. Schetselaar, E.; Chung, C.-J.F.; Kim, K.E. Integration of Landsat TM, Gamma-Ray, Magnetic, and Field Data to Discriminate Lithological Units in Vegetated Granite–Gneiss Terrain. *Remote Sens. Environ.* **2000**, *71*, 89–105. [CrossRef]
63. Hakku Data Service. Available online: https://hakku.gtk.fi/en (accessed on 15 May 2019).
64. Morton, C. Geochemical Anomaly Mapping along the Southern Flank of the Carnmenellis Granite, Cornwall, UK. Master's Thesis, Camborne School of Mines, University of Exeter, Exeter, UK, 2017. Unpublished. 45p.
65. Knights, K.V.; Heath, P.J. Quality control statistical summaries of Tellus stream sediment regional geochemical data. In *Tellus Project Report*; Geological Survey of Ireland: Dublin, Ireland, 2016; 252p.
66. O'Connor, P.J.; Reimann, C. Multielement regional geochemical reconnaissance as an aid to target selection in Irish Caledonian terrains. *J. Geochem. Expl.* **1993**, *47*, 63–87. [CrossRef]
67. Faria, C.; Gomes, C.L. Structure of the Granitic Pegmatite Field of the Northern Coast of Portugal—Inner Pegmatite Structures and Mineralogical Fabrics. *Heritage* **2019**, *2*, 315–330. [CrossRef]
68. Breiter, K.; Škoda, R.; Uher, P. Nb-Ta-Ti-W-Sn-oxide minerals as indicators of a peraluminous P- and F-rich granitic system evolution: Podlesí, Czech Republic. *Mineral. Petrol.* **2007**, *91*, 225–248. [CrossRef]
69. Knésl, I.; Jandova, T.; Rambousek, P.; Breiter, K. Calibration of portable XRF spectrometer in Sn-W ore-bearing granites: Application in the Cínovec deposit (Erzgebirge/Krušně Hory Mts., Czech Republic). *Inżynieria Mineral.* **2015**, *16*, 67–72.
70. Dehaine, Q.; Filippov, L.O.; Glass, H.J.; Rollinson, G. Rare-metal granites as a potential source of critical metals: A geometallurgical case study. *Ore Geol. Rev.* **2019**, *104*, 384–402. [CrossRef]
71. Hoal, K.O.; Stammer, J.G.; Appleby, S.K.; Botha, J.; Ross, J.K.; Botha, P.W. Research in quantitative mineralogy: Examples from diverse applications. *Mineral. Eng.* **2009**, *22*, 402–408. [CrossRef]
72. Sandmann, D.; Gutzmer, J. Use of Mineral Liberation Analysis (MLA) in the Characterization of Lithium-Bearing Micas. *J. Mineral. Mater. Charact. Eng.* **2013**, *1*, 285–292. [CrossRef]
73. Sousa, R.; Simons, B.; Bru, K.; Botelho de Sousa, A.; Rollinson, G.; Andersen, J.; Martin, M.; Machado Leite, M. Use of mineral liberation quantitative data to assess separation efficiency in mineral processing—Some case studies. *Mineral. Eng.* **2018**, *127*, 134–142. [CrossRef]
74. Simons, B.; Rollinson, G.K.; Andersen, J.C.Ø. Characterisation of lithium minerals in granite-related pegmatites and greisens by SEM-based automated mineralogy. Poster Presentation. In Proceedings of the Mineral Deposits Study Group Winter Meeting, Brighton, UK, 4 January 2018.
75. Hale, M.; Plant, J.A. (Eds.) Drainage geochemistry. In *Handbook of Exploration Geochemistry*; Elsevier: Amsterdam, The Netherlands, 1994; Volume 6.
76. McMartin, I.; McClenaghan, M.B. Till geochemistry and sampling techniques in glaciated shield terrain: A review. In *Drift Exploration in Glaciated Terrain*; McClenaghan, M.B., Bobrowsky, P.T., Hall, G.E.M., Cook, S.J., Eds.; Special Publications; Geological Society of London: London, UK, 2001; Volume 185, pp. 19–43.
77. Turner, D.J.; Young, I. *Geological Assessment Report on the SELWYN 1–10 Claims, Victoria, British Columbia, Canada*; War Eagle Mining Company: Vancouver, BC, Canada, 2008. Available online: http://yma.gov.yk.ca/095100.pdf (accessed on 18 July 2019).
78. Joint Ore Reserves Committee (JORC). The JORC Code. 2012 Edition. Available online: http://jorc.org/docs/jorc_code2012.pdf (accessed on 19 August 2019).

79. BRGM. InfoTerre. Available online: http://www.infoterre.brgm.fr (accessed on 14 August 2019).
80. Korsman, K.; Koistinen, T.; Kohonen, J.; Wennerström, M.; Ekdahl, E.; Honkamo, M.; Idman, H.; Pekkala, Y. (Eds.) *Suomen Kallioperäkartta—Berggrundskarta över Finland-Bedrock Map of Finland 1:1 000 000*; Geological Survey of Finland: Espoo, Finland, 1997.